KU-360-343

Construction Planning, Programming and Control

Third Edition

Brian Cooke and Peter Williams

WILEY-BLACKWELL

A John Wiley & Sons, Ltd., Publication

This edition first published 2009
© 2009 Brian Cooke and Peter Williams

Blackwell Publishing was acquired by John Wiley & Sons in February 2007. Blackwell's publishing programme has been merged with Wiley's global Scientific, Technical, and Medical business to form Wiley-Blackwell.

Registered office
John Wiley & Sons Ltd, The Atrium, Southern Gate, Chichester, West Sussex, PO19 8SQ, United Kingdom

Editorial offices
9600 Garsington Road, Oxford, OX4 2DQ, United Kingdom
2121 State Avenue, Ames, Iowa 50014-8300, USA

For details of our global editorial offices, for customer services and for information about how to apply for permission to reuse the copyright material in this book please see our website at www.wiley.com/wiley-blackwell.

The right of the author to be identified as the author of this work has been asserted in accordance with the Copyright, Designs and Patents Act 1988.

All rights reserved. No part of this publication may be reproduced, stored in a retrieval system, or transmitted, in any form or by any means, electronic, mechanical, photocopying, recording or otherwise, except as permitted by the UK Copyright, Designs and Patents Act 1988, without the prior permission of the publisher.

Wiley also publishes its books in a variety of electronic formats. Some content that appears in print may not be available in electronic books.

Designations used by companies to distinguish their products are often claimed as trademarks. All brand names and product names used in this book are trade names, service marks, trademarks or registered trademarks of their respective owners. The publisher is not associated with any product or vendor mentioned in this book. This publication is designed to provide accurate and authoritative information in regard to the subject matter covered. It is sold on the understanding that the publisher is not engaged in rendering professional services. If professional advice or other expert assistance is required, the services of a competent professional should be sought.

Library of Congress Cataloging-in-Publication Data

Cooke, B. (Brian)
 Construction planning, programming and control / Brian Cooke and Peter Williams. – 3rd. ed.
  p. cm.
 Includes bibliographical references and index.
 ISBN 978-1-4051-8380-2 (pbk. : alk. paper) 1. Building–Planning. I. Williams, Peter. II. Title.
 TH153.C597 2008
 690.068–dc22

                                                               2008022171

A catalogue record for this book is available from the British Library.

Set in 10/12pt Minion by Aptara® Inc., New Delhi, India
Printed in Singapore by Utopia Press Pte Ltd

1 2009

Contents

PART A PLANNING

Preface

The aims of the first edition have not changed. This is a book for students of construction-related professional and degree courses – construction management, building, quantity surveying and building surveying alike. Students of all these topics need a basic understanding of planning, procurement and programming at some time in their career development.

The text of the second edition was extensively rewritten and restructured to reflect changes in the construction industry, and this edition has been updated and extended to include new contracts, legislation and working practices.

In this edition, special emphasis has been placed on practical aspects of construction planning and programming. New chapters have been included dealing with construction sequences, method statements and planning for safety. Extensive guidance has been provided on the identification of hazards, the management of risk and the thought process required for the effective planning of construction projects. There is a new chapter dealing with delay and disruption and this includes an introduction to the theory and practice of delay analysis.

The book has been updated for changes in commonly used construction contracts including the 2005 versions of the JCT family and the NEC 3rd Edition. Important changes to the Construction (Design and Management) Regulations have also been included. A brand new case study has been introduced which includes new topics such as vertical access planning and materials handling. A new section on construction waste has been included which introduces the statutory requirement for Site Waste Management Plans.

Our special thanks go to the directors and staff of Countryside Properties PLC and Alfred McAlpine Special Projects for their permission to use case study material from their sites and to the Totty Construction Group who helped us so generously with the second edition. The invaluable assistance of our friend and former colleague Paul Hodgkinson is gratefully acknowledged. Paul has again produced the excellent diagrams and managed to keep track of the 224 figures now in the book!

Introduction

1 The construction industry

1.1 Introduction

The effective planning and control of construction projects requires the application of systematic and logical methods and tried and tested techniques aimed at ensuring successful project outcomes for the client, the contractor and all other project participants. Control is not possible without a plan, and without a programme there is no effective means of exercising control.

Successful projects cannot happen in a vacuum, however, and no one can effectively plan and control a construction project without understanding the culture and methodologies of the industry that organises and carries out the work as well as the impact of the various procurement strategies open to the client.

1.2 Industry structure

Construction is a large and complex industry comprising many types and sizes of organisations and a diverse range of professional and other representative bodies. Clients, professional practices, contractors and specialist firms all have their own 'agenda' and allegiances; Walker (2007) explains the complexities of the interrelationships that can occur in construction projects between clients and their professional advisors and between the contractors and specialists they engage to construct the project.

The annual output of the UK construction industry is approximately £114 billion, which represents almost 9% of gross domestic product. The industry is also a large employer with around 1.9 million people, of which 0.7 million are self-employed. Construction has a unique structure with a small number of very large firms carrying out the majority of its turnover. It also has a very large number of relatively small firms with few criteria limiting entry to the industry.

There is a wide disparity in the standards of competence across the industry: a largely semi-skilled and itinerant workforce and a generally low standard of education and qualifications in the managers employed in construction.

The Department for Business, Enterprise and Regulatory Reform (BERR) (2007) reports that the industry is made up of some 186 000 firms of which:

- 1% are large firms employing over 80 people
- 6% are medium-sized employing between 14 and 79 people
- 93% are small firms employing less than 14 people

The very large contractors (1200+ employees) have considerable influence on the industry as they enjoy a large share of the construction market. Consequently, the Major Contractors Group (MCG) have lined up behind the Construction Skills Certification Scheme (CSCS) in order to help promote greater competence and attention to health and safety risks. CSCS is a voluntary registration scheme whereby trainees, workers, managers, supervisors, visiting professionals, etc., can obtain the relevant colour card which enables them to gain access to MCG sites. A health and safety awareness test must be passed in order to obtain the card.

The industry has a diverse range of suppliers as well as contractors, including manufacturers of materials and components, suppliers of quarry products and ready-mixed concrete, builders' merchants and plant hire firms. Some of the industry's suppliers are larger than the largest contractors working in the industry and they also have considerable influence on the way the industry operates.

Morton and Ross (2007) examine the structure of the industry in detail and discuss issues such as safety, efficiency, employment and industry practices.

1.3 Industry culture

The scale and diversity of the clients, contractors, professionals and others who are involved in construction is enormous. A contractor may be a small 'sole trader' with an annual turnover of £100 000 or a large public company with a yearly workload of £1 billion. A client could be a domestic householder, private sector corporation or government department. The project 'architect' could be a single practitioner, a local authority or a large public company.

This means that there is nothing 'typical' about the customs and practices in construction. There is, however, one common 'cultural' feature at all levels of the industry and that is the 'project-based' nature of the industry. The whole focus is on the project; the time, cost, quality, resources, problems and solutions are all geared to the project. This brings enormous pressures on project teams, however big or small they are, to ensure that the project is completed 'on time, on budget and to the correct quality standards'. However, more or less every project has its own individuality and peculiarities depending on the site and location, the design and type of construction, the business arrangements between the parties and the hopes and expectations of all those involved; and this 'one-off' nature of construction creates additional pressures.

One of the big problems in construction is the extent to which the industry separates design from production to a far greater extent than in other industries. This particular feature of the industry is still common despite the deficiencies of traditional procurement and the benefits offered by newer and more flexible approaches.

The industry is a highly fragmented project-based industry and leadership in construction comes largely from clients and not from the contractors and specialists who carry out the work. In 1998, the Construction Clients' Forum (now the Confederation of Construction Clients) established the Clients' Charter which sets out the minimum standards in procurement expected by clients. By registering for the charter, clients commit themselves to establishing a modern business culture with their suppliers, steady improvement of standards measured against nationally accepted criteria and the exchange of best practice experience.

Consequently, it is the clients and their professional advisors who dictate the procurement methods used in the industry and it is the contractors who have to react to the latest 'flavour of the month' in the context of the organisation and management of the project.

1.4 Construction industry reports

In order to stimulate debate about construction industry practices and procedures, a number of well-known reports have been published over the years, including those shown in Table 1.1.

Several of the reports published prior to Latham and Egan raised similar criticisms about the customs and practices of the industry. Banwell even suggested that a common form of contract should be adopted for use on all construction projects.

These problems, and many others, have been recognised since World War II but despite identifying the problems and proposing solutions, most of the reports have had little influence on either government or the industry over the years.

Murray and Langford (2003) discuss these reports in detail and the extent to which government has tried to shape the performance and attitudes of the industry. Brief summaries of some of the more influential recent reports on the construction industry are given below.

Table 1.1 Construction industry reports.

Report	Title	Year
Simon Report	The Placing and Management of Building Contracts	1944
Emmerson Report	Survey of Problems Before the Construction Industries	1962
Banwell Report	The Placing and Management of Contracts for Building and Civil Engineering Work	1964
National Economic Development Office	Action on Banwell	1967
Tavistock Report	Interdependence and Uncertainty	1966
Latham 1	Interim Report – Trust and Monies	1993
Latham 2	Final Report – Constructing the Team	1994
Levene Efficiency Scrutiny	Construction Procurement by Government	1995
Egan Report 1	Rethinking Construction	1998
National Audit Office	Modernising Construction	2001
Egan Report 2	Accelerating Change	2002
National Audit Office	Improving Public Services through better construction	2005

1.5 The Latham reports

'Trust and Monies'

Perhaps the most influential of all the reports concerning the industry and its problems is 'Constructing the Team' written by Sir Michael Latham (1994) who was commissioned by both government and the industry to review the procurement and contractual arrangements in the UK construction industry.

Prior to final publication of his report in July 1994, Sir Michael produced an interim report in December 1993 called 'Trust and Monies'. This report raised concerns about the extent of mistrust between professionals and contractors and between contractors and subcontractors in construction, and also flagged up the endemic culture of late and conditional payments operating in the industry. The prevailing atmosphere of mistrust and slow payments was reported to result in disharmony in project teams, poor standards of work and poor client satisfaction.

'Constructing the Team'

'Constructing the Team' is better known as the 'Latham Report' and its purpose was to find ways to 'reduce conflict and litigation and encourage the industry's productivity and competitiveness'.

The specific terms of reference for the review were to consider:

- Current procurement and contractual arrangements
- Current roles, responsibilities and performance of the participants, including the client

The report took account of the structure of the industry and the need for fairness, accountability, quality and efficiency and paid particular regard to:

- Client briefing
- Procurement methods
- The design process
- The construction process
- Contractual issues
- Dispute resolution

The report runs to some 130 pages and contains 30 main observations and recommendations, with the principal emphasis being on 'teamwork' in order to achieve 'win-win' solutions.

Latham noted several issues which influence the ability of the construction industry to respond effectively to its customers' requirements, and these can be summarised briefly as:

- Sensitivity to changes in government spending patterns
- Intense competition for work
- Inability to respond to increased demand
- Lack of competency testing of firms/workers entering the industry
- Lack of training
- Mistrust between the participants in construction projects
- Inadequate capital base (i.e. most contractors are undercapitalised)
- Adversarial attitudes
- Claims-conscious contractors
- High levels of insolvency

Some of the main points made by Latham clearly have important consequences for the planning, production and control of construction and are therefore directly relevant to this book. These include:

- The need for a set of basic principles for modern contracts
- Greater use of the New Engineering Contract, which could become a common contract for the whole industry
- Improved tendering arrangements and more advice on partnering arrangements
- Evaluation of tenders on quality as well as price
- Fairer treatment of subcontractors, with particular regard to tendering and teamwork on site
- A real cost reduction target in construction of 30% by the year 2000
- Pay when paid contract terms to be outlawed
- Adjudication to be the normal method of dispute resolution
- Fair contract terms backed up by a legislation
- Insolvency protection by means of trust funds

One of the key issues considered by Latham was the productivity of the industry, and Latham clearly considered that this is linked to the quality of design preparation and information. Inefficiency creeps in where designs are incomplete or information given to the contractor is conflicting or too late to allow proper planning of production.

An issue of major importance is conflict in the industry both between clients and contractors and between contractors and their subcontractors. Latham suggested that considerable efficiencies can be gained by making changes in 'procurement practice, contract conditions, tighter restrictions over set-off and the introduction of adjudicators as a normal procedure for settling disputes'. He also concluded that the 'most effective form of contract in modern conditions should include:

- A specific duty for all parties to deal fairly with each other, and with their subcontractors, specialists and suppliers, in an atmosphere of mutual cooperation
- Taking all reasonable steps to avoid changes to pre-planned works information. But, where variations do occur, they should be priced in advance, with provision for independent adjudication if agreement cannot be reached'

and that 'subcontractors should undertake that, in the spirit of teamwork, they will coordinate their activities effectively with each other, and thereby assist the achievement of the main contractor's overall programme. They may need to price for such interface work.'

The conclusions of the review were clearly extensive and led to the formation of the Construction Industry Board (subsequently replaced by the Strategic Forum) and an extensive programme of initiatives including:

- Steps to improve productivity
- Better quality design and improved briefing of designers
- Changes to trade and professional training and education
- Improving the image of the industry
- Encouraging fewer disputes by encouraging partnering between contracting parties
- Improved quality of construction professionals, contractors and subcontractors
- The publication of reports from 12 working groups concerning subjects as diverse as client briefing, education and partnering

Some of the Latham recommendations were included in the Housing Grants, Construction and Regeneration Act 1996.

1.6 The Egan reports

'Rethinking Construction'

'Rethinking Construction' was published in July 1998 (Construction Task Force 1998) and represents the work of a special task force which was set up by the government to identify the scope for improving quality and efficiency in construction. The task force was chaired by Sir John Egan, hence the popular title for the report – the Egan Report.

The Egan Report at 40 pages is certainly not as comprehensive as its predecessor, the Latham Report, but is no less searching and probably considerably more controversial. It contains many 'home truths' but may also be said to contain unfair criticisms, particularly with respect to comparisons with factory-based manufacturing industries, such as the motor industry.

Latham looked at designing an infrastructure for the industry aimed at removing the inefficiencies and inconsistencies, especially in terms of client briefing, better design management and more coherent project strategies. In 'Rethinking Construction', there is no industry 'blueprint' for change but the Construction Task Force, which produced the report, took the lead on a number of new initiatives including:

- Movement for Innovation (known as m4i) – a board of members whose task is to coordinate a number of demonstration projects, to disseminate best practice information and to oversee industry-wide benchmarking
- The Construction Best Practice Programme provides information for firms wanting to improve their performance
- The idea behind Inside UK Enterprise was for top-performing companies to have an 'open day' where other firms could visit and find out how things are done by the 'host' company

The Egan Report has probably had more publicity than the Latham Report and certainly there has been plenty of action as a result of the report, including the introduction of key performance indicators and demonstration projects exemplifying best practice.

The Egan Report undoubtedly recognises both the good and bad in construction and seeks to build on those aspects of the industry which are excellent in a worldwide context. However, on balance, the conclusion of the report is that the industry as a whole is underachieving and that there should be radical change in key areas of its performance. These include quality, productivity, cost and time certainty, and health and safety.

In the Executive Summary, the Egan Report makes the following observations:

- The UK construction industry at its best is excellent. Its capability to deliver the most difficult and innovative projects matches that of any other construction industry in the world.
- Nonetheless, there is deep concern that the industry as a whole is underachieving. It has low profitability and invests too little in capital, research and development and training. Too many of the industry's clients are dissatisfied with its overall performance.

- If the industry is to achieve its full potential, substantial changes in its culture and structure are also required to support improvement. The industry must provide 'decent and safe working conditions and improve management and supervisory skills' at all levels. The industry must design projects for ease of construction, making maximum use of standard components and processes.
- The industry must replace competitive tendering with 'long-term relationships based on clear measurement of performance and sustained improvements in quality and efficiency'.

The Egan Report identified five key drivers of change needed to set the agenda for the industry:

(1)   Committed leadership
(2)   A focus on the customer
(3)   Integrated processes and teams
(4)   A quality-driven agenda
(5)   Commitment to people

Among the year-on-year targets proposed by Egan were:

- 10% reduction in construction time from client approval to practical completion
- 10% increase in productivity
- 20% reduction in the number of reportable accidents
- 10% increase in turnover and profits of construction firms

One of the problems with the Egan Report is that the emphasis is placed on the 'top-end' of the industry, whereas Latham looked at the fundamental problems of the entire industry. So while Egan has led to the development of several good ideas and worthwhile aims, the concepts may take some time to filter down to the lower echelons of the industry.

'Accelerating Change'

This report – the second Egan Report – presents the first year's work of the Strategic Forum for Construction (2002), which replaced the now defunct Construction Industry Board that was set up following the Latham Report.

'Accelerating Change' identifies ways of increasing the pace of change following the recommendations in 'Rethinking Construction', reports on the progress made to date and sets out a strategic direction with targets. The report identifies three main drivers to accelerate change in construction and introduce a culture of continuous improvement in the industry:

- The need for client leadership
- The need for integrated teams and supply chains
- The need to address 'people issues', especially health and safety

The vision and aspirations set out in 'Accelerating Change' emphasise the need for collaboration between the whole supply team, including clients and manufacturers. The report represents a 'manifesto for change' for all involved in construction, including government, schools and further/higher education and professional bodies.

The strategic targets identified in 'Accelerating Change' include:

- By the end of 2004
  - 20% of construction projects by value to be undertaken by integrated teams and supply chains
  - 20% of clients to adopt the principles of the Clients' Charter
  - 10% annual improvement by adopting the Clients' Charter
- By the end of 2007
  - These figures rising to 50%
- By the end of 2006
  - 300 000 qualified people to be recruited to the industry
- By 2007
  - 50% increase in applications for built environment courses
- No later than 2010
  - A certificated fully trained, qualified and competent workforce

The report suggests six key steps that clients must consider when considering whether or not to build:

(1) Verification of need
(2) Assessment of options
(3) Develop procurement strategy
(4) Implement procurement strategy
(5) Project delivery
(6) Post-project review

The effectiveness of post-Egan initiatives is examined by Morton and Ross (2007).

1.7 The National Audit Office Report

'Modernising Construction'

This 103-page report by the National Audit Office (2001) recognises that there are inefficiencies in the traditional methods of procuring and managing major projects and, especially, in the practice of awarding contracts on the basis of lowest price. The report suggests that this does not provide value for monies in terms of either initial costs or through-life and operational costs.

According to the report, 73% of construction projects carried out by government departments and agencies were overbudget and 70% were delivered late. It is also reported that a benchmarking study of 66 central government construction projects (total value of £500 million) carried out in 1999 showed that three-quarters of projects exceeded their budgets by up to 50% and two-thirds exceeded their original completion date by 63%.

The report suggests that a major contributory factor to this poor performance is the adversarial relationships that exist between construction firms, consultants and clients and between contractors, subcontractors and suppliers. It is further suggested that the entire supply chain must be integrated and that risk and value must be managed in order to improve buildability, reduce accidents and drive out waste.

'Modernising Construction' identifies that government departments and agencies have adopted three main strategies for ensuring more collaboration, integration and value for monies in the procurement of construction services, including:

- Private Finance Initiative (projects over £20 million)
- Prime contracting
- Design and build

The report emphasises the benefits of long-term collaborative relationships between clients and contractors (partnering). It is emphasised that partners should be appointed competitively and that a hands-off but eyes-on approach is needed. A key factor in the success of partnering is continuous improvement and open-book accounting where cost and efficiency gains are auditable through access to the contractor's records.

References

Construction Task Force (1998) *Rethinking Construction.* Department of Trade and Industry.

Department for Business, Enterprise and Regulatory Reform (BERR) (2007) *Construction Statistics Annual.* The Stationery Office.

Latham, M., Sir (1994) *Constructing the Team.* HMSO.

Morton, R. & Ross, A. (2007) *Construction UK: Introduction to the Industry.* Blackwell Publishing.

Murray, M. & Langford, D. (2003) *Construction Reports 1944–1998.* Blackwell Publishing.

National Audit Office (2001) *Modernising Construction.* Report by the Comptroller and Auditor General HC 87. The Stationery Office.

Strategic Forum for Construction (2002) *Accelerating Change.* Rethinking Construction.

Walker, A. (2007) *Project Management in Construction,* 5th edn. Blackwell Publishing.

2 Construction contracts

2.1 Contracts and planning

A contract is a means of formalising the relationship between the contracting parties in which the rights and obligations of the parties are agreed and the balance of risk between the parties established. Formal contracts establish how administrative procedures and the serving of formal notices are to be conducted and they establish mechanisms for dealing with contract payments, delays, compensation and disputes. The choice of contract has an important bearing on the planning of a construction project for a number of reasons:

- The contract terms usually dictate what sort of programme the contractor has to provide
- Some contracts specify what information must be included in the programme
- Some contracts specify how the programme is to be presented and how it is to be used as a management control document during the execution of the works

From a financial planning point of view, the form of contract is important because payment terms and payment periods are specified and the retention percentage is stated. During the project, the contract explains procedures for dealing with delay, how contract variations are to be dealt with and what entitlements the parties have to damages for delay.

A wide variety of contracts are used for construction projects, including a large number of standard forms. The best-known main 'families' of standard contracts are the JCT (Joint Contracts Tribunal), ICE (Institution of Civil Engineers) and the ECC (New Engineering Contract (NEC) Engineering and Construction Contract) families, which together produce a bewildering variety of contracts for use in different circumstances. Other 'families' of contracts include:

- The GC/works forms for government contracts
- The PC/works forms for local authorities, National Health Service (NHS) Trusts and the private sector, etc.
- The FIDIC series – the international forms of contract
- The Association of Consultant Architects (ACA) forms

The Architects and Surveyors Institute (ACI), the Joint Council for Landscape Industries (JCLI), the Federation of Property Services, Defence Estates and NHS Estates are other producers of standard contracts.

2.2 Contracts in use

Every two years since 1985, the Royal Institution of Chartered Surveyors (RICS) has commissioned a survey to determine trends in the use of standard forms of contract and methods of procurement in construction. The survey is a representative sample of new build and refurbishment work carried out in the UK but it excludes civil engineering and routine maintenance work.

'Contracts in use' (RICS 2006) found that a standard form of contract was used on 92% of building projects (97% by value of contracts). JCT standard forms were by far the most popular, accounting for 71% of contracts by value. There appears to have been a marked increase in the uptake of the NEC Engineering and Construction Contract for building work (13% by value), which was heavily promoted by the Latham and Egan reports. It appears that the NEC is the most commonly used contract for civil engineering work in the UK and the NEC User Group (www.neccontract.co.uk) now has nearly 500 members. The NEC is also widely used in a number of countries abroad.

The RICS survey indicates that lump sum contracts are popular for smaller projects and that the use of drawings and specification dominates this market. Design and build predominates in the mid-sized project market but traditional bills of quantities have, surprisingly, enjoyed a resurgence in popularity. There has been a significant increase in the use of contracts on the basis of approximate quantities. For larger projects, construction management or some sort of design and build arrangement remains popular but management contracting is reported to be in terminal decline.

Design and build still represents over 40% of the market by value but partnering and target contracts have gained significantly in popularity. Since the last survey, it seems that formal partnering arrangements have gained in popularity and use of the ACA PPC2000 contract has sharply increased. Partnering appears to be most popular in the £2–20 million contract value range and approximately 24% of contracts between £10 and 20 million employed the PPC2000 contract. Figure 2.1 illustrates the relative popularity of different standard forms of contact for building work.

2.3 The Housing Grants, Construction and Regeneration Act 1996

Whilst not entirely devoted to construction, the so-called 'construction act' is recognised as an important landmark for the industry because it contains specific provisions that relate to the way the industry conducts its business. The 'construction' provisions included in the Act are generally credited to the efforts of Sir Michael Latham and others and they impose, for the first time, statutory requirements that relate to the way construction contracts are conducted.

The Act is in five parts and the Latham provisions are in Part II – Construction Contracts under Sections 104–117. This legislation applies to all construction contracts whether main or subcontracts and also to contracts for design or other professional services.

The Act confers certain statutory obligations and entitlements on parties to such construction contracts and where suitable clauses are not present in a contract, they are provided by regulations called the Scheme for Construction Contracts. The scheme represents the 'back

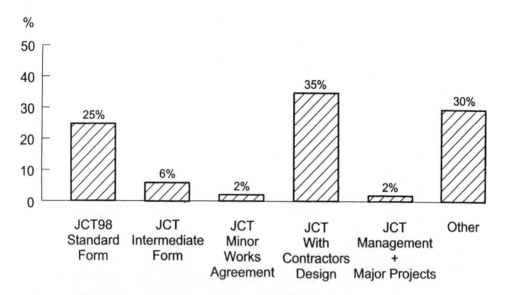

By value of turnover
excluding civil engineering, maintenance and repair work
Source: RICS

CONTRACTS IN USE

Figure 2.1

stop' where perhaps standard conditions of contract are not employed or where contracts are entered into on the basis of an exchange of letters.

The issues covered by the act include:

- The right to refer disputes on construction contracts to immediate adjudication rather than having to wait for arbitration or litigation
- The right to stage payments for work carried out under a construction contract
- The right to proper arrangements for payment including amount, dates and how it was calculated
- The obligation to give notice of intention to withhold payment including how much and why
- The right to suspend performance of the contract pending payment
- The prohibition of 'pay when paid' provisions (e.g. main contractor delaying payment to a subcontractor pending payment from the employer)

Most standard forms of contract have been amended to bring them into line with the act. In particular, the JCT forms were changed to include the right of interest on late payments and the right to suspend work, subject to notice, if payment is not received on time.

2.4 JCT Contracts

JCT 2005 Standard Building Contract

The JCT 2005 Standard Building Contract With Quantities (SBC/Q) is a lump sum contract based on bills of quantities. A remeasurement contract is possible using either the Approximate Quantities (SBC/AQ) or Without Quantities (SBC/XQ) versions. The contract assumes separation of design and construction and the architect undertakes the roles of designer, client's agent and contract administrator.

Partial contractor design is possible using the Contractor's Designed Portion provisions of the contract, in which case the contractor provides an analysis of that part of the contract sum which relates to the contractor's designed portion. Sectional completion provisions are available for completion of the work in phased sections (as opposed to partial handovers at the contractor's discretion).

The architect or contract administrator determines extensions of time under this contract and certifies monthly payments normally on the basis of interim valuations by the client's quantity surveyor who is named in the contract. Interim payments include materials on site and there are complex rules for the valuation of variations which are also paid as the work proceeds. Interest is payable on late payments.

Where the employer/architect wishes to exercise some control over the choice of specialist subcontractors for the project, a short list of named subcontractors may be included in the contract documents, but they can no longer be nominated. Consequently, the final choice of all subcontractors, whether named or otherwise, will be made by the main contractor.

The contractor is to provide a full-time competent person in charge and, when given possession of the site, is to commence the works and proceed regularly and diligently in order to complete on or before the completion date.

The contractor is to supply copies of his master programme but the programme does not impose any additional obligations on the contractor other than those determined by the contract documents. As the programme is not a contract document, failure to observe the programme is not a breach of contract. Under the contract, an information release schedule may be provided stating what information the architect will release and when (this is optional for the client).

In practice, even the best-laid plans rarely work out exactly but the only obligation to submit a revised programme under JCT 05 is where an extension of time is granted under the contract. The architect/contract administrator has no powers to require the contractor to accelerate the works but he may issue a notice of determination if the contractor persistently fails to progress satisfactorily in accordance with the intended programme.

JCT 2005 Intermediate Building Contract

This is a popular form of contract for work which is straightforward, where there are no complex services and where the design is to be carried out by, or on behalf of, the employer. Subcontractors can be named by the contract administrator. In most respects, the contractual provisions are similar to the JCT Standard Building Contract.

There is no contractual requirement for a programme under the contract and, therefore, where the contractor is required to produce a programme for the project, perhaps in a specified format, this should be stated in the bills of quantities, specification or preambles.

JCT 2005 Design and Build Contract

This contract is used for lump sum design and build contracts where the basis of the contract is employer's requirements and contractor's proposals. The contractor is to supply a contract sum analysis of his price. The employer's agent is named in the contract but there is no named quantity surveyor.

Interim payments are based on the contractor's application either at monthly intervals or on completion of prescribed stages of work. Materials on site are not included in interim valuations in circumstances where certificates are based on stage payments.

Unlike the Standard Building Contract, the Design and Build form contains no provision for a master programme. However, provision is made for determination of the contractor's employment under the contract for failure to proceed regularly and diligently with the works.

It would therefore be prudent for the client's advisors to have a copy of the contractor's up-to-date programme so as to monitor performance. This could be provided for in the contract preliminaries or in the instructions to tenderers. Alternatively, a requirement for a programme could be included in the employer's requirements which are incorporated in Article 4 of the articles of Agreement to the Conditions of Contract.

The contractor's programme may well be an important document as payment in design and build may be linked either to construction stages or to activities on the programme.

JCT 2005 Major Project Contract

This relatively new standard form of contract represents a significant departure from conventional JCT thinking and contains a number of novel features more reminiscent of the NEC style of contracts. It is, for instance, written in a much shorter and simpler style and avoids the extensive procedural provisions associated with conventional JCT contracts.

As its name suggests, the Major Project Construction Contract is intended for use with experienced clients, contractors and subcontractors on large commercial projects where the client would otherwise wish to draft a bespoke contract or extensively amend an existing standard form. 'Major Project' is not defined, but it seems the intention is to exclude 'run of the mill' projects for inexperienced clients.

The contractor undertakes more risk and responsibility than under conventional JCT contracts and it is essential that the contractor is well versed in the principles and application of risk management. The contract assumes that the client will have in-house procedures and protocols dealing with day-to-day matters such as instructions, record keeping, delays, updating the programme and health and safety monitoring that will be familiar to the contractor. Such detailed procedures, familiar to users of the main JCT forms of contract, have been much simplified in the Major Project Construction Contract and a certain degree of reliance is placed on the contractor having worked for the client on previous projects.

Some of the key features of the contract include:

- The employer describes the project in requirements
- Consultants, subcontractors and specialists may be named in the employer's requirements, but there is no provision for nomination
- Certain provisions of the contract may be exercised by funders, tenants or purchasers via the third-party rights schedule
- The contractor's proposals show how he intends to meet the requirements
- The pricing document sets out the means by which the contractor will be paid
- The contractor undertakes any further design beyond that given in the requirements
- The contract includes a design submission procedure
- Payments for work in progress may be by interim valuation, stage payments or may be related to progress or, alternatively, may be another method (e.g. by S curve or formula)
- There is no provision for deduction of retention from interim payments
- The contractor undertakes the roles of construction design and management coordinator and principal contractor under the Construction (Design and Management) Regulations 2007
- The contractor is given access to the site rather than possession of it as is conventionally the case on the assumption that the client may require others to carry out work on the site
- Unexpected ground conditions are dealt with in a similar manner to Clause 12 in the ICE Conditions of Contract.
- The contractor is encouraged to suggest changes in order to make cost savings or value improvements
- There is no requirement for the contractor to provide a master programme
- Provisions are included to allow acceleration of the project by agreement between the parties
- Provision is made for bonus payments to be made for early completion
- The project may be carried out in stages without the need for a sectional completion supplement

The Major Project Construction Contract has been drafted by a firm of consultants, rather than the JCT itself, under the direction of a steering group mainly representing the Construction Confederation and the British Property Federation. It is a much shorter form of contract than the JCT 2005 Design and Build Contract, which it is likely to replace in due course.

JCT 2005 Construction Management Agreement

Under this standard form, the contract is between the client and the construction manager who is effectively on the client's team and is intended to manage the project on the client's behalf. The client appoints a consultant team and consultant team leader and the construction manager has a limited role as client's agent with limited powers to spend money. The construction manager has a contractual obligation to consult with the consultant team on relevant matters.

The works are carried out by trade contractors, directly engaged by the employer, under the Construction Management Trade Contract.

The construction manager's obligations include:

- Recommend economies in buildability, methodology, cost and time during design development
- Prepare the project programme showing critical path, lead times and key milestones agreed with the client
- Prepare a detailed week-by-week programme listing of all trade contracts and showing how the project is to be achieved
- Prepare a tender events schedule for each trade element
- Prepare trade contractors' detailed programmes and any acceleration proposals
- Prepare drawing and information release schedules showing procurement periods
- Advise on the division of the project into trade contracts
- Prepare and agree a suitable tender list and tendering arrangements
- Interview all tenderers before and during tender period
- Analyse tenders and recommend contractors
- Advise on the pre-ordering of materials and plant
- Submit monthly progress reports to the client
- Instruct, direct and manage trade contractors
- Check applications for payment from trade contractors and prepare valuations as necessary
- Make recommendations for payment
- Manage and coordinate the work of trade contractors
- Expand, update and adapt the project programme
- Monitor and report on actual expenditure against the project cost plan
- Organise and manage the site
- Maintain complete and accurate records

JCT 2005 Management Building Contract

Under these conditions, the contractor tenders on the basis of a management fee and works contracts are placed later for various 'packages' as and when the design is sufficiently developed. The works contractors are engaged and managed by the management contractor under the Management Works Contract.

The intention of this method of procurement is for the client to appoint professional advisors for the architectural and engineering design and quantity surveying (the professional team) and to engage a contractor in a 'professional' capacity to advise on design, technical and buildability issues.

The contract requires the management contractor to cooperate with the professional team during the design stages and in the planning, programming and cost estimating for the project. This involves the contractor in preparing the project programme which has to be agreed on by the client's professional team.

Additionally, the management contractor is required to prepare all necessary programmes for the execution of the project, which includes the preparation of a detailed construction programme. The management contractor's duties include maintaining and regularly updating the detailed construction programme.

2.5 ICE Conditions

The ICE Conditions of Contract 7th Edition

This form of contract is used for works of a civil engineering nature or where building work is a minor part of the project. The contractor is required to submit his programme for the approval of the engineer, showing the order in which he intends to proceed with the works. This programme is commonly referred to as the Clause 14 Programme as it is required under Clause 14(1) of the contract.

In addition to this obligation, the contractor is required to submit a written description of his proposed arrangements and methods for the project and, if requested by the engineer, detailed information on construction methods, temporary works and contractor's equipment. Additionally, the contractor is required to explain his intended methods of working and resourcing of the works to the reasonable satisfaction of the engineer, through Clauses 13 and 14.

Submission of this information puts the engineer under an obligation to respond within a limited time by either accepting or rejecting the contractor's proposals or requesting him to submit further details. The contractor's programme is not usually a contract document in civil engineering but, nevertheless, far greater emphasis is placed on the programme than in building works.

While a programme is not a contract document in its own right, the submission of a programme is a contractual requirement under Clause 14. The engineer plays a significant role in its final production and he has the power to reject the proposed programme until such time as he is satisfied with the contractor's order of working.

Under ICE7, an up-to-date programme is a contractual requirement and Clause 14(4) entitles the engineer to ask the contractor to submit a revised programme where actual progress is not as intended. Under Clause 46(1), if the engineer considers that the contractor is not progressing in accordance with the programme, he is entitled to ask the contractor to expedite or speed up his work so as to complete the works on time. This does not, however, give the engineer the power to ask the contractor to accelerate the works so as to finish earlier than the contract completion date.

The provision for acceleration in Clause 46(3) does not have the same meaning as 'expedite'. Acceleration provides a means for the client (employer) to ask the contractor to finish the job before the contract completion date (or the extended date) when perhaps an opening or commissioning deadline is important. In such circumstances, the parties can agree on special arrangements for earlier completion where mutually acceptable, but such an agreement will be outside the original contract.

The ICE Design and Construct Conditions of Contract 2nd Edition

This contract is used for civil engineering works where the contractor is to design the works.

To all intents and purposes the contract is similar to the 7th Edition, with the obvious exceptions that apply to design and build arrangements. For example, the contract price is based on the employer's requirements and the contractor's submission. There is no engineer to the contract but there is an employer's representative instead.

The contractual obligation to provide a programme is similar to the 7th Edition.

2.6 The Engineering and Construction Contract (NEC)

The NEC was first published in 1991 and is now in its third edition following initial amendments recommended by the Latham Review and, more recently, by user feedback. The NEC was considered by the Latham Report to be capable of being a common contract for the whole industry both in the public and in private sectors, but whilst the contract is gaining in popularity, this aspiration is a long way from fruition.

The main form of contract is the ECC within the NEC family of documents, which also includes standard forms of subcontract and professional services contracts. The fundamental concept of the NEC is a contract designed to be flexible in its use and applicable to a variety of types of construction work including:

- Building
- Civil engineering
- Process and other engineering
- Refurbishment work
- Maintenance work

The contract can be used for both major and minor works and its flexibility extends to employing a variety of contractual arrangements with different types of project documentation.

A project can be set up on the basis of the main options shown in Table 2.1.

There is no architect or engineer under the ECC. Instead, these traditional appointments are separated into the functional roles of project manager, supervisor and adjudicator although, of course, they could be undertaken by one individual if preferred. The flexibility of this approach can be exemplified, for example, where a construction management approach is required. Here, the client would enter into direct contracts with each of the works contractors using the main

Table 2.1 Main options of ECC.

Option	Brief description
A	A priced contract using an activity schedule (a list of activities linked to the programme)
B	A priced contract using bills of quantities (e.g. SMM7 or CESMM)
C	A target contract using an activity schedule
D	A target contract using a bill of quantities
E	A cost-reimbursable contract using schedules of actual cost (similar in principle to daywork)
F	A management contract (with works package contractors engaged directly by the management contractor)

ECC form and appoint the management contractor as 'project manager' under the same form of contract.

One of the basic concepts of the NEC is that any contract requires certain fundamental clauses, such as provision for time, quality, payment and dispute resolution. Other clauses, on the other hand, can be considered as 'optional extras'. These might deal with liquidated damages, bonds and guarantees, retentions, early completion bonuses, etc. Consequently, the ECC has 9 'core clauses' which would be in every contract, and 15 'secondary option clauses' (plus a partnering option) from which to choose to bespoke the contract to the client's requirements. The flexible approach offered by the NEC facilitates consultant or contractor design and also contracts based on conventional lump sum, remeasurement, management arrangements and target cost or other reimbursable contracts.

The programme is an important document under this form of contract and there are contractual obligations placed on the contractor under Clause 30 prescribing information to be shown on the programme and to make sure it is kept up-to-date. The programme must show:

- Critical dates
- Operations by the contractor and others
- Equipment and resources for each operation
- Order and timing of the works
- Float and time risk allowances
- Health and safety requirements
- Other general information and dates

2.7 ACA Standard Form of Contract for Project Partnering PPC2000

This form of contract has been specifically designed by the Association of Consultant Architects for use on projects where a partnering arrangement is intended. It may be used in conjunction with a partnering charter, which is an agreed statement of values, goals and priorities signed by the partnering team members. The key features of the contract are:

- A partnering team who are the signatories to the partnering contract
- A contractual undertaking by the partnering team to work together in a spirit of trust, fairness and mutual cooperation to the benefit of the project
- A client representative acting as fully authorised client's agent (except membership of the core group)
- A partnering advisor
- A core group, drawn from the partnering team, which meets regularly to arrive at consensus decisions which are binding on the partnering team
- A list of six partnering objectives and ten partnering targets
- An early warning system whereby partnering team members notify others of potential problems
- A partnering timetable setting out the activities of the partnering team prior to commencement on site
- A project timetable for implementing the project post-commencement agreement

- Implementation of an open-book approach to supply chain relationships
- Risk management exercises aimed at identifying, eliminating or reducing risks and their costs and apportioning risks to the partnering team members best able to manage them
- Sixteen provisions enabling the constructor to be granted extensions of time subject to his best endeavours to minimise delay and increased cost
- Monthly payment intervals, unless stated otherwise
- Payments subject to valuation by the client's representative
- Payment 15 days from valuation or 10 days from receipt of VAT invoice
- Payment under a pre-possession agreement for work done before commencement
- Payment in accordance with the price framework subject to an agreed maximum price

Reference

RICS (2006) *Contracts in Use.* Royal Institution of Chartered Surveyors.

3 Procurement methods

3.1 Procurement criteria

The construction industry attracts a wide variety of clients all of whom will have their own objectives and priorities for their particular project. Consequently, when a client is considering the decision to build, a number of important decisions need to be made in order for the project to have a good chance of a successful outcome. The CIB (1997a) suggests developing a strategy for the project on the basis of the client's needs and that the first priority is to decide how to procure the skills and resources needed for the design and construction of the project. Turner (1997) suggests that an appropriate procurement route will develop from an understanding of the objectives and priorities of the client and that a number of procurement assessment criteria will need to be considered.

These criteria were originally noted in the NEDO report 'Thinking about Building' (NEDO 1985) which suggested eight procurement factors to be considered by the client at the procurement assessment stage. Of these, the three main considerations are:

- Time – the overall timing of the project from inception to completion
- Quality – the client's required standards of design and workmanship as expressed in the specification
- Cost – certainty of the market price at the bid submission and final account stages

Five other client considerations were also highlighted as being influential on the choice of procurement route. These are:

- Complexity – Complexity of building design, layout and services provision
- Controllable variation – How sure is the client of his requirements? The cost of variations to the contract may prove difficult to agree and assess
- Degree of competition – Does the client wish to create competition at the design and construction stages of the project?
- Client responsibility – Does the client wish to be directly involved in decision-making during the project or does he simply want to return from the South of France and cut the tape on occupation of the building?
- Risk in the project – Commercial risk, occupation risk, design and construction risk

Taking these factors into account, decisions will have to be made in relation to:

- Choosing an appropriate procurement strategy for managing the project
- Selecting the client's principal advisor or lead consultant, that is architect, engineer, project manager, etc.
- The appointment of other consultants
- Establishing the client's brief and assessing the project's feasibility
- Obtaining the required project finance
- Choosing how to procure and manage the design and construction
- Establishing a realistic time period for the overall project

A detailed overview of procurement assessment criteria is outlined by Turner (1997) and Seeley (1997) and both Morledge et al. (2006) and Masterman (2002) discuss a variety of procurement strategies for different clients and different circumstances. Morledge notes that most projects are needed within a time frame or by a specific date, but this might mean that the client may want the project to be:

- completed as quickly as possible (because time is money)
- completed by a specific date (that must be achieved)
- completed on time once it commences (with no time slippage)

The choice of procurement route for each of these situations will be quite different notwithstanding the additional complexity of other criteria that might be important to the client (such as price certainty, quality, etc.).

3.2 The design process

The procurement of design services is a crucial consideration when choosing the procurement route for a project. Decisions have to be made such as who will prepare the design, who will bear the design risk and when, if at all, the contractor will be involved in the design. The design process for construction projects is complex and evolves through several stages and iterations with numerous different professional disciplines involved.

The RIBA Outline Plan of Work (RIBA 2000) is a model of the design process which explains in detail what happens at each stage and how the various engineers, the quantity surveyor and others contribute. An overview of the Plan of Work can be seen in Figure 3.1. This shows the work stages that the architect follows from initial briefing through to completion of the project using a traditional procurement route. The work stages are broadly classified under three main headings:

- Feasibility – stages A–B
- Pre-construction – stages C–H
- Construction – stages J–L

The Plan of Work indicates that the architect has two roles – design/design leader and contract administrator – and also that the design will be complete before tenders are invited

RIBA OUTLINE PLAN OF WORK
TRADITIONAL CONTRACT

Procedures undertaken by an architect
during the designing and administration of a building project

STAGE	COMMENTS	
A – APPRAISAL	Confirmation of client's requirements, identify procedures. Engagement of cousultant and select procurement route.	**FEASIBILITY**
B – STRATEGIC BRIEFING	Preparation of client's brief. Confirm key requirements. Identify procedures. Organise structure of design team.	
C – OUTLINE PROPOSALS	Develop full brief. Prepare outline proposals. Estimate of cost. Review procurement route.	**PRE-CONSTRUCTION PERIOD**
D – DETAILED PROPOSALS	Complete brief. Prepare detailed proposals. Apply for development control approval.	
E – FINAL PROPOSALS	Preparation of final proposals, scheme plans, elevations, Coordination of all components and elements,	
F – PRODUCTION INFORMATIOAN	Preparation of production information to enable tender documents to be prepared. Application for statutory approvals.	
G – TENDER DOCUMENTATION	Preparation of tender documentation. Tenders to be invited for project.	
H – TENDER ACTION	Evaluation of tenders received. Tender comparisons. Appraising tenders and recommendation to client.	
J – MOBILISATIOAN	Award of building contract, appoint contractor. Issue production information. Occupation of site by contractor.	**CONSTRUCTION PERIOD**
K – CONSTRUCTION STAGE	Administration of contract to practical completion stage. Provision of contract information during project.	
L – AFTER COMPLETION	Administration of contract after completion. Final inspections, settle final account.	

Figure 3.1

at Stage H. The stages where the architect's main design input is required are shown below:

- Stage C – Appraise design options and develop the client's brief
- Stage D – Prepare the site plan, building layouts and schematics

- Stage E – Develop the design with floor plans, elevations, sections and external works layout
- Stage F – Complete the design including construction details, schedules and specifications

When the client is considering if, or when, to involve the contractor in the design, and to what extent, he will have to judge the most suitable stage for this to happen bearing in mind his own requirements for the project in hand and the degree of risk that he is most comfortable with.

It should be noted that the RIBA Plan of Work does not completely represent reality because it is rare to have a fully developed design at tender stage in a traditional lump sum contract. In practice, design Stage F often overlaps construction and it is also quite common on traditional contracts for the contractor to be asked to design part of the permanent works. This option is available under the JCT 2005 Standard Building Contract (Contractor's Designed Portion) and under ICE7 Conditions (Clauses 7 and 8 refer).

A model of the design process for design and build procurement is also included in the RIBA Outline Plan of Work. The work stages are the same as for a traditional procurement route but provisions are included for the development of the employer's requirements (Part 3) and the contractor's proposals (Part 4).

3.3 Procurement options

The procurement of a construction project requires a balance between time, quality and cost constraints. A critical eye needs to be kept on the client's requirements and the forecast budget and this necessitates effective briefing, design and cost control in relation to the final choice of procurement route.

Procurement methodology concerns not only the contractual arrangement for the construction work but also the sourcing of professional services for design, project administration and health and safety management. Turner (1997) considers who should carry the risk and how the right choice of procurement route can ensure that the risk is allocated to whichever party is best placed to manage it.

Procurement methods are constantly under review and whilst there may be a significant trend towards design and build arrangements (both client-led and contractor-led design), other procurement options and partnering arrangements are coming to the fore. For large contractors operating in the public sector, the decision to tender for Private Finance Initiative (PFI) contracts requires major decisions at senior management level because of the tendering costs and risks involved.

Common procurement options available to clients are outlined in the rest of this chapter.

3.4 Traditional procurement

Introduction

Despite procurement trends and fashions, traditional procurement routes remain popular and, in many cases, the client's lead consultant will be an architect/designer. The popularity of traditional procurement is shown in Figure 3.2, which indicates a current market share of 37% based on turnover.

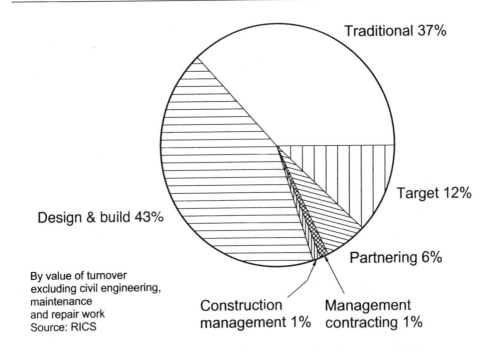

Traditional 37%

Target 12%

Design & build 43%

Partnering 6%

By value of turnover
excluding civil engineering,
maintenance
and repair work
Source: RICS

Construction
management 1%

Management
contracting 1%

POPULAR PROCUREMENT METHODS

Figure 3.2

Sometimes referred to as the 'end-on' or 'design-bid-build' method of procurement, the traditional route separates design from construction with an intervening 'tendering' period. For price certainty, the project should be fully designed before tender but this is frequently not the case. A variety of contractual arrangements are available with traditional procurement:

- a lump sum contract based on:
  - drawings and specification where the builder prepares the quantities
  - full documentation with bills of quantities
- a remeasurement contract based on approximate/notional quantities
  - the contract price is determined at the end of the contract by measuring the actual work carried out.

The traditional method of procurement may be accelerated by using a **two-stage tender approach** which enables the preferred contractor to be selected at an earlier stage and thereby be in a position to influence the cost efficiency and practicality of the design. Alternatively, the contract may be **negotiated** with a single contractor, or a small select list of firms, favoured by the client or architect. Two-stage tendering and negotiation are not limited to traditional procurement and can equally well be used with other methods including design and build.

Traditional projects are usually architect-led and the procedures followed are modelled in the RIBA Outline Plan of Work (RIBA 2000). As an alternative to an architect-led project, a project manager may be engaged as the client's representative who then manages the design consultants. The project manager may then decide to follow the project development procedures recommended in the CIOB Code of Practice for Project Management (CIOB 2002). In any event, the lead consultant will interpret and develop the client's brief, prepare cost and feasibility studies, arrange for scheme drawings to be finalised and organise the tenders.

Forms of contract

For traditional procurement, the JCT Standard Building Contract (with or without quantities) may be used but other forms such as the JCT Intermediate Building Contract or the JCT Minor Works Building Contract may be applicable for less complex or small projects. The Engineering and Construction Contract (ECC) may also be considered, with the client representative acting as 'project manager', that is both lead designer and contract administrator. For civil engineering works, the ICE Conditions would normally be used with the engineer in the role of client representative.

An alternative is to appoint a project manager to represent the client as lead consultant, with the design team, including the architect, under his direction. This would require careful consideration, especially in terms of the form of contract used, as only the ECC gives the opportunity to separate the lead design and contract supervisor roles in the contract. Other contracts would need to be amended because, for instance, the JCT Standard Building Contract and the Intermediate Building Contract both refer to the 'architect/contract administrator' and there is no provision for the separation of the two roles.

The JCT Standard Building Contract has provisions which cater for variations, extensions of time and loss and expense, but it is arguable whether contractors recover the true cost of delay and disruption to the work should this happen. These difficulties are usually handled by most contractors with good humour and a ready acceptance of a reasonable reimbursement, but frustration sets in when the client refuses to agree or pay for the extra work. Litigation often results and brings to an end a reasonable relationship which the parties worked hard to establish during the project.

Project relationships

Figure 3.3 illustrates the relationships between the client, client's design team and the main contractor in traditional procurement. The architect is normally regarded as the leader of the design team for building projects and he or she coordinates the input of specialist designers and cost consultants etc. as well as undertaking duties regarding the administration of the client's contract with the builder. The architect may be assisted by a clerk of works who acts as his 'eyes and ears' on site to make sure the design and specification are followed by the contractor.

Figure 3.4 indicates the relationships between the parties on a civil engineering project using the ICE Conditions of Contract 7th Edition. The client's lead consultant is normally a consulting engineer who will be responsible for coordinating the design and construction process. Works on site are usually under the direction of a resident engineer assisted by an inspector of works to ensure quality and workmanship.

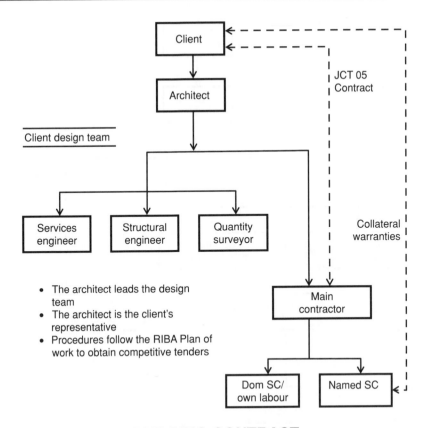

BUILDING CONTRACT
(JCT 05 with Quantities)
TRADITIONAL PROCUREMENT RELATIONSHIPS

Figure 3.3

Advantages of traditional procurement

Despite the criticisms, traditional procurement has the advantage of price competition from competing contractors at the tender stage and should work successfully provided that:

- The design is complete before the tender stage (ensuring price certainty for the client)
- The designer understands how the building 'fits together', ensuring buildability
- The design does not substantially change during construction, so contract variations are kept to a minimum

Summary

- Client knows the lump sum cost before being committed
- Firm contractual date for completion

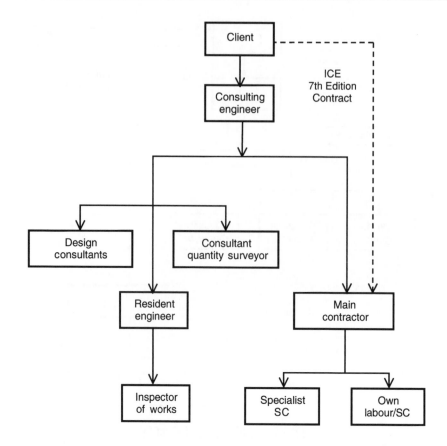

CIVIL ENGINEERING PROJECT
TRADITIONAL PROCUREMENT RELATIONSHIPS

Figure 3.4

- Client retains control over the design team and quality can be assured
- Capable of obtaining the best contract price for the full scope of the works
- Variations and contract changes are relatively easy to handle (but often not so easy to agree!)

Disadvantages of traditional procurement

At its best, traditional procurement is ideal for many clients, but all too frequently:

- The end product is not finished on time
- The budget is overspent as the final account far exceeds the client's expectation
- Disputes arise
- Adjudication, arbitration or litigation results with the only winner being the legal profession

Summary

- Commitment to a lump sum price is often undermined by a lack of information or client changes
- Tenders can only be finalised when the design is complete, which may lead to an extended programme
- Responsibility for specialists lies with the client
- All design risk is ultimately carried by the client
- High pre-contract design fees likely

3.5 Design and build

Introduction

The development of design and build procurement is well documented (Janssens 1991, Masterman 2002, Morledge et al. 2006, etc.) and has grown in popularity since the early 1970s to a market share of some 43% (see Figure 3.2). Whilst there are several varieties of design and build, the basic idea is that there is a single point of responsibility for both the design and construction of the project. The client benefits from the contractor's expertise early in the project and there is also a common law fitness for purpose liability for contractor's design. This benefit is often lost under some standard contract conditions which limit the contractor's design liability to that of an architect (i.e. reasonable skill and care).

Research (Gidado and Arshi 2004) suggests that there may be as many as six versions of design and build but new ideas are never in short supply in the construction industry. For instance, Rawlinson (2008) reports on the recent successful use of develop and construct with a two-stage tender and partnering arrangement, a novated design, open-book accounting and with the risk and value benefits of a collaborative approach – all on the same project! Two extremes will be considered here in relation to design and build procurement, Extreme A and Extreme Z, and a brief explanation of other design and build variants will then follow.

Extreme A – client-led design and build

With Extreme A, the client can be fully involved in influencing the design development. Prior to tender action (Work Stage H), it may be that the design will be virtually complete possibly with full bills of quantities or notional bills. There may also be prescriptive specifications and possibly performance specifications for specialist installations such as heating, ventilating and air conditioning. At this stage, a small number of contractors will be invited to tender for the project. Here, the design risk is taken fully by the client because tenderers will have little to add to the fundamental design.

The client will be responsible for all design fees from commencement to completion of the project. The architect may be appointed lead consultant or the client may decide to appoint a project manager in the lead consultant's role. Figure 3.5 illustrates the possible relationships between the client, design team and contractor.

This method is referred to as **develop and construct** because the contractor inherits the design and develops the detail required for the production stage of the project. It is preferable

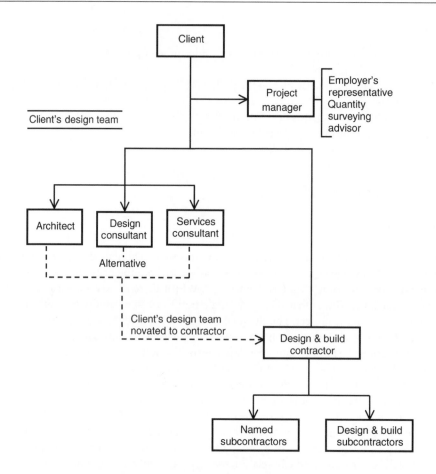

DESIGN AND BUILD
CLIENT-LED DESIGN RELATIONSHIP

Figure 3.5

to invite tenders at RIBA Plan of Work Stage D because this way the client retains some of his control over the design but also benefits from imaginative design submissions from contractors.

An alternative twist on develop and construct, which is popular with many clients, is to novate the architect or design team to the contractor once the contract has been awarded. This variant is known as **novation design and build** because the client's architect, and maybe other members of the design team, are legally passed over to the contractor in order to produce the detailed aspects of the design. The novation arrangement should be agreed with the architect when he is first appointed.

In this way, the client will maintain an interest in the design but the contractor will pay the continuing design fees through to contract completion and become responsible for the entire design. Where novation is chosen, it is advisable that the client's project manager has a consultant quantity surveyor available to give cost management advice. In some cases, once practical completion has been achieved, the architect will revert back to work for the client in order to undertake snagging and defects correction duties.

Extreme Z – contractor-led design and build

In Extreme Z, the client may wish to provide the contractor with minimal information in the form of an outline brief (RIBA Stage B – Strategic Briefing). This will leave the design and build contractor fully responsible for the conceptual and detailed design in order to meet the employer's requirements.

The contractor will then submit a bid on the basis of the Contractor's Proposals. Full design responsibility will be taken by the contractor to produce a building in respect of the client's time, quality and cost requirements.

This variant is known as **traditional design and build** and Figure 3.6 illustrates the relationship between the parties. The contractor may provide an in-house design facility or independent design teams may be used. With a team arrangement, it is normal for the design and build contractor to employ a design team coordinator as a key member of the team to ensure the flow of information between the design team and the project team in order that key design and construction dates are adhered to.

Other design and build options (somewhere between extremes A and Z)

It is most probable when using design and build procurement that procedures will fall somewhere between the two extremes described above. Definitions of alternative arrangements are outlined by Janssens (1991) (see Table 3.1), Masterman (2002) and Morledge et al. (2006) and these indicate the wide range of options available.

The main difference between the options in Table 3.1 relates to the proportion of design work undertaken by the employer's consultants. Consequently, the design may simply be a conceptual design or something more developed. The choice will largely be governed by the degree of control that the client wishes to have over the design and specification. More client control will probably mean less contractor design input and therefore less buildability and more cost. It is this design that is included in the tender enquiry to the contractor from which the contractor's proposals are developed.

Design and manage is very similar to management contracting except that both the construction 'works contractors' and the design team are engaged directly by the contractor. The design–manage contractor is often selected in competition usually on the basis of a design and management fee and, possibly, the contract preliminaries. Since there is very little firm information at tender stage, the client has very little control over eventual design quality. As the contractor's design emerges, packages of work can be prepared and prices obtained under the scrutiny of the client's cost advisor. The contractor undertakes the design risk in this method. **Design, manage and construct** is a similar method where the contractor undertakes some of the work packages as well as coordinating the other package contractors on site.

One of the very first methods of design and build was the **package deal** where contractors offered standard building designs which were then adapted to suit individual clients and site

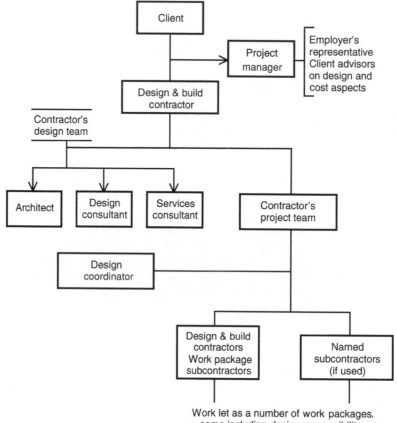

DESIGN AND BUILD
CONTRACTOR–LED DESIGN

Figure 3.6

Table 3.1 Design and build arrangements.

Extreme	Variety
A Employer-led design	Develop and construct Design and build (single-stage tender)
Z Contractor-led design	Design and build (two-stage tender) Negotiated design and build Design and manage Turnkey

conditions. Some of the designs were system buildings. This method of design and build still suits some clients. Package deals are sometimes called **turnkey contracts,** but true turnkey projects provide the client with a fully functioning building, including the provision of staff working inside, and the method is mainly used for overseas projects.

Forms of contract

A number of standard forms of contract may be used for design and build procurement but the choice will depend on the extent of contractor design required. For instance, where only partial contractor design is wanted, the Contractor's Designed Portion provisions (Clause 2.2) of the JCT 2005 Standard Building Contract may be used or, for smaller projects, the JCT 2005 Intermediate and Minor Works Building Contracts have similar provisions. For civil engineering works, Clauses 7 and 8 of the ICE Conditions of Contract 7th Edition would apply for partial contractor design of the permanent works.

For full contractor design, the JCT 2005 Design and Build Contract and the ICE Design and Construct Conditions 2nd Edition are available. Alternatively, the ECC is flexible enough to allow for contractor design.

Advantages of design and build

- Price certainty is secured early in the project
- Contractual completion dates are fixed early in the design process
- Less risk of price changes during the design development
- Main contractor provides single-point responsibility for design and construction
- Major risks lie with the contractor

Disadvantages of design and build

- The contractor assumes greater financial risk and this is often reflected in the price
- Difficulties arise for the employer in matching like-for-like prices at tender stage. For example, the client's requirements may have indicated that he or she wanted a lemon tree but some contractors priced for an orange tree and others simply priced for a tree!
- The tender period and negotiation tends to be much longer
- It is more problematic to control design and quality
- Changes can prove to be expensive and disruptive to the contractor
- The project becomes price driven at the expense of quality

3.6 Management contracting

As an alternative to the traditional approach, the client may decide to appoint a management contractor to engage and manage a number of work package subcontractors.

Management contracting enables the client to create a competitive situation between management contractors at the appointment stage of the project. For instance, four management contractors may be invited to submit quotations for managing the project and would then present their proposals to the client and his design team. It is important that the design team feel that they can 'gel' with the management contractor and work as a project team. The presentations would indicate the management contractor's approach to:

- The programme
- The procurement of work packages
- The control of information flow
- Buildabilty considerations

Bids are based on the contractor's assessment of management and preliminaries costs associated with managing the work packages. Further competition is created during the letting of the work packages as each work package subcontractor bids competitively to obtain the work. Certain work packages may contain a design element. Figure 3.7 illustrates the relationships between the parties involved in the project.

The work package approach allows maximum overlap between design and construction activities. As each section of work is designed, the work package subcontractor may commence the works on site. This tends to result in a reduced overall contract period. If only it worked consistently in practice, it would be wonderful.

Strict control of the work package budget is essential for the success of the project and many prestigious projects have shown the tendency to overspend. Problems in practice have also resulted from the management contractor failing to pay the work package subcontractors in due time. When valuation payments are sometimes in the order of £10 million per month, a 1-week delay in paying the subcontractors will certainly help to improve the management contractor's cash flow. Management contractors have even been known to use the 'payment threat' as a means of getting the work package subcontractor to perform.

3.7 Construction management

Figure 3.8 illustrates the relationship between the parties in construction management procurement. In this situation, the construction management consultant or construction manager joins the design team early in the project.

The construction manager's services are based on a negotiated fee with the client simply to supervise and plan the work to be undertaken by the work package contractors. Normally, this method of procurement is reserved for major projects but on a small factory/office block project, the construction manager could simply be a self-employed person offering a management service to supervise and coordinate a number of subcontractors.

Each work package contractor has a separate contract with the client and is paid directly by the client each month.

Advantages of construction management and management contracting

- Maximum overlap between design and construction. As each package design is completed work may commence on the project

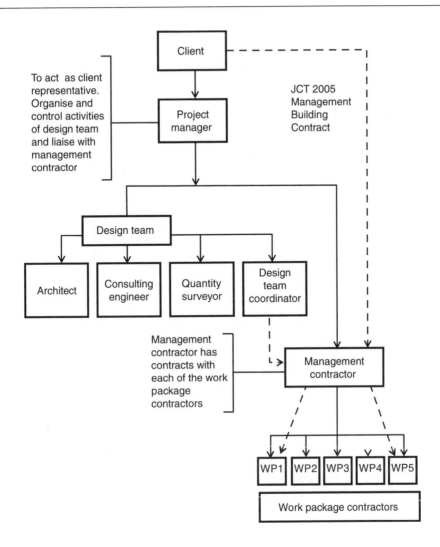

**MANAGEMENT CONTRACTING
ARRANGEMENT**

Figure 3.7

- Quality can be assured – a high level of supervision and quality control can be achieved at site level
- The construction expertise of the management contractor aids buildability and value engineering
- Variations can be kept under control as the price of changes can be agreed before commencing a variation

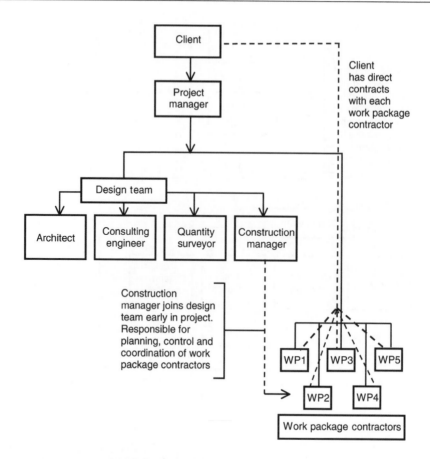

CONSTRUCTION MANAGEMENT ARRANGEMENT

Figure 3.8

- The contractor takes a less adversarial approach, being part of the client's team (he has learnt all the 'dodges' by practical experience)
- Suits complex projects where the design can be developed in stages

Disadvantages of construction management and management contracting

- The client has no commitment from the contractor on price certainty. Strict control of the work package budget is essential
- Total cost of the project is not usually known until the project is well into the construction programme
- Blue-chip work package subcontractors are often chosen with no incentive to reduce costs
- Early letting of packages may lead to extensive design changes

- Damages for delay are difficult to pin on one subcontractor and they are expensive to negotiate
- The client takes all the risk, particularly in the construction management arrangement

3.8 The role of the project manager

Definitions

Walker (2007) suggests that:

- Management is the dynamic input that makes the organisation work
- Organisation is the pattern of the interrelationships, authority and responsibility that is established between the contributors to it

More pragmatically, the Chartered Institute of Building Code of Practice for Project Management (CIOB 2002) provides clients and all members of the construction team with a comprehensive management strategy for any project from inception to completion.

The Code defines project management as 'the overall planning, control and coordination of a project from inception to completion aimed at meeting a client's requirements in order to produce a functionally and financially viable project that will be completed on time within authorised cost and to the required quality standards'. Time, cost and quality control are the keystones to the success of a project and many project managers find that this is a difficult balance to achieve.

Under the direction and supervision of a project manager, management of the project becomes a team effort, as shown in Figure 3.9. The duties of the project manager, and his formal agreement and conditions of engagement with the client, are indicated in the CIOB Code.

Walker suggests that the client's project manager may be classed as an 'executive' or 'non-executive' project manager. The **non-executive** project manager is effectively a team coordinator who may have another role within the team (such as design). The **executive** project manager, on the other hand, is independent from the other members of the team and is purely concerned with managing their activities.

Confusion often arises in the use of the title 'project manager' because the contractor may also use the title 'project manager'. This title is used to describe his senior construction representative on site and this often implies that he or she is a 'super-duper' site agent.

Confusion also arises with the use of the term 'client's agent', which may imply either an executive or non-executive role.

Project management stages

Figure 3.10 illustrates the main stages involved during a project from inception to completion. This process somewhat mirrors the RIBA Plan of Work where the architect would take the role of lead consultant (non-executive).

Figure 3.11 is based on the project management handbook of a local authority which has established procedures for taking a project management approach to managing new build

PROJECT MANAGEMENT – TEAM STRUCTURE

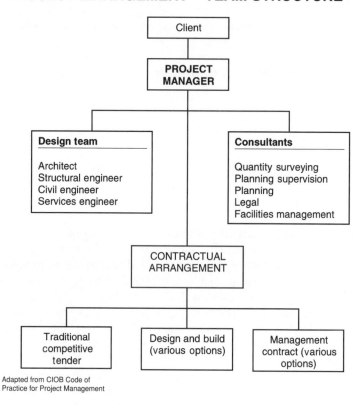

Adapted from CIOB Code of
Practice for Project Management

Figure 3.9

projects. This model has proved extremely successful with the authority due to the simplicity of its application to projects.

3.9 Private finance initiative

PFI is an arrangement where public sector assets and services are acquired through private sector funding with the idea of reducing government/public sector borrowing. PFI is one of several types of public–private partnerships and typical projects undertaken include roads, bridges, schools, hospitals and prisons. Projects are typically over £20 million in value.

Once the public sector sponsor has established a business case strategy, the project will be advertised in the *Official Journal of the European Community* (OJEC) and pre-qualified bidders will be shortlisted. Bidders for PFI contracts will invariably be a joint venture or consortium of firms (contractors, or contractors and design consultants) who will set up a Special Purpose Company (SPC) especially for the project.

The SPC will bid for the project in competition and, following a costly and lengthy process, a preferred bidder will be chosen. This point is reached through a series of 'gateways' which punctuate the process at key stages. The contract will then be awarded and notified in the OJEC.

DEVELOPMENT PROCESS AND THE PROJECT MANAGER

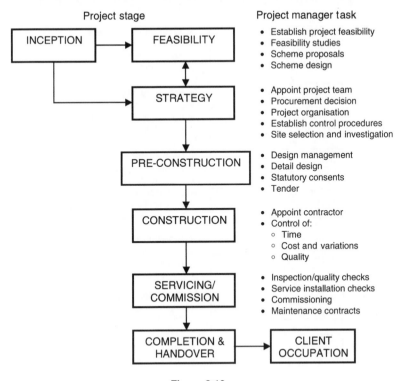

Figure 3.10

The project will normally be financed by borrowings from a bank or investment company at a high rate of gearing such that, typically, the SPC members will provide equity of 10% and the remainder will be borrowed. This is risky for the project sponsor (client) and therefore a 'step-in' agreement may be arranged which will enable the project to continue if the SPC fails. Subcontractors on site are particularly at risk as they are effectively unsecured creditors of the SPC whose only assets are the project and any equity funds remaining. The complex contractual and funding relationships are simplified in Figure 3.12.

Most PFI projects are design–build–operate schemes with capital invested coming back over a 25–30 year concession period during which tolls (e.g. roads) or service charges (e.g. schools) are made. Some schemes are partly financed by the public sector.

Examples of PFI projects include the Dartford and Skye Bridge crossings, the Channel Tunnel and the Second Severn crossing, which operate under a DBOOT (design–build–own–operate–transfer) arrangement.

Advantages of PFI

- Potential for high returns
- Continuity of work
- Involvement in design

LOCAL AUTHORITY MODEL
IMPLEMENTATION OF PROJECT MANAGEMENT

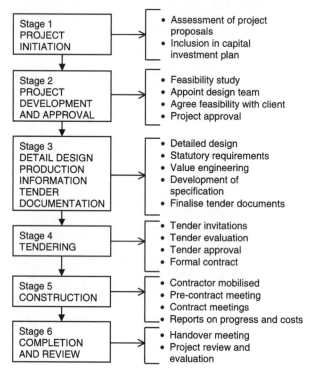

Figure 3.11

- Buildability input
- More control over the programme

Disadvantages of PFI

- Very high bidding costs
- Competitive market
- Long bidding process
- High level of resources required
- Complex and demanding
- Tough contract terms
- High level of liquidated and ascertained damages
- Fixed price risk to contractor

3.10 Framework contracting

In recent years 'traditional' approaches to tendering and contractor selection have been found to be defective, as reported by Latham and Egan for example. In this respect, final delivery of

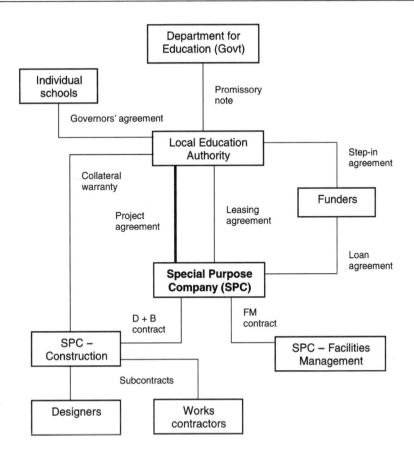

PFI RELATIONSHIPS FOR A SCHOOL'S PROJECT
WITH PARTNERING

Figure 3.12

projects based on the criteria of selection by price has been found to result in poor quality and unreliable outcomes and this is why other approaches have been devised.

One of these is framework contracting. This idea is similar to the traditional standing tender list in that approved suppliers are chosen or 'preferred' to go on the list for future contracts. The framework suppliers include contractors, specialist contractors, suppliers of goods and suppliers of professional services (such as architects and surveyors) who are invited to tender for packages of work.

The big difference, though, is that with a framework contract there will be no 'main contractor' as such. Instead, there will be a number of 'first tier' framework suppliers tendering for the main packages of work, and possibly 'second tier' suppliers tendering for the smaller or specialist packages.

Another difference is that the choice of successful bidder is not made solely on the basis of price. In this way, the framework of supervision, project management and construction is built up for a specific project that is then implemented on a sort of 'partnering' basis.

Some big clients use this system and their projects are run by the framework team and not one single 'managing' contractor as is traditional in the building industry.

3.11 Prime contracting

Prime contracting runs on similar lines to the standing approved lists but the prime contractors are chosen very carefully, usually via several stages of selection. The system is used by some government departments such as Defence Estates and the National Health Service.

The idea is to have 'prime contractors' who can provide all the project deliverables for capital projects, including design, planning and cost control. There are also prime contractors or 'one-stop shops' who are able to deliver all property maintenance and capital works in particular regions. Once a prime contractor is chosen, they become part of the client's project team for perhaps up to 7 years. The client then takes a 'hands-off–eyes-on' approach to their projects but with 'partnering' very much as the focus.

Selection is based on 'hard' and 'soft' criteria. Hard criteria are the usual ones of price, competence and standing, and the soft criteria focus largely on ideas, attitudes, willingness to share risk, flexibility, etc.

Features of prime contracting include:

* The point when a price is agreed with the prime contractor can vary
* The price could be fixed price, guaranteed maximum price, target price, etc.
* The form of contract could vary
* The period of facilities management is flexible
* Long-term commercial agreements could be built in for facilities management, other contracts, etc.

3.12 Partnering

Definition of partnering

Partnering may be defined as 'a structured methodology for organisations to set up mutually advantageous commercial arrangements, either for single projects or in long-term strategic relationships, which help their people work together more effectively' (CIB 1997b).

It should be stressed that partnering is entirely a voluntary arrangement which operates within the chosen procurement method, but it may become formalised in the contract should the parties so desire. The essential components of partnering are illustrated in Figure 3.13 and these components are fundamentally geared to establishing a commitment from participants to work together for mutual success where:

* All participants seek 'win–win' solutions
* Value is placed on a long-term relationship

PARTNERING

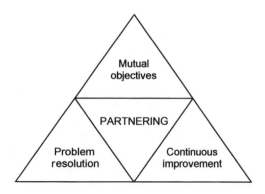

Source: The Reading Construction Forum – 'Trusting the Team'

Figure 3.13

- Trust and openness are norms
- An environment for long-term profitability exists
- All are encouraged to openly address any problems
- All understand that no one benefits from exploitation of the others
- Innovation is encouraged
- Each partner is aware of the other's needs, concerns and objectives and is interested in helping their partner achieve this
- Overall performance is improved

Partnering arrangements

Partnering arrangements are quite normal in construction and a number of close client/ contractor relationships have developed over the years. One of the best known 'alliances' is between Marks and Spencer and Bovis, although this has never developed into true or exclusive 'partnering' and other contractors also tender for and work for this particular client.

There are two types of partnering arrangement:

- Strategic partnering
- Project partnering

Strategic partnering is where a relationship is developed over an indefinite period and where there is a long-term commitment to the partnering approach. However, partnering may be for a particular project or length of time, such as for an individual contract or for a 3-year maintenance contract. This would be a 'project partnering' approach.

Forms of contract

Partnering arrangements may be based on a formal contractual arrangement between the partners, such as the PPC2000 Form of Contract for Project Partnering or the ECC using secondary Option X12. Also, the JCT 2006 Constructing Excellence Contract has been specifically developed for use in partnering and collaborative working and can be used for a target or lump sum contract whether or not design services are required.

Morledge et al. (2006) observes that *some project teams do not want their partnering arrangements to become legally enforceable* and suggests that legal formality may even be considered by some to be contrary to the very principles of partnering. As an alternative to a contract, there can be a non-legally binding relationship based on a charter with a mission statement and a set of common objectives. The charter could be drawn up specifically for the project or an 'off-the-shelf' document, such as the JCT 2005 Non-binding Partnering Charter, could be used. Whether or not there is a formal agreement, there needs to be a basis for agreeing prices and conditions of contract that are fair to both parties and this is left to the discretion of the contracting parties as normal.

Morledge suggests that conventional lump sum contracts may not be the best arrangement for partnering because of the need to 'incentivise' the contract. He also observes that many partnering contracts are based on a target cost with a guaranteed maximum price arrangement which is believed to give a 'fair price' for the job. This is a type of 'cost plus' reimbursement contract where the percentage added to the actual recorded cost acts as the 'competitive' element of the contract.

Partnering charters are common and a charter may even be included in a formal partnering contract (see PPC2000, Clause 5.6). The contents of partnering charters vary but might contain some or all of the following objectives and agreements:

- Budget certainty for the client
- Agreed maximum price
- Positive cash flow for the contractor
- Reasonable profitability for the contractor
- Completion on time
- A good/exemplary safety record on site
- Good relationships
- Enhanced reputation
- Trust and fairness
- Encouragement of ideas and innovation
- Sharing of cost savings
- Early warning of problems (two-way)
- Joint problem solving
- Use of conciliation/alternative dispute resolution where all else fails

A partnering culture includes:

- Open communication
- Top-down commitment
- Honest dealings (the difficult bit!)

EXAMPLE OF CLIENT/CONTRACTOR PARTNERING USING
TRADITIONAL PROCUREMENT METHODS

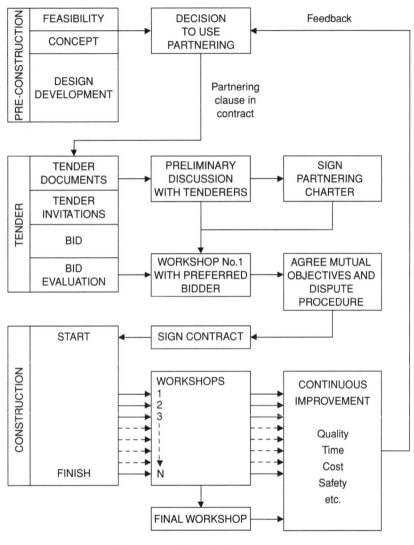

Figure 3.14

- Elimination of confrontation
- A problem-solving approach
- Proactive teamwork

The partnering team will consist of all supply-chain participants in the project and will include the customer and end-user.

Partnering in practice

Project partnering is most common in the UK but both strategic and project specific relationships need to be based on trust, dedication to common goals and an understanding of each other's individual expectations and values. Expected benefits include improved efficiency and cost-effectiveness, increased opportunity for innovation and the continuous improvement of quality products and services.

Partnering arrangements may be preferred where the client has repeat business for the industry or perhaps wishes to adopt a Latham/Egan or team-building approach to a specific project. This can be achieved by entering into a relationship over a specified period or for a longer period, where both parties agree to work together to achieve particular goals, for mutual benefit. However, any arrangement that does not have the triangle of features in Figure 3.13 is an 'alliance' and not true partnering.

Figure 3.14 illustrates the partnering process using a traditional procurement route but the principles of open discussion and consensus using workshops or 'shirt-sleeved' meetings are common whatever the preferred procurement arrangements.

The benefits of building up long-term relationships were recognised in the Latham Report in terms of improving quality and timeliness of completion while reducing costs. However, Latham also recognised that the benefits need to be mutual for the system to work and, whilst contractors may be assured of continuity of work, clients must be assured that prices remain competitive within a relationship of trust.

Feedback from completed projects suggests that typical benefits from using a partnering approach include:

- Cost savings through value engineering
- Fewer interest charges due to no retention
- Cost savings through buildability
- Cost savings by eliminating formal tendering
- Reduced site correspondence
- Good relationships leading to repeat business
- Effective cost management of variations
- Zero accidents

References

CIB (1997a) Construction Industry Board – Working Group 2. *Constructing Success*. Thomas Telford.

CIB (1997b) Construction Industry Board. *Partnering the Team*. Thomas Telford.

CIOB (Chartered Institute of Building) (2002) *Code of Practice for Project Management*, 3rd edn. Blackwell Publishing.

Gidado, K. & Arshi, S. (2004) Suitability of Different Design and Build Configurations for Procurement of Buildings, COBRA.

Janssens, D.E.L. (1991) *Design–Build Explained*. Macmillan Education.

Masterman J.W.E. (2002) *Building Procurement Systems*, 2nd edn. Spon Press.

Morledge, R., Smith, A. & Kashiwagi, D. (2006) *Building Procurement*. Blackwell Publishing.

NEDO (1985) *Thinking about Building*. National Economic Development Office. HMSO.

Rawlinson, S. (2008) *Procurement – Develop and Construct*, Building.
RIBA (2000) The Architect's Plan of Work, Royal Institute of British Architects. RIBA Enterprises.
Seeley, I.H. (1997) *Quantity Surveying Practice*, 2nd edn. Macmillan Press.
Turner, A.E. (1997) *Building Procurement*, 2nd edn. Macmillan Press.
Walker, A. (2007) *Project Management in Construction*, 5th edn. Blackwell Publishing.

Part A **Planning**

4 Management and organisation

4.1 Management principles

The principles of management established by Henri Fayol in the early twentieth century are as applicable today as they were then. The seven main principles are outlined by many management writers including Brech, Denyer, Drucker, Calvert, Cole and Clutterbuck.

Clutterbuck and Crainer (1990) provide a particularly useful overview of individual contributions to management development in a chronological order from 1841 to date.

The seven principles of management

Fayol developed seven basic principles which are generally applicable to a wide range of business organisations. The seven principles comprise:

- Forecasting and planning
- Organisation
- Commanding or directing
- Controlling
- Coordination
- Motivation
- Communicating, which encompasses them all

Each of these principles is now considered below as applied to the management of a construction organisation.

Forecasting and planning

Forecasting is looking into the future and planning involves the making of decisions on the basis of these forecasts. Types of planning at management level include considerations of strategy (strategic planning), business planning and long- and short-term business planning. Setting business objectives and policymaking form an integral part of forecasting and planning.

Planning first involves consideration of the objectives of the business. A contractor's objective is to make a profit in relation to the amount of capital invested – no profit, no business. Second, policies must be established in relation to the planned rate of growth and expansion.

Forecasting involves the preparation of:

- Financial forecasts
  - company annual turnover
  - company cash funding requirements
  - individual project cash funding
- Construction workload forecasts and estimating workload
- Resource forecasts
  - Staffing and key labour
  - Subcontractor resource

Organisation

Organisation is the grouping of work and the allocation of duties, responsibilities and authority.

- Organisation structure of company
- Organisation of individual functions (departments)
- Organisation structure of projects
- Defining roles and responsibilities within the organisation
- Providing job descriptions for individual staff

One of the steps in organising is to divide the business into sections which will be of most help in the efficient administration of the business. These sections may be related to functions, product or location. Many construction firms are organised functionally, that is surveying, buying, administration, planning and construction activities.

Examples of the overall organisation structure of a construction firm and the organisation of a single project are given later in this chapter.

Commanding or directing

Commanding is the giving of instructions to ensure that the agreed policies are carried out. It also involves the granting of authority before commands can be issued.

Control

Control involves the continuous checking of performance with the plan and taking corrective action. The principles of control in relation to the control cycle are outlined in Chapter 15, where control procedures have been applied to a project in relation to contract time (programme), money and cost (resources). Examples of control in construction include:

- Establishing budgetary control procedures within the organisation
- Control of cash flow (the movement of money)
- Control of project cost – monthly cost reporting
- Control of progress

Coordination

Coordination is the unification of effort between the company's personnel to ensure that the declared policies are fully implemented. Coordination is necessary between personnel in different departments to ensure the smooth running of the business.

Motivation

Motivation theories were developed by McGregor (1960) and Maslow (1954). Maslow's hygiene factors may be related to a modern contracting organisation by posing the question 'what motivates management personnel?' Motivation factors include:

- Pay
- Job security
- Recognition, promotion
- The working environment and working conditions
- Office 'perks' such as company car, holidays, use of the company boat or flat
- Balance between work, pleasure and home life
- Creating the 'no blame' culture
- Company training culture and quality of training provision

Communicating

Communications are necessary to enable the business to function effectively. This involves the three-way transfer of information between client, contractor and staff.

Modern communication systems ensure that data can be transmitted rapidly by e-mail and fax, and electronic tendering procedures ensure that less paperwork is transferred between client and contractor at the bid stage of a project (Cartlidge (2002) explains how e-tendering works).

The office notice board is a useful means of communicating company policy statements, contract awards and staff promotions and appointments, and a company newsletter may help to boost morale and highlight regional activities and achievements.

4.2 Leadership styles

The Association for Project Management Body of Knowledge (2006) defines 'leadership' as *the ability to empower and inspire people to achieve project success*, whereas Partington (2003) maintains that leadership *comes from the involvement, participation and empowerment of followers*. As businesses expand, the leadership style of the principal or managing director becomes apparent. The various leadership styles are summarised in Figure 4.1 as being autocratic or democratic.

Traditionally leadership has tended to be associated with autocratic command, especially within the small-sized organisation. Many still see leadership mainly in terms of issuing orders which are obeyed by subordinates without question. Drucker (1989) asserts that leadership is the lifting of man's vision to higher sights and the raising of man's performance to a higher

LEADERSHIP STYLES

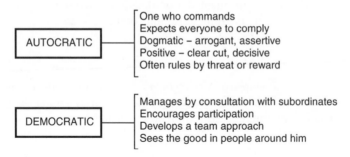

Figure 4.1

standard. Management can only create the leadership under which potential leadership qualities become effective.

Clutterbuck and Goldsmith (1984) indicate that to be effective 'leaders must be seen'. Perhaps the question should be posed: 'how many times in the last 12 months have you had personal contact with your company chairman or chief executive?' Perhaps not at all?

Dixon (1991) defines leadership as the process of directing and influencing the work of team members. Leadership is concerned with guiding and directing others. The style adopted depends in part on the manager's view of human nature in general and the ability of his/her subordinates in particular. The manager's attitude to his subordinates may be depicted by McGregor's Theory X and Theory Y, as outlined in Figure 4.2. The Theory X manager favours the autocratic approach while the Theory Y manager favours democracy.

McGREGOR's THEORY X AND THEORY Y

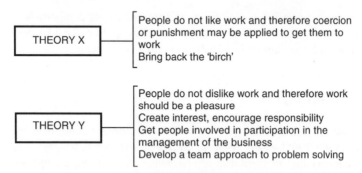

Figure 4.2

4.3 Project teams

It is commonly accepted that construction is a project-based industry requiring multi-disciplined teams of people to work together. Teams of architects, surveyors and engineers frequently represent the client and conversely teams of managers, engineers and subcontractors carry out the work on site. Nowadays client and contractor teams work collaboratively together due to more enlightened approaches to procurement and project management.

Burke (2003) makes the point that 'teamwork' should aim to bring individuals together so as to increase their effectiveness without the loss of their individuality. This must be achieved without stifling creativity or the generation of ideas however. This fits with the definition offered by the Association for Project Management (APM) Body of Knowledge that 'teamwork' *is when people work collaboratively towards a common goal.*

Burke emphasises that a team *implies a number of people working together to achieve results while a group of people does not.* He also explains that a group of people may be working on the same project but that they do not necessarily interact with each other. Dysfunctional teams, where there may be clashes of personality or unreliable subcontractors, contribute to unsuccessful projects. Winch (2002) points out that *a team adds value to the group because it is a team.*

In construction, the numbers of people working together on a project can be considerable and the numbers on site may vary between 20 and 200 people consisting of managers, engineers, surveyors, operatives and subcontractors etc. This does not square easily with the research findings referred to by Burke, Winch and others that the ideal team size is between 5 and 10 people. Burke further explains, however, that *teams tend to grow in size until some magic number is reached – then they subdivide.*

This is certainly the case on many projects where, for instance, there may be several subcontractors working on site. Even on a medium-sized project there may be 20 or 30 subcontractors involved and in such cases it is common for the project team to include a subcontract coordinator. It is this person who works with the subcontractors' representatives and then reports back to the project team thereby subdividing an otherwise large group.

4.4 Team building

Winch argues that unless the team building takes place in the context of a partnering-type arrangement, there is a danger that all that will result is a pseudo team which performs less well – more akin to a representative group than a team.

Lavender (1996) also makes a valid point that team building is difficult because individual members may not be involved in the project all the time. This is true in construction where it is commonly the case that there will be a visiting quantity surveyor or engineer and the health and safety advisor, whilst integral to the team, will almost certainly be head office based and only visit the site infrequently. The same could be said of subcontractors who come and go as the project evolves.

Construction project teams are at best temporary – they break up when the project is finished – and more often than not are transitory. There may be a small 'core team' leading the project but the other members come and go as the project develops – some may come back again later whilst others move on to the next job. Even members of the 'core team' may change as the site engineer is no longer required or the general foreman is replaced by a finishing foreman etc.

Teams are nevertheless important because they serve as a means of building self-esteem and a sense of belonging among team members but, conversely, some people are happier working on their own. This is quite understandable and it is perfectly acceptable for a project to benefit from ideas generated outside the team as well as within it. Part of the role of the project manager is to work with all sorts of talented individuals and to get the best out of them for the benefit of the project as a whole.

Winch (2002) reminds us that many team-building programmes or events involve the participants in getting cold, wet and muddy. Perhaps this is why so many project teams in construction are successful!

4.5 The size of construction firms

'Modernising Construction' (National Audit Office 2001) reported that there are over 160 000 firms operating in the construction industry and many of these are small contractors. In order to give a scale to the industry 'pyramid', the current annual turnover of the top 100 contractors in terms of annual turnover is shown in Table 4.1.

Table 4.1 represents less than 1% of the firms in the industry and indicates that the vast majority are either medium-sized or small.

Table 4.1 Top 100 contractors and housebuilders.

Turnover (£ million)	Number of firms
Less than 100	0
100–249	37
250–499	26
500–999	20
1000–1999	10
2000–2999	1
3000–3999	4
4000–4999	0
5000+	2

Source: Building

4.6 The characteristics of firms

It is interesting to compare the changes in the organisational structure of companies as company expansion or change takes place. In many construction situations, the management is not able to cope with the management of changing size.

Drucker (1989) states that the biggest problem in business is growth, that is the problem of changing from one size to another. Many principals or owners of construction firms face this problem as business expansion takes place. Often they cannot cope with the new situation facing them. This is due to their lack of vision and competence to manage people around them. Many directors cannot delegate responsibility to subordinates due to a lack of trust and

Drucker indicates that a change in behaviour, attitude, competence and vision is needed by people at the top.

Success in business often results from a company providing a good service to clients and doing a good job. The business can only service its customers by becoming bigger.

An interesting approach is taken by Drucker (1989) in respect of defining the four stages of business growth. These have been summarised as:

- The **small** business which is distinguished from the one-man proprietorship by requiring a level of management between the man at the top and the workers. Also, small businesses tend to be organised functionally
- The **fair-sized** business (later referred to as medium-sized) – in this size of organisation, the role of the managing director has become a full-time position. He is required to concentrate his efforts on guiding and managing the affairs of the business. The company needs some formal organisation structure which focuses the vision and efforts of managers directly on business performance and results. In the majority of construction firms in this category, the tendency is to develop a departmental approach to the various sections of the organisation, such as construction, surveying, estimating, plant and administration
- The **large** business – at this stage of development the setting of overall objectives becomes far too big for one person and becomes a shared responsibility of the management team. A large construction organisation may be managed by a main board of directors, supported by departmental directors and regional and technical directors
- The **very large** business – this is characterised by the fact that the overall business objectives and resulting actions must be organised on a team basis. Each position in the organisation requires the full-time services of several people

4.7 The small firm

National statistics categorise the small firm as a business with between 1 and 24 directly employed staff (or staff and operatives). This represents some 93% of companies in the UK Construction Industry. The number of construction firms in the European Union is in the order of 1 100 000 with 91% employing less than 10 people.

Business profile of a small company

Company background

The business is managed by the principal or owner with three directly employed staff. Sixty percent of the work is obtained by negotiation and the rest from competitive tenders. Quotations are based mainly on a drawings and specification basis, which places extensive risk on the business enterprise. Turnover after 2 years of trading is in the order of £800 000 and the largest project undertaken to date is £250 000.

The policy of the company is to use a mixture of labour-only and directly employed subcontractors. Figure 4.3 indicates the organisation structure of the company, if a structure can be said to exist, and Figure 4.4 illustrates how the management approach the control of projects.

ORGANISATION STRUCTURE – SMALL COMPANY

PRINCIPAL OWNER

RESPONSIBLE FOR:

Contact with clients / architects etc.
Preparation of tenders / obtaining work
Submission and adjudication decisions
Pre-contract arrangements
Letting of subcontracts
Planning - programming of work
Interim certificates / final accounts
Payment to subcontractors / suppliers
Establishing policy
All decision making

Figure 4.3

Business control procedures

Control is maintained by constant visits to projects by the principal in order to coordinate materials, plant and subcontractors. The principal's major concern is providing a personal service to clients and ensuring that projects are completed on time at a reasonable profit. A non-confrontational approach is adopted in the settlement of contract final accounts. The business relies extensively on negotiating further contracts with satisfied clients. Maintaining client contact during a project is considered essential to the success of the business.

Management checks are made on project profitability at interim valuation stages on a somewhat ad hoc basis. Monthly site meetings on all projects are attended by the principal. The majority of the principal's time is spent on preparing tenders for new work and chasing round his existing projects.

PROJECT ORGANISATION – SMALL COMPANY

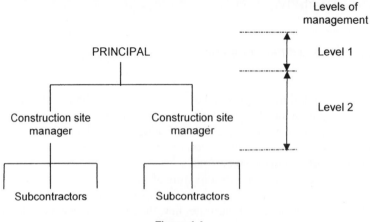

Figure 4.4

Commentary on the role of the principal

The principal tends to make all decisions, and tends to do everything himself, which is the typical small company approach to managing a construction enterprise. He does not appear to be able to delegate responsibility to people around him but with further business expansion the principal considers that he will lose control. Unless he overcomes this problem, the business will fail to expand and the principal will finish up having a heart attack!

A large number of small companies in this situation may, however, be content with the profit returns they are achieving and may not wish to become any bigger. Perhaps the attractions of working hard for 9 months and spending 3 months lying in the sun is too much to give up for the sake of success!

4.8 The medium-sized firm

The medium-sized firm has developed from expansion of the small business, brought about by the increase in workload, turnover and business diversification – possibly away from the core business. This may have developed from a need for the company to consider broadening its work base. The core business may have been refurbishment work, but the company may now be offering a design and build or a work package service to clients. Alternatively, the business may have moved into its own speculative refurbishment projects. Many such options are available as a business expands.

A team approach to the management of the company has to be established by the appointment of directors responsible for contracts, surveying, estimating, office organisation and financial aspects of the business. As Drucker (1989) states, 'the role of the managing director now becomes a full time position'. A strategic plan and clear policy objectives must be established. Policy decisions must be communicated throughout the organisation, and the development of teamwork is necessary at all levels of management.

Policy must be clarified in such areas as:

- Safety, health and welfare
- Training and recruitment
- Company control and reporting procedures
- Planning procedures to be utilised on projects

Delegation of responsibility at all levels of management must be established. Control and reporting procedures must be established for reporting on contract performance during the progress of construction projects. Channels of communication must be established throughout the company in order that staff know the person they are responsible for reporting to.

Business profile of a medium-sized company

Company background

The company was established in the early 1980s as a two-partner business undertaking public house and club refurbishment work. The company has now expanded into a major specialist

ORGANISATION STRUCTURE – MEDIUM – SIZED COMPANY

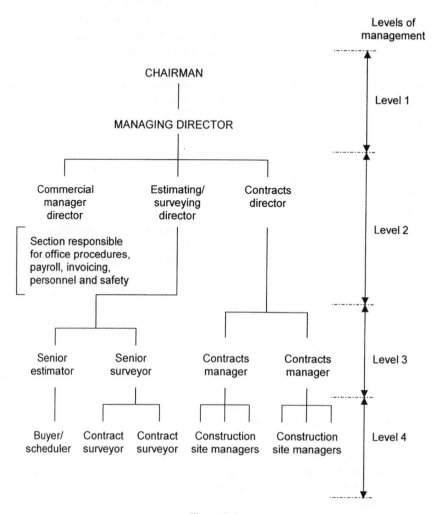

Figure 4.5

contractor servicing the leisure and hotel industry and, in 2006 turnover approximated to some £72 million – a remarkable growth. Recently, the company has expanded to offer a specialist design and build service to the leisure and hotel industry. Contract work is organised on a work package basis and the policy is to use labour-only and established domestic subcontractors. The company employs between 150 and 180 office, site staff and key operatives.

Figure 4.5 illustrates the current organisation structure showing the business divided into three functional areas, that is estimating/surveying, contracts and office/commercial management, and Figure 4.6 indicates the approach taken to the organisation of a major project. This simply consists of a construction manager who manages subcontractors. Head office

PROJECT ORGANISATION - MEDIUM-SIZED COMPANY

RESPONSIBILITY LIES WITH CONTRACTS DIVISION

SITES ARE SERVICED BY HEAD OFFICE - BUYING / SURVEYING

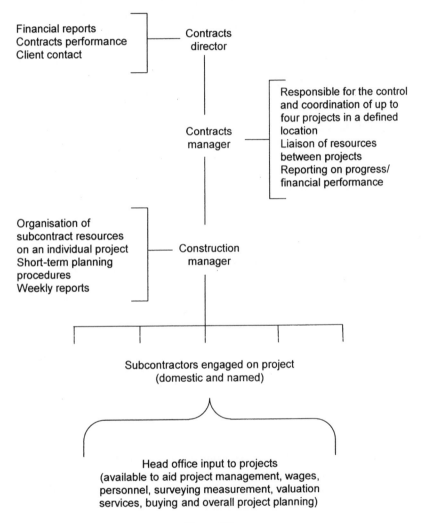

Financial reports
Contracts performance
Client contact

Contracts
director

Responsible for the control
and coordination of up to
four projects in a defined
location
Liaison of resources
between projects
Reporting on progress/
financial performance

Contracts
manager

Organisation of
subcontract resources
on an individual project
Short-term planning
procedures
Weekly reports

Construction
manager

Subcontractors engaged on project
(domestic and named)

Head office input to projects
(available to aid project management, wages,
personnel, surveying measurement, valuation
services, buying and overall project planning)

Figure 4.6

support services are provided as indicated. Powerproject is used for programming within the organisation.

This company is in a very competitive market and pricing risks are taken due to the nature of the enquiries received. In order to reduce the risk, the company is moving towards offering a design and build service on the basis of the client's scheme drawings or in-house design services. The largest single project undertaken to date was worth approximately £2–2.5 million.

Business control procedures

Control of the various functions of the business is delegated to a team of directors who cover construction management, estimating and surveying. The company is very surveying-orientated due to the extensive amount of subcontract work involved in the contracts. Cash flow is considered to be the lifeblood of the company and there is a very strong emphasis on the control of time and money. Strict guide rules have been established for reporting on the cost and value situation at monthly intervals on all projects and an effective cost–value reconciliation system has been established. This provides data on the performance of each contract in the short term and provides the degree of control and reporting needed for overview by the directors.

Programming and progress reviewing is the responsibility of the contracts manager, whereas short-term planning and the coordination of subcontractors is the responsibility of the construction manager. Strict control of subcontract orders is maintained by the link between the buyer and surveyors. Extensive pressure is applied to subcontractors to perform to programme and the success of projects relies largely upon subcontractor coordination and control.

Good client–contractor relationships are established early in the project by the appointment of a project director. This allows links between the client and contractor to be maintained at the top of the organisation and often leads to further negotiated contracts. All work is currently obtained by negotiation and, in this company at least, competitive tendering is now considered a thing of the past.

4.9 A business unit approach in a medium–large firm

A medium–large northwest-based contractor has taken an innovative approach to managing the business, involving establishing six business units within the organisation. This approach is illustrated in Figure 4.7, which also shows how head office services and individual projects are managed.

The business units are not based on the allocation of work within a region, but countrywide. Each business unit aims for a turnover of some £20–30 million each year. Certain units may be set up to service a single client (say on a partnering arrangement).

The aim of the business unit managers is to develop a team approach to managing projects that aids both company morale and clients alike. The role of the business manager is taken by a senior construction manager or it may be a shared position between a construction manager and senior quantity surveyor. Monthly reporting on progress and profit performance is made directly to the managing director and construction director.

The development of construction management project teams results in harmony between construction managers, site engineers, quantity surveyors and clients' representatives.

4.10 The large firm

Most large contracting organisations develop originally from the expansion of a medium-sized business. Sometimes the growth is by acquisition when separate businesses join together via either a merger or a takeover. Large firms typically employ between 300 and 1200 people and therefore the organisation structure must be designed to meet the specific needs of the company. The structure is often developed to suit the market in which the contractor obtains

BUSINESS UNIT APPROACH TO PROJECT MANAGEMENT

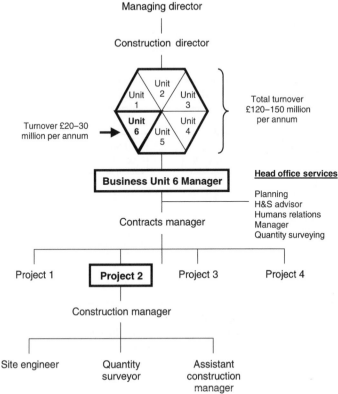

Figure 4.7

his work but no two contractors operating in the same field are likely to be organised in the same way. Organisation structures tend to be based on the individual preferences of the owners or directors and are very much personal to each type of business. This is illustrated in Figures 4.8 and 4.9 and in the case study below.

The organisational needs and services to be provided by each company differ widely due to market orientation and a contracting business operating predominantly in the competitive tendering market will need a different structure to that of a speculative developer. For instance, the organisation structure of Laing O' Rourke, which is a major contractor in the competitive market, will differ to that of a major speculative developer and house builder such as Barratt Developments. The nature and source of their respective workloads and the expertise available within the firm are possible explanations for this difference.

Laing O'Rourke has four main divisions based on the products and services they provide but some areas of their business are organised geographically and others have a customer or product focus:

- Construction north and south, infrastructure and value-added services
- Specialist services such as piling, building services and plant hire, etc.

ORGANISATION STRUCTURE – LARGE COMPANY

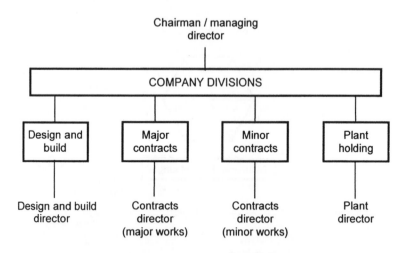

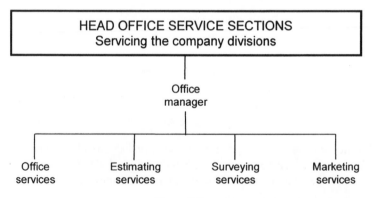

Figure 4.8

- Residential and commercial development
- International business

Barratt Developments, on the other hand, also has four divisions but they are focused on four distinct brands targeting specific markets or customers:

- Housebuilding, urban regeneration and social housing
- Prestige housing
- Quality homes in the South-East of England
- Commercial property development including retail, leisure, industrial, office and mixed-use developments and regeneration schemes.

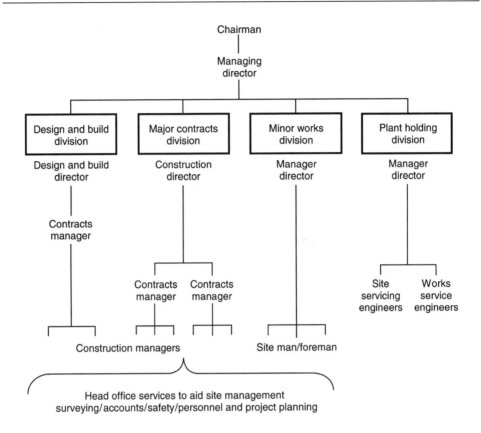

ORGANISATION OF CONSTRUCTION ACTIVITIES IN A LARGE COMPANY

Figure 4.9

Large firms tend to be involved in a wide variety of activities and this often results in an organisation structure based on some sort of 'grouping' arrangement. Lavender (1996) suggests several ways of grouping business activities but emphasises that some companies may have a structure that combines different methods:

- **Functional** – this is where the firm is divided into departments such as marketing and human resources and the production department is responsible for running the projects.
- **Product/Service** – this grouping may arise where the company has special expertise in a number of markets such as design and build, major projects, civil engineering.
- **Geographical** – some firms prefer a regional structure aimed at creating a local identity. An example may be a large national contractor with several regional offices. Alternatively, when large national firms acquire smaller local companies they often retain the name to establish the local identity.

- **Customer** – this type of grouping may arise perhaps where the company offers services aimed at particular customers such as design services, project management, PFI, rail infrastructure.
- **Capital** – such a grouping may arise where the company has invested heavily in specialist equipment such as slip-forming, tunnelling, cranage and lifting services.
- **Project** – this grouping typifies the traditional 'building contractor' who undertakes a series of discrete projects with head office support for estimating, purchasing and administration, etc. Small firms are often organised this way too.

Figure 4.8 illustrates a large company divided into four divisions each being responsible for a particular type of construction activity. This would be a 'product' grouping with each one managed by a main board director. If the product grouping was also organised regionally, there would be a regional director in charge of each product group reporting to the main board of directors. Figure 4.9 shows how each of the four divisions organises the construction projects within its responsibility.

Head office is responsible for providing services to all projects in the form of office administration, estimating, surveying, project planning, safety and marketing. Head office support may be provided from a single head office or may be regionalised in order to serve its customers' needs better.

Figure 4.8 illustrates typical service departments within a large firm, although these can vary enormously. Some construction companies do not have any quantity surveyors for instance! The service departments may be organised under an office manager who reports to the managing director (as shown in Figure 4.8) or there may be a finance director or human relations director etc. on the main board who deals with the general office functions.

Business profile of a large residential development company

Company background

The company is a large privately owned developer/contractor with a turnover of approximately £500 million and 800 employees. The head office is in London and the company is organised into two main divisions – Special Projects and Residential Housing. There are three regional offices – two in London and one in the North-West of England – and two land purchase offices in Bristol and Birmingham.

The company's main work sources are:

- New build multi-storey flats and low-rise private housing
- Refurbishment schemes
- Strategic projects/high class commercial projects

Organisation Structure

Residential development work is undertaken in three regions, each region under the direct control of a regional operations director.

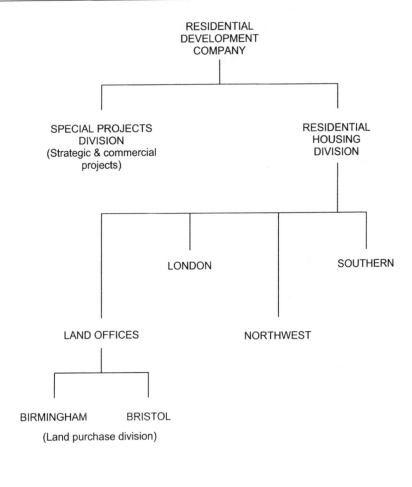

LARGE COMPANY ORGANISATION

Figure 4.10

Figure 4.10 gives a broad overview of the company's main activities, and Figure 4.11 shows the organisation structure for the north-west region. Strategically, the region has been divided into two main geographical areas as shown. Turnover in each of the area varies from £100 to 150 million per annum. Construction activity is managed by two construction directors. Central head office services are provided to each contract as work progresses from scheme design stage through to contract completion and handover.

REGIONAL ORGANISATION

NORTHERN REGION ORGANISATION

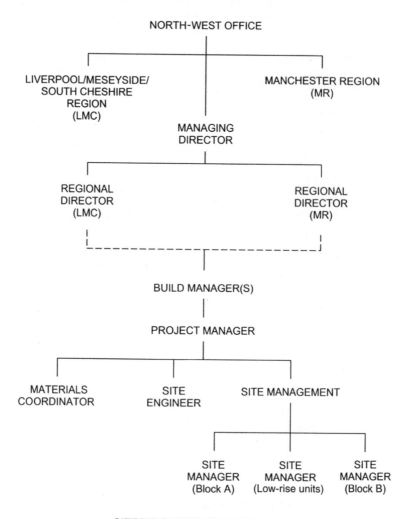

SITE/PROJECT ORGANISATION

Figure 4.11

An important aspect of the organisation structure is the interface between the construction director and the build managers. They are the main motivators for controlling the projects and are in direct contact with the build managers on each project. Emphasis is placed on achieving key programme dates which are closely related to the project handover schedule. Towards the end of each financial year, more emphasis is placed on projected completions because this has a significant impact on the annual accounts, which measure the success/failure of the business in each accounting period. The company operates a financial incentive system which is paid to site management personnel on the basis of project performance in meeting completions and handovers and building within budget.

References

Association for Project Management (2006) *APM Body of Knowledge*. APM Publishing.

Burke, R. (2003) *Project Management*, 4th edn. John Wiley & Sons.

Cartlidge, D. (2002) *New Aspects of Quantity Surveying Practice*. Elsevier Butterworth-Heinemann.

Clutterbuck, D. & Crainer, S. (1990) *Makers of Management*. Macmillan Press.

Clutterbuck, D. & Goldsmith, W. (1984) *The Winning Streak*. Weidenfeld & Nicholson.

Dixon, R. (1991) *Management Theory & Practice*. Management Made Simple Series, Butterworth-Heinemann.

Drucker, P.F. (1989) *The Practice of Management*. Butterworth-Heinemann.

Lavender, S. (1996) *Management for the Construction Industry*. Pearson Education.

Maslow, A.H. (1954) *Motivation and Personality*. Harper.

McGregor, D. (1960) *The Human Side of the Enterprise*. McGraw-Hill.

National Audit Office (2001) *Modernising Construction*. Report by the Comptroller and Auditor General HC 87. The Stationery Office.

Partington, D. (2003) *People in Project Management*, J.R. Turner (ed.). Gower Publishing.

Winch, G.M. (2002) *Managing Construction Projects*. Blackwell Publishing.

5 Tendering procedures

5.1 Introduction

Some companies in the construction sector do not actually tender in the competitive market because they negotiate most or all of their work. Others create their own workload by finding land and building speculatively. However, negotiated work is much sought after and land is at a premium and there are always other companies chasing the same opportunities. Consequently, the process of finding the right sort of work to suit the skills of the particular company starts well before the contractor is invited to tender for a contract or before a land deal is made.

Construction companies are in business like those in any other industry and competition for work is fierce, especially in the competitive tendering market. Consequently, contractors need to market their skills and convince prospective customers that they can offer the range of services required to satisfy the client's needs and expectations. A top-ten contractor will typically enjoy a market share of 2–3% of the £114 billion annual output of the industry (BERR 2007) and the largest of these will undertake 5–6% of the total work available.

To win this level of work requires a great deal of effort and therefore the contractor needs to be alive to business opportunities by checking local planning registers, reading the *Official Journal of the European Communities* (OJEC) and being included on the tender lists of all the major clients, consultants, local authorities and central government departments and agencies.

The pre-tender planning process is described below under the following headings:

- Establishing contract leads
- Tender pre-qualification
- The tendering process
- Traditional competitive tendering

5.2 Competitive tendering

In the competitive construction market, tendering arrangements for construction projects were traditionally conducted according to the procedures laid down in various National Joint Consultative Committee (NJCC) Codes of Procedure for tendering. The NJCC no longer exists, but the codes of procedure are still available and used for a variety of procurement arrangements, including single-stage, two-stage and design and build (CIS 2008). These codes are not mandatory but are widely observed in the industry and they provide guidance to good practice with respect to suitable tender periods, the number of contractors invited to tender and the general conduct of the tender process.

Further guidance on how to conduct the tendering process is available in the CIB (1997) publication *Code of Practice for the Selection of Main Contractors*, which sets out good practice guidelines for the pre-qualification of contractors, compilation of the tender list, tender periods, tender submission and assessment and what to do when a suitable tender has been accepted.

For public sector projects over certain value thresholds, UK tendering regulations, which implement EC directives, have to be followed. The rules vary according to the value of the project. Open, restrictive (selective) and negotiated tendering is allowed and there are proposals to include framework arrangements in a future consolidated EC public sector directive. Public sector projects over a certain value have to be advertised in the OJEC.

The effect of the public sector tendering regulations is to impose time constraints on tendering periods, and 'lead' times have to be allowed for receiving requests to tender (minimum 36 days) and for receipt of tenders (up to 40 days minimum). These have to be taken into account when deciding on the overall timescale for public sector projects.

Twort and Rees (2004) give a good overview of tendering procedures and legislation.

5.3 Establishing contract leads

A marketing approach

All construction firms need to develop contacts in order to secure opportunities to tender for work or to be in the right place at the right time when the chance to negotiate contracts presents itself. The opportunity to tender for a factory extension for Acme Widgets is more likely to arise in the private boxes at Old Trafford or the Cheltenham Festival than in a back street public house bar!

Within medium- and large-sized companies, however, a formalised marketing approach is common so that contract leads can be developed strategically at an early stage so as not to miss potential opportunities to be considered for inclusion on the final tender list.

Tendering costs money and therefore contractors cannot afford to tender for contracts willy-nilly without any thought as to the risk and profit potential or whether the type of work or contractual arrangements are suitable.

It is clearly important then for the contractor to be on the tender lists for jobs he is keen to win and not to waste time and money tendering for unattractive contracts. Indeed, a more selective tendering strategy may increase the contractor's tender success rate from, say, 1 in 8 bids to 1 in 6 or better.

Public sector projects

In the public sector, major construction works or supply contracts are advertised in the OJEC. This publication covers tendering opportunities over £300 000 in the public and government sectors for projects such as motorways, hospitals, community buildings and projects involving European funding.

These advertisements follow a standard procedure and public works projects and supply contracts are notified in the local language under standard headings. A typical tender opportunity for a hospital project in Sunderland would include the following list of information:

- Awarding authority
- Award procedure and contract type
- Site location and design information available
- Completion deadline
- Legal arrangements for joint venture or consortia bidders
- Deadline, address and language for submitting tender list applications
- Final date for dispatch of tender invitations
- Deposits, bonds or guarantees required
- Financing and payment (e.g. monthly payment against invoice)
- Qualifications such as past experience, technical expertise and financial stability
- Award criteria (e.g. price, technical merit, quality, value for money)
- Variants, which would include permissibility of tender qualifications or alternative bids
- Other information about the project in hand
- Various official dates and notices

Due to the extensive numbers of advertisements and the language difficulty, most contractors use the services of the local Euro Information Centre or perhaps a firm of contract leads/contract data consultants in order to find out what is happening in the market place. A selective approach to following up potential leads can be undertaken through these sources on the basis of searches using, say, five preferred criteria, for example:

- Project type
- Project size
- Contractual arrangements
- Form of contract and contract terms
- Location

Private sector projects

Contract leads may also be obtained from other sources:

- Local authority planning applications
- Planning committee meetings
- Minutes of local authority meetings
- Public press announcements
- Trade journals and magazines
- Trades representatives (they are often 'in-the-know')
- Keeping your eyes and ears open (don't forget Old Trafford!)

Whatever the source, a lead can take a considerable time to come to fruition. For instance, an average contract lead can take 10 months to track from an initial planning application to inclusion on the tender list.

5.4 Tender pre-qualification

More often than not, contractors are required to undergo some form of pre-qualification in order to be selected for inclusion on a tender list. This is a formal process which usually requires prospective tenderers to answer a standard questionnaire and perhaps attend a formal interview and make a presentation.

On most projects, the main contractor will usually be appointed as the principal contractor under the Construction (Design and Management) Regulations 2007, and therefore pre-qualification will also include questions about their health and safety record, health and safety training and the qualifications and experience of their staff and operatives. In all cases, contractors must be able to demonstrate that they have appropriate procedures in place to comply with health and safety law, as well as possessing the usual qualities and resources expected of a competent contractor.

Consideration of the contractor's financial standing would involve scrutiny of the contractor's accounts to determine profitability, solvency, liquidity, asset strength and payment record, as well as taking out references from previous clients, bankers and the trade, for example suppliers and builders' merchants.

Where sensitive information is concerned, trade and other references may be subject to the Data Protection Act 1998. For instance, normal trade information concerning a contractor's financial stability, the length of time taken to settle invoices or the contractor's credit limit with suppliers and builders' merchants might be caught by the Act.

The legislation protects individuals from the improper use and dissemination of sensitive data by data controllers but, whilst the Act does not apply to incorporated companies, little distinction seems to be made in practice between sole traders, partnerships and limited companies.

In a limited survey conducted by the authors:

- Local authorities were reluctant to provide financial references for contractors on their approved lists
- Some materials suppliers were cautious about the sort of 'sensitive' information they were prepared to give
- Some firms of quantity surveyors and architects establish a contractor's financial standing by asking how much they pay for providing contract bonds or insurances, some carry out company searches through Companies House and some use credit reference agencies which have special status under the Act

In any event it seems that the days of the 'trade reference' may be numbered.

As the construction industry moves towards a less adversarial approach to contracting, a factor of particular importance to some clients is the contractor's attitude towards claims. This may be seen as reflecting their ability to engage in partnering and more open methods of doing business.

Examples of issues which should be considered when choosing prospective tenderers are as follows.

General criteria

- Company details
- Completed contract record

- Technical expertise
- Plant and resources
- Financial accounts
- References

Competence

- Past experience
- Response to pre-construction health and safety information
- Organisation
- Safety advice
- Training provision
- Management procedures
- Accident statistics and prosecutions
- Vetting and control of subcontractors
- Safety management system

Resources

- Provision for health and safety
- Plant and equipment
- Technical support
- Trained personnel
- Adequate time

As well as the general information listed above, pre-qualification may also involve submitting a company video and giving a formal PowerPoint-type presentation. Many companies produce formal pre-qualification documents or brochures, the typical contents of which might be:

- Company organisation and structure
- Regional/national/international offices
- Key statistics, for example registered office, bankers, capital structure, annual turnover, profit progression, 5-year summary of accounts
- Audited annual accounts for the last 3 years
- Current projects
- Relevant completed projects
- Special expertise or experience
- Résumés of key personnel
- Project organisation structures and management approach
- Typical outline method statement
- Typical outline programme
- Cost planning approach for management or target contracts
- Details of insurances
- Evidence of quality assurance certification
- Health and safety policy and safety record statistics

Company brochures can be produced impressively and professionally in-house using modern computer software such as desktop publishing packages and photo-image software for presenting completed projects in colour. Project management software such as Power Project can be used to present typical contract programmes in colour format.

An average cost of around £500 per brochure is a relatively small investment for the opportunity to be included on a tender list for an £8 million warehouse project.

5.5 The tendering process

The pre-tender planning process begins with the tender enquiry to the contractor and ends with the contractor's tender submission. The steps in the process vary according to the procurement method chosen by the client or his advisors.

A central feature of pre-tender planning is the tendering process which is described in detail in the CIOB Code of Estimating Practice (CIOB 2008). This provides an authoritative guide to good practice and is based on procedures developed by a wide range of construction organisations. The Code distinguishes between estimating and tendering, which are separate though closely interrelated processes.

The estimating stage of a bid is simply putting together the numbers. It is the 'technical' process of arriving at the cost of a proposed project by taking into account all factors that contribute to the costs of construction. Estimating involves:

- Making enquiries to suppliers for the cost of materials
- Asking for specialist quotations from subcontractors
- Arriving at the cost of employing different types of labour, including the on-costs such as national insurance and holidays with pay contributions
- Pricing the items in the bills of quantities provided in the tender enquiry from the client
- Planning the timescale for the project (usually with other colleagues)
- Pricing the cost of running the project on site (preliminaries)
- Producing a summary of the 'tender' for directors

Tendering, on the other hand, is converting the numbers into a competitive bid after consideration of market factors and risk. It is carried out by the directors and the senior managers of the company, taking into account such factors as who the client is, the level of competition at the time of tendering, the technical and commercial risks involved and whether the company needs the work or not. This is essentially a commercial decision which may result in making alterations to the estimate by adding or deducting sums from the total arrived at by the estimator(s). A decision is then made on the final figure to be submitted to the client.

The CIOB Code contains many interesting and practice-based pro formas which a contractor may consider using to collate data and build up his estimate. A series of flow charts illustrates the various stages of the estimating process. Further explanation, together with worked examples, can be found in a number of textbooks including Brook (1993), Smith (1995) and Bentley (1987).

The following description of pre-tender planning procedures is typical of those used by many medium and large-sized contractors. Procedures for a traditional competitive tender are

SMALL/MEDIUM–SIZED COMPANY ORGANISATION STRUCTURE

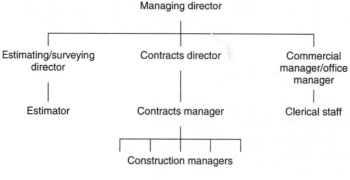

Figure 5.1

described first, followed by a comparative description of procedures for a design and build tender.

5.6 Procedures for a traditional competitive tender

The procedures outlined below are those undertaken within the medium/large construction organisation. Within larger firms, a more structured functional approach to the estimating and tendering process is often developed. These approaches are illustrated in Figures 5.1 and 5.2.

LARGE COMPANY – FUNCTIONAL ORGANISATION STRUCTURE

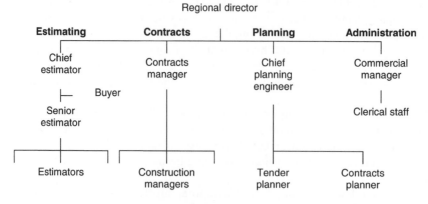

Figure 5.2

In either case, the contractor will have a lot to consider during the preparation of the estimate and its subsequent conversion into a tender bid submission. The following stages summarise the pre-tender planning sequence:

- Decision to tender
- Pre-tender arrangements
- Site visit and report
- Enquiries to subcontractors and suppliers
- Tender method statement
- Build-up of estimate
- Pre-tender programme assessment
- Build-up of contract preliminaries
- Response to pre-construction health and safety information
- Management adjudication and risk assessment
- Analysis of tender performance

Within the medium-sized company, it is likely that there would be no separate buying facility available and that the company would not have any specialist planning staff to call upon. Responsibility for estimating and tendering would lie with three principal directors. Their involvement in the bidding process is shown in Table 5.1.

Table 5.1 Principal directors' tendering responsibilities.

Personnel	Involvement
Managing director	**Decision to tender** Organisation of bid Monitoring progress **Tender adjudication and risk assessment** **Analysis of tender performance**
Estimating/surveying director	**Decision to tender** Site visit and report Enquiries to subcontractors and suppliers Tender method statement Build-up of estimate Build-up of contract preliminaries **Tender adjudication and risk assessment** **Analysis of tender performance**
Contracts manager/director	**Decision to tender** Site visit (with estimator) Advice on construction methods Advice on plant and preliminaries Pre-tender programme Response to pre-tender health and safety plan **Tender adjudication and risk assessment** **Analysis of tender performance**

Within larger organisations, a full tender team would be involved in the tender planning process. This would be based on a functional organisation structure from which a dedicated estimating team would be developed. The team would have full access to:

- A buying section to deal with subcontractor and material suppliers' enquiries
- A planning service usually provided by the planning department
- Full administration support staff for dealing with tender enquiries and correspondence
- Senior construction personnel for advice on methods of construction, plant and preliminaries

Figure 5.3 shows a programme of tender activities for a traditional competitive tender, highlighting the involvement of key personnel within the overall estimating and tendering period. A minimum period of 4 weeks is normally allowed from tender enquiry to tender submission.

5.7 Procedures for a design and build tender

With a design and build arrangement, the contract may be based on either a partial or full contractor's design. In either case, the tendering contractor will need time to arrange for design work to be prepared in sufficient detail to enable quantities to be prepared for pricing purposes. This is an added risk factor for the contractor. The overall stages involved in pre-tender planning will vary according to the extent of design input by the client or contractor at the tender stage.

The pre-tender planning stage will involve input from either the contractor's in-house design team or independent design consultants appointed by the contractor. With many design and build projects, extensive liaison with subcontractors forms a necessary part of the tendering process.

In the larger organisations, a bid manager (senior estimator) may be appointed on major tenders to coordinate information flow during the estimate preparation and tender stage. The role of the bid manager would involve responsibility for:

- Receipt of tender documentation
- Attendance at the tender clarification meeting
- Organisation and distribution of documents and correspondence received during the tender
- Managing the tender launch and decision to tender meeting
- Updating and reporting to senior management on the progress position during the tender preparation
- Attending tender adjudication meeting and any settlement meetings during estimate
- Arranging for bid submission on appropriate date

Figure 5.4 shows the various stages involved in the organisation of a design and build tender in bar chart form.

Project success often relies on the ties developed at tender stage between the subcontractors and the main contractor, and many of the subcontract packages will involve responsibility for design. The design and build contractor often has only one chance of getting it right and

PROGRAMME OF TENDER ACTIVITIES (Traditional contract)

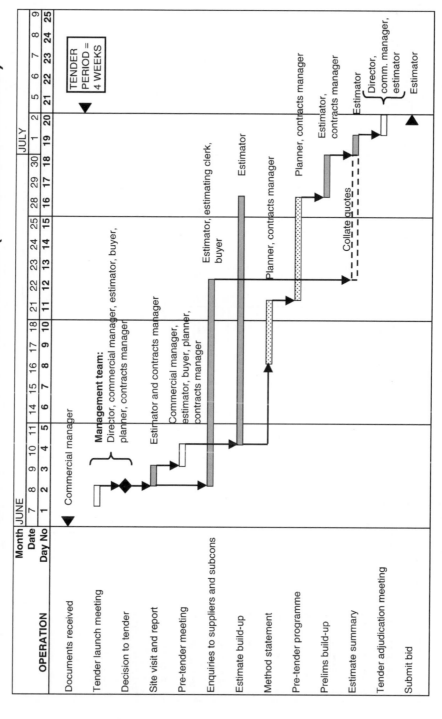

Figure 5.3

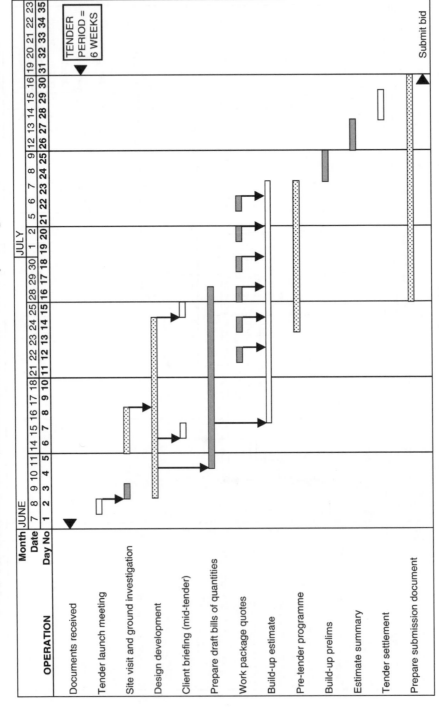

Figure 5.4

whether or not he does largely depends on his ability to manage his subcontractors during the design as well as the construction period of the project. Trust between both parties is essential for a successful project.

Figure 5.4 identifies the design and build contractor's responsibility for both quantity assessment and design and also shows the requirements for close liaison with work package subcontractors during the bid preparation.

5.8 Traditional competitive tendering

Competitive tendering is still the most widely used method of procurement in construction despite the popularity of design and build and partnering (see Chapter 3). Irrespective of the procurement method used, however, the principles which apply to traditional tendering are generally applicable as in most cases the client will require a price for the job and the contractor will have to be sure that he is giving the right price to enable the contract to make a profit.

The remainder of this chapter is devoted to the procedures inherent in preparing a 'traditional' tender, which comprise the following stages:

- Decision to tender
- Pre-tender arrangements
- Site visit report
- Tender enquiries to subcontractors and suppliers
- Build-up of the estimate
- Pre-tender method statement
- Preparation of pre-tender programme
- Build-up of contract preliminaries
- Management adjudication of the estimate
- Analysis of tender performance

5.9 Decision to tender

The decision to tender needs to be carefully considered by the contractor as tendering costs money. Several factors have to be weighed up, as listed below. Not least of these factors is the 'quality' of the enquiry in terms of both the status of the client and the prestige of the project.

Some contractors grade the tender enquiries received from clients at the decision to tender stage. Status grades from 1–4 may be applied. However, there appears to be little to be gained from this practice especially in terms of how much effort the contractor puts into a grade 2 tender as opposed to one ranked at grade 4.

A more sensible approach might be for contractors to consider each tender on the basis of submitting a winning bid, in which case the ultimate price should be influenced only by risk and market factors. Where circumstances change during the bid process – perhaps the contractor wins another contract unexpectedly – this can be taken into account at the adjudication stage.

A company's tender success rate may be influenced more by being selective at the decision to tender stage than by developing some esoteric tender grading system.

Checklist

The following factors will influence tendering decisions in general, together with specific market conditions at the time of tendering.

General tendering policy

- Is it our kind of work – does it fit into our strategic plan?
- What is the current workload in both the contracts division and estimating section?
- Do we have the financial and management resources to undertake the work?

Working capital

- Is sufficient working capital available to fund the project?
- What will be the effect on company financial resources?
- The working capital required to fund a £500 000 project will be approximately 15–20% of the monetary value at the peak funding month (say £100 000–150 000)

Availability of resources

- General management personnel (e.g. contracts managers, planning engineers, quantity surveyors)
- Site management (e.g. site agents, foremen/gangers, site engineers)
- Labour and plant
- Subcontractors – are suitable subcontractors available and what is their resource situation?

Location

- Is the project located within our trading area?
- What management and control problems will there be with a contract located some miles from head office?

Size and type of work

- What is the monetary value of the project?
- Is the contract too big for the company to undertake?
- Taking on a project which is too big could be damaging to future planning and growth
- What impact will there be on the viability of the business if the contract fails to make an adequate margin?
- If a contractor with an annual turnover of £10 million wins a £4 million contract and this project makes a loss, the whole business could be put at risk
- A major project could give the company severe liquidity problems
- How did the company perform on similar types of work in the past?

Subcontract element

- What is the extent and value of the contractor's work in the project compared with the subcontract element?
- Is the main contractor simply being asked to manage a number of subcontractors?
- Is a reasonable markup on subcontractors likely?
- Is there a risk of incurring liabilities from subcontractor claims?
- What future liabilities may be incurred as a result of defective work by subcontractors?

Degree of competition

- How many contractors have been invited to tender?
- Which contractors are on the tender list?
- Obtain details of competitors by making contact with material suppliers or specialist sub-contractors listed in the contract documents
- Do we want the work?

Tender period

- Have the recommendations of the NJCC Code of Procedure for Selective Tendering been followed?
- For a contract value of £500 000, a period of 4–6 weeks should be allowed for tendering

Terms and conditions of contract

- Which form of contract is to be used?
- Is there any contractor design element?
- Are there any amendments to the standard form?
- Is the contract period stated in the documents?
- What is the rate of liquidated and ascertained damages?
- Is retention applicable?
- What is the defects liability period?
- Is a bond to be provided by the contractor?
- What are the insurance requirements?
- What are the payment conditions and terms?

Tender documentation

- Is there any quantity risk placed on the contractor (e.g. drawings and specification tender)?
- What is the quality and accuracy of the documents provided?
- Is there potential for contractual risk in the contract documentation (e.g. liability for ground conditions)?
- If the tender documentation is poor, what problems or potential claims could arise?
- Potential for claims is not always attractive to contractors as there is no guarantee of success, large sums of money can be outstanding for long periods and the costs of compiling claims can be high

Client and other participants

- What is the name of the client?
- What is the name of the architect, engineer, quantity surveyor, etc.?
- Have there been any previous difficult experiences working with these people?
- Do we need to allow a contingency factor for this?
- What is the financial standing of the client?

Market factors

- What is government policy (now and future) and its effect on the construction sector?
- What are the general market conditions in relation to the availability of work?
- What is the current level of bank interest rates?
- Is our current borrowing facility adequate?
- Is there a feel-good factor in the business economy?

The decision to tender is the responsibility of senior management. In order to aid this decision, project particulars may be set out on a Preliminary/Tender Enquiry Form as shown in the CIOB Code of Estimating Practice (CIOB 2008).

Tendering costs

From a limited survey of contractors, Hughes (2004) has found that contractors can spend in the order of 3% of turnover on winning work. This includes not only tendering costs but also the costs of marketing and following up enquiries.

Bidding for design and build work is intrinsically more expensive than for architect-designed projects because of the costs of design work and the preparation of bills of quantities. However, Hughes's research suggests that traditional competitive tendering is actually up to four times more expensive than design and build. This is probably due to the higher levels of competition and lower tender success rates in traditional tendering.

Hughes indicates that tendering success rates are normally around 1 in 5 within larger contracting organisations but this will vary according to the type of work tendered for, the balance of competitive and negotiated work, the nature of the procurement arrangements for projects and the market circumstances at the time of tendering.

The average time spent on a major enquiry was found to be in the order of 400–500 man-hours per bid. Taking an average 5-week tendering period, this would suggest that a contractor employs between two and three people full-time on each tender.

Contractors' head office overheads vary of course but can be as low as 3–5% in a large company and thus tendering costs represent a considerable proportion of the total. However, without this overhead expense, the contractor cannot win any work.

5.10 Pre-tender arrangements

Where the contractor has decided to submit a tender, the estimating, buying and construction team will need to be motivated into action.

In small- and medium-sized companies, the senior personnel involved in the tendering process will normally be the managing director, the director responsible for estimating and surveying and the contracts manager. The main responsibility for putting the tender together will lie with the estimator/surveyor in conjunction with the contracts manager. A commercial/office manager will normally be responsible for the distribution of tender documentation from the client.

A typical organisation structure for such a business is shown in Figure 5.1 where it is assumed that the company has no separate buying or project planning facilities.

In medium and large companies, a functional organisation structure is common with separate departments for the estimating/surveying function, contracts and planning function and the administration of the business, as shown in Figure 5.2. In this type of organisation structure, the personnel involved in the estimating and tendering process will be:

- Senior estimator/buyer/estimator
- Contracts manager
- Tender planning engineer
- Commercial manager/office manager and clerical staff

Advice on contractual matters and commercial risk issues will also be sought from the company's quantity surveying department.

A large company may employ up to five or six estimators under the direction of a chief estimator. Standard procedures and estimating pro formas are often developed on the basis of computer spreadsheets. Information flow is controlled by the chief estimator or commercial manager.

An important aspect of the chief estimator's job is the management of the tendering process, which includes considering the workload of the estimating team. Figure 5.5 shows a simple bar chart which might be used to indicate the workload and key-estimating dates over a monthly period.

Tender submission deadlines are invariably tight and the tendering process needs careful management because mistakes are likely to be costly. If procedures are not followed, the contractor might miss the submission date or lose the contract on a technicality. Worse still, the contractor might lose money should the tender be successful. Most competent contractors will have a system of pre-tender procedures and tasks will be delegated by the chief estimator to make sure things happen as they should. These tasks will include:

- Receipt of tender documents
- Registering the tender and allocating a tender reference number
- Circulating the tender enquiry summary
- Setting up the decision to tender meeting (tender launch meeting)
- Distributing documents to the tender team
- Arranging review meetings during tender
- Arranging adjudication meeting and submission of final bid

The chief estimator will also delegate responsibility for the preparation of the estimate and tender summary. This will be coordinated by the estimator but several other staff will also be involved. Their input will be as shown in Table 5.2.

ESTIMATING WORKLOAD

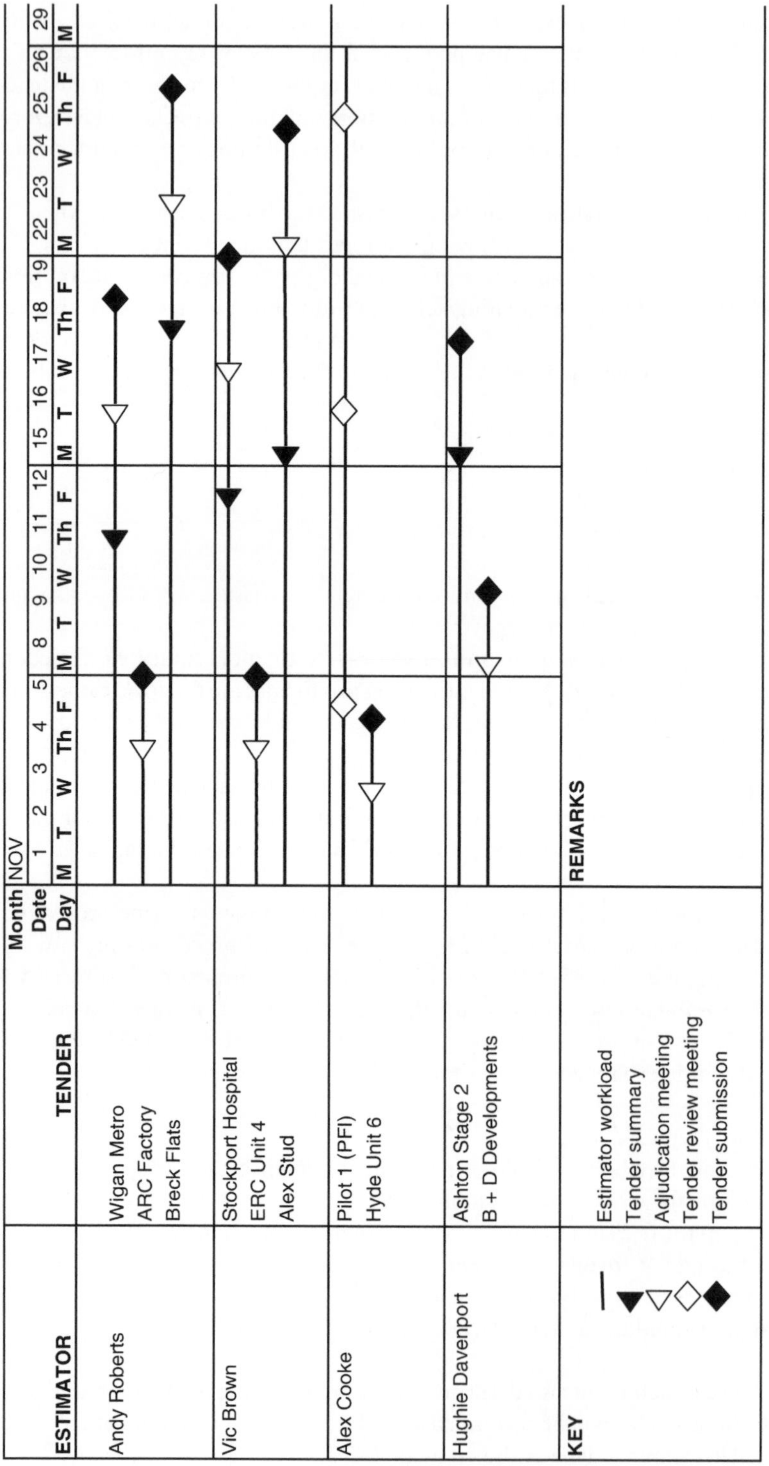

Figure 5.5

Table 5.2 Estimating responsibilities.

Personnel	Responsibility
Estimator	Dealing with the management and pricing of the bills of quantities Dealing with subcontract and material price enquiries when there is no company buyer available
Contracts manager	Arranging to visit the site with the estimator Preparing the site visit report Preparing the method statement and discussing it with the estimator Preparing an assessment of the project plant requirements Advising the estimator on the requirements of the contract preliminaries Preparing an assessment of the pre-tender programme, in conjunction with the company planning engineer Preparing an assessment of the project safety requirements in response to health and safety legislation Arranging to view the project drawings, if not included with the tender documentation
Company planning engineer	Preparation of an assessment of the pre-tender programme Assistance in the assessment of the construction method statement
Senior management	Overseeing all stages of the estimate Liaison with the contracts manager with regard to major decisions Chairing the tender adjudication meeting and taking all necessary decisions on markup and profit additions

5.11 The site visit report

Before starting to price the tender, the contractor should visit the site in order to satisfy himself of the prevailing conditions. The visit will usually be made by the estimator and perhaps a colleague such as the contracts manager.

As well as being a sensible precaution, the site visit is usually a condition of tendering and there is normally a 'sufficiency of tender' provision in the contract preliminaries. This means that the contractor is deemed to have satisfied himself, as far as he is able, with regard to the condition of the site and that the tender takes this into account. The following factors will directly influence the contractor's tender price:

- Access to the works – effect on construction methods
- Access restrictions affecting the utilisation of plant
- Site topography, ground conditions and groundwater levels
- Distance to local tips for the disposal of material
- Provision of site security
- Restrictions imposed by adjacent buildings and services

Many of the large construction firms make use of a standard site visit report pro forma, which acts as a checklist during the site visit and ensures that essential data is not overlooked. A model site visit report is provided in the CIOB Code of Estimating Practice (CIOB 2008).

Checklist

Points to be noted when preparing a site visit report include the following.

Names of parties involved in the contract

- Particulars of local authorities and planning and building control, etc.
- Details of statutory undertakings (water, electricity, telephones)

Site access

- Low bridges affecting the movement of plant and equipment to site
- Busy roads and potential offloading problems
- Traffic management on and around the site
- Specific site requirements for temporary works including access roads

Nature of site

- Topography – flat/sloping/extent of bushes and trees, etc.
- Site location relative to adjacent roads
- Other contracts in the immediate area
- Existing services locations

Existing buildings

- Extent of buildings to be demolished – site photographs prove useful when pricing
- Consider recording the condition of buildings on a digital camera
- Location of and dangers from buildings adjacent to the works – this may influence the cost of temporary works and excavation methods
- Details of any fly-tipped materials on site

General ground conditions

- Site surface conditions – adjacent excavations may provide evidence of ground conditions and water table levels
- Details of any watercourses crossing the site should be noted

Disposal of excavated material and waste

- Distance to local tips
- Tipping charges and landfill taxes
- Types of materials to be removed from the site – this will influence disposal costs, particularly if 'special waste' is involved
- Local skip-hire firms

Existing services

- Evidence of water and electricity supplies, drains and manholes
- Presence of telegraph poles and overhead cables

Site security

- Possible requirements for hoarding and site compounds
- The need for the provision of secure containers especially on inner city sites
- Likely requirements for security patrols and dogs

Local labour

- Assessment of the labour availability in the area
- Availability and quality of local subcontractors
- Telephone number and address of local Jobcentre Plus agency

Material suppliers

- Names of local builders' merchants and suppliers
- Local plant hire firms and size of plant fleet

Working space

- Space available for cranes and hard standings
- Site boundary constraints requiring crane 'easements'
- Lay down areas for material storage
- Site parking for staff and operatives
- Vacant land nearby for rent if required

As a note of caution, many a contractor has lost a project because of the inability to carefully assess risk. Tenders are often won on the basis of items that the contractor does not include in the price, either missing them altogether or simply taking a 'flier'. Site ground conditions are particularly important and the effect on pricing of temporary earth support needs careful consideration. This particular item may be treated as an item of contractual risk within the tender.

The site visit provides vital clues to the presence of these risk factors and the estimator needs to pick them up and make proper provision in the pricing. Whether the directors decide to put them down to 'contractor risk' is another matter!

5.12 Tender enquiries to subcontractors and suppliers

Subcontractors

One of the key decisions facing the estimator is how to 'package' the work when sending out enquiries to subcontractors at tender stage. The problem is that the bills of quantities will normally have been prepared using one of the standard methods of measurement used in the industry, but the bills are not necessarily arranged conveniently for subcontract packaging (or 'parcelling' as it is sometimes called).

If work is packaged by following the bills of quantities work sections, enquiries may contain work activities which are outside the scope of the work normally carried out by a particular subcontractor. Conversely, where there are interfaces between packages, for example demolition and excavation, or groundworks and reinforced concrete work, there may be uncertainty as to what each of the subcontractors have priced into their tenders. Such interfaces can also result in disputes where none of the package contractors has included for a particular item of work or, alternatively, duplication may occur where several firms are responsible for the same item. Mistakes can easily be made here and the estimator must ensure that all bill items are priced one way or the other.

It is usually best to parcel work using normal trade demarcations of the industry, thus ensuring that each subcontractor will receive an enquiry for work with which he is familiar. Alternatively, the main contractor will send enquiries to firms which specialise in, say, formwork, rebar and concreting or excavation and drainage, by selecting appropriate parts of the bills of quantities to send to them.

An important consideration here is the need to obviate further subletting of the package to sub-subcontractors or even sub-sub-subcontractors which can lead to multi-level interface problems resulting in coordination difficulties and uncertainty as to responsibilities and liabilities on site. If well thought out and planned, packages can be let so as to avoid this situation, thereby reducing the potential for confusion, mistakes and accidents.

Partnerships with preferred subcontractors are now common in the industry and Chapter 6 deals specifically with procedures for managing the supply chain at the tender stage and during the contract on both traditional and design and build projects.

Suppliers

Part of the estimating process concerns sending enquiries to material suppliers so that the estimator can price the measured items in the bills of quantities. Main contractors tend to use subcontractors for most site operations nowadays, but for certain trades the contractor will engage his own labour and price the work himself.

At tender stage, quantities for the main materials have to be extracted from the bills of quantities and this is done either by the estimator or by an estimating clerk. This is not straightforward as the bills of quantities contain items of finished work measured in m, m^2 or m^3, but materials are often supplied in different units. For instance, brickwork is measured in m^2 but supplied in packs and priced per 1000 bricks. The CIOB Code of Estimating Practice (CIOB 2008) indicates the information to be included in enquiries to material suppliers.

When the estimator receives prices from the various suppliers these have to be compared to see which is the most appropriate price to use. A comparison spreadsheet is usually used for this. The CIOB Code provides useful schedules for guidance.

5.13 Build-up of the estimate

Pricing is the sole responsibility of the estimator although on major projects there may be an estimating team involving several estimators, planners and contracts staff. Assistance may also be available from the company buyer, if the company possesses such expertise. Within the medium-sized organisation, it is usual for the estimator to be responsible for sending enquiries to subcontractors and suppliers. The various sources of data available to the estimator are summarised in Figure 5.6.

The build-up of the estimate will involve the estimator in the processes described below.

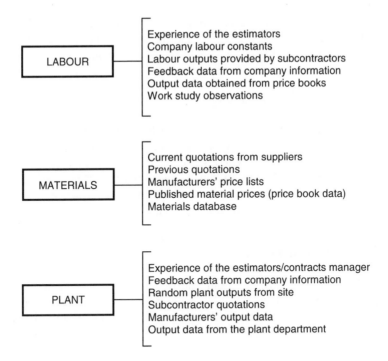

SOURCES OF ESTIMATING DATA

Figure 5.6

Build-up 'all-in rates' for labour

The 'all-in rate' for craft operatives and labourers varies from contract to contract due to varying conditions on site and the degree of supervision allowed in the build-up. The CIOB Code of Estimating Practice (CIOB 2008) illustrates a typical build-up.

Pricing the bill sections

Bill sections may be built up in analytical form with a separate labour, material and plant inclusion for each bill item. Rates are usually established net with the percentage for overheads and profit added at the adjudication stage. Alternatively, the estimator may use his own method of building up rates, based on his experience and knowledge of labour-only rates currently applicable.

Analysis of the quotations

This is a comparative analysis of prices received from various subcontractors and materials suppliers. Adjustments need to be made for differences in the quotations received to enable like-for-like comparisons to be made. For instance, allowances need to be made for items not priced, qualifications to the price, discounts allowed and facilities required from the main contractor for offloading, hardstandings, access, etc. (called attendances).

Build-up of the contract preliminaries

Preliminaries are the contractor's general cost items which do not form part of the permanent works. They include such items as site supervision, provision of temporary accommodation, on-site services and power, temporary access, hoardings, etc. Preliminaries items generally fall into two main categories: fixed costs (one-off charges) and time-related costs (those that vary with time). The estimator will often receive advice from the contracts manager and senior management when building up the preliminaries items.

Preparation of tender summary

This involves collating the estimate, analysing the overall price into labour, materials, plant and subcontractor elements and identifying other major items of cost such as preliminaries, prime cost and provisional sums, insurances and bonds. A tender summary sheet will be prepared by the estimator.

Attending the tender adjudication meeting

The estimator will present a report to management summarising the approach taken to the pricing of the tender and identifying the key risk issues and allowances which will need to be considered by the adjudication panel.

The estimator is usually the one who gets the blame for winning tenders that ultimately finish up losing money. It is so easy for site management to blame their own inefficiencies on bad estimating. The estimator always gets the negative feedback but is never told when his rates make money. Estimating is a no-win situation!

5.14 Pre-tender method statement

The pre-tender method statement outlines the sequence and method of construction upon which the tender is to be based and supports the pre-tender programme which forms the basis for calculating the cost of preliminaries.

For civil engineering tenders, greater emphasis is placed on construction method at tender stage because different methods of measurement are used in civil engineering and the methods of pricing are also different.

On building projects, there is normally a detailed bill of quantities, usually prepared on the basis of the Standard Method of Measurement (SMM7), and the 'unit-rate' method of estimating is mostly used, i.e. a price per unit of measured work (m/m^2/m^3 etc).

Bills of quantities for civil engineering works are usually based on either the Civil Engineering Standard Method of Measurement (CESMM) or the Highways Method of Measurement and the items are mostly 'composite'. (Each measured item includes several items of work that would be measured separately under SMM7.) For instance, the item coverage for a deep excavation would include excavation, disposal, earthwork support, trimming and preparation, etc., whereas these items would be measured separately under SMM7.

Consequently, 'operational' pricing is often used for civil engineering work where the estimator considers the time to carry out the operation, the type of plant required, how many excavators and waggons will be needed and what the labour gang size will be. He will calculate the total quantity of materials required, including waste, and the type and extent of temporary works such as sheet piling. A total amount of money is thereby determined for the operation. This may then be divided by the total quantity to give a rate per unit of measured work.

5.15 Preparation of pre-tender programme

In the majority of construction contracts, it is usual to have an agreed period within which the works are to be completed. This gives certainty to the contract and allows key events to be triggered. These include extensions of time for completion and the charging of liquidated and ascertained damages for delay.

The period for completion will either be stipulated within the tender documents or stated in the contractor's tender offer. On occasion there will be a stipulated period and also an opportunity for tendering contractors to state the period within which they propose to carry out the works as an alternative for consideration by the client.

As part of the tendering process, the contractor will have to think about how the job will be built so that the tender figure is realistic and technically achievable. The contractor cannot simply rely on the contract period stated in the tender documents for building up the costs of preliminaries and major items of plant.

A pre-tender programme is essential in order that the contractor may be reasonably sure that he will be able to complete the project on time, as there could be severe financial consequences for failure to do so. For instance, time overruns can result in additional costs and there may also be liquidated and ascertained damages to pay to the employer for failure to hand over the building on time.

The pre-tender programme will be presented in sufficient detail to enable the estimator to price the time-related elements of the tender and enable the adjudication panel to judge the time- and method-related risks attached to the project.

When the programme has been thought through, it may be possible to tender on a reduced timescale to that which is stipulated in the tender documents. This will facilitate a reduction in the time-related elements of the preliminaries and thus give the contractor a competitive edge over other tenderers.

On the other hand, the contractor may find that the stipulated period is insufficient to complete the job safely and to the desired quality. It is no use tendering for a contract on the basis of a 30-week programme if the contractor considers that he needs 38 weeks to construct the project. In such circumstances, the contractor may qualify his tender to this effect. This is done by enclosing a letter of qualification with the tender submission.

However, it might be a condition of tendering that qualifications are not allowed. In this event, the contractor will have to make an allowance in his bid for the cost of overrunning the programme (i.e. extra preliminaries) and he would also be wise to make an allowance for any liquidated and ascertained damages that might be charged to him for failure to complete on time.

Where the contract period is specified, it is necessary for the contractor to assess whether this period is realistic or not.

At tender stage, the pre-tender programme is normally presented in bar chart format. The programme will cover the major stages of the work to be undertaken on the basis of the information available from the tender drawings. The use of linked bar chart techniques based on project management software will help the contractor to present a professional-looking programme. A well-presented programme may be important because it is now becoming common practice for the pre-tender programme and method statements to be submitted with the priced bid. This is because some clients like to assess bids on the basis of programme and quality as well as price.

5.16 Build-up of contract preliminaries

The purpose of the contract preliminaries is to include for the pricing of items which cannot be reasonably included in the measured work. Contract preliminaries are priced on a fixed and time-related basis. Certain preliminary items relate to a one-off or fixed charge to the project, that is the erection of cabins and site accommodation, while other items are directly related to the contract duration, that is the utilisation and weekly hire costs of using a tower crane.

The division of the preliminaries costs into fixed and time-related elements is directly reflected in additional charges to the client in relation to extensions of time and claims for loss and expense. It has been shown that tender success rates improve where management involvement is evident in the build-up of the contract preliminaries.

The CIOB Code of Estimating Practice (CIOB 2008) includes an extensive project overheads schedule to assist in the build-up of preliminaries costs. This ensures a logical approach to the assessment of project overheads.

Preliminaries checklist

Employer's requirements

- Time-related costs include allowances for site accommodation and telephone facilities for the use of the architect and other consultants; also, hire costs of a computer facility if specified in the documents.
- Fixed costs include the erection and dismantling of accommodation, fitting out and decorating, furniture and fittings, technical and surveying equipment, telephone installation and site notice boards.

Contractor's management and site staffing

- Time-related costs include site staff allocated to the project, including site manager, trades supervision, site engineering staff and site support staff. In certain organisations, it may be policy to treat staff servicing the site as part of the head office overheads – contracts managers, surveying and planning personnel, etc. This depends on the policy within the organisation and their approach to the pricing of project overheads.

Site accommodation costs

- Time-related costs include hire charges for all required site accommodation, including the servicing of such facilities.
- Fixed costs include the delivery to site, erection, fitting out and the dismantling and removal on completion.

Attendance on domestic and nominated subcontractors

- Time-related costs include unloading and distribution of materials, attendance on domestic subcontractors and progressive clearing of building waste.
- Fixed costs include the cleaning out of the building on completion of the works.

Facilities and services associated with the works

- Time-related costs include weekly costs of telephones, office equipment and stationery, safety, health and welfare provision, power and lighting for offices and site security costs per week. This may also include such items as the removal of site rubbish, drying out buildings and office cleaning services.
- Fixed costs include installation and connection charges relative to the above. The majority of the costs will be involved in temporary service connections.

Temporary works costs

- Time-related costs include maintaining, cleaning and servicing access roads, hard standings, compounds and hoardings. Monies may also be included for complying with traffic regulations and work in connection with attendance on pumping and de-watering.
- Fixed costs include the initial installation and removal of items such as access roads, hard standings, compounds and hoardings. Items in relation to pumping and de-watering may be included separately.

Mechanical plant

- The selection of plant for a project will have a direct effect on preliminaries costs and ultimately the competitiveness of the tender. The pre-tender method statement requires careful review in the light of proposed alternatives, and cost exercises may need to be undertaken. Installing a track for a tower crane may prove expensive and must not be considered lightly at estimate stage. Consideration must also be given to the provision of power supplies to large items of plant. All these factors may force the contractor into taking more risk at the tender stage.
- Time-related costs include the hire charge per week of cranes, hoists, mixers, vibrators, compactors, dumpers, access equipment and mobile elevating working platforms (MEWPs), tele-handlers and forklift trucks and site vehicles for the collection of goods. Plant items

such as excavators may be included in the built-up rates. The contractor's approach to the inclusion of mobile concrete pumps is debatable, as this may be treated as a preliminary cost in the mechanical plant section or included in the unit rate for concrete work.

- Fixed costs include the transport of the plant to the site, cost of receiving and installing and removal on completion. This should include any additional costs in providing temporary works such as the track or base for tower cranes.

Non-mechanical plant (or small tools and equipment)

- This includes such plant items as bar bending machines and provision of saw benches, ladders, wheelbarrows, road barriers and lamps, slings and chains, skips and surveying equipment. The main item included here is scaffolding, which will carry the bulk of the cost in this section.
- Quotations will normally be obtained for scaffolding as part of the subcontract enquiry process. The contractor may also include as a fixed-cost items such as power tools, including the transportation to and from site. Many contractors tend to price non-mechanical plant as a percentage addition on the labour cost (typically 1–2%).

The preliminaries are an important element of the contractor's tender and may represent between 5 and 15% of the tender sum. However, if the contractor included every conceivable item in the contract preliminaries section, he would never win a competitive tender. Clearly, the contractor needs to be competitive and it is common for directors to shave off money from the preliminaries to help the contractor win the contract. This is often done in conjunction with considerations of the project programme, especially where the contractor decides to tender on a shorter duration than stipulated in the tender documents.

Contractual risk must, however, come into the equation somewhere. As previously stated, the contractor tends to win the tender on items he has not included rather than those he has.

5.17 Management adjudication of the estimate

When the estimator has completed the pricing, a summary sheet will be prepared which will indicate a bottom-line figure. This is the total estimate for the project. Prior to submitting the tender at the appointed time, the directors will have to consider the estimator's calculations and make some decisions, including:

- Is the estimate accurate?
- Have the items been priced sensibly?
- Are the subcontract prices used achievable?
- Can the prices used be bettered if we win the contract?
- Is the pre-tender programme realistic?
- Can the programme be shortened?
- What are the risk issues related to this job?
- How much does the company need the work?
- What is the level of competition for the work?
- Does the tender documentation offer any commercial opportunities?

When these questions have been answered, the directors will be in a position to:

- Agree with the estimator's price
- Add money to the price or
- Reduce the price

Whatever the decision, this will be the tender sum which will be offered to the employer via the form of tender. If the tender is accepted, the contractor will have to live with the decisions made and get on with building the project for the price offered. The pressure will then be on the site team to work within the budget and return the desired margin (profit).

The CIOB Code of Estimating Practice proposes a model form for the estimate summary, analysis and report.

Adjudication checklist

The following checklist provides an aide-memoire to assist the tender adjudication process and make sure that key issues are not overlooked. All information generated in the build-up of the tender figure needs to be included. For a decent-sized tender, the amount of supporting documentation may be substantial, as follows.

General information

- Tender enquiry form including details of parties involved in the contract
- Site visit report
- The estimator's report
- The quantity surveyor's commercial risk assessment
- Knowledge of competitors
- Details of current and future contractual commitments

Pricing data

- Net cost analysis including the value of work to be sublet
- Analysis of subcontract and materials prices
- Schedule of prime cost and provisional sums
- Schedule of project overheads
- Tender summary

Pre-tender planning data

- Overview of construction methods allowed in estimate
- Planning report from the planner/contracts manager comprising comments on the contract period given by the client and the feasibility of the pre-tender programme
- Pre-tender programme (linked bar chart)
- Pre-tender method statements

Risk issues

- Ground conditions report
- Temporary works allowances
- Design liability (if any)
- Summary of contract conditions including any amendments to the standard form
- Details of the damages for late completion
- Details of insurance requirements and bonds
- Working capital implications for the contractor's borrowings
- Any cash funding requirements required

Finally, the contractor must consider the desired markup in the context of the quality of tender information, the technical and contractual risks identified in the tendering process and the construction time and methods allowed. Profit percentages could be anything from 1 to 20% depending on how keen the contractor is to win the contract.

In a very keen market, the contractor may look for ways of taking money out of the tender by anticipating discounts and buying savings from subcontractors and suppliers or by gambling on possible claims and other commercial opportunities that might arise during the contract. At this stage, it may be better to hire a crystal ball!

A percentage addition is normally applied to cover head office overheads. This might be in the range of 5–10% depending on the size of the organisation.

A typical agenda for the adjudication meeting is given below:

AGENDA FOR TENDER ADJUDICATION MEETING

Project title
Tender no.
Date

Personnel present
Project details
- Client
- Description of works
- Budget (if known)
- Form of contract
- Liquidated and ascertained damages
- Completion period

Project strategy
- Construction methods
- Programme
- Site establishment
- Technical issues
- Health, safety and environmental issues

Tender strategy
- Market conditions
- Risk issues

Tender analysis
- Net cost of labour, plant and materials
- Subcontractors and discounts

- Preliminaries
- Temporary works
- Markup

Risk management
- Risk register
- Payment terms
- Insurances
- Onerous contract conditions
- Programme

Submission documents
- Priced bills of quantities
- Contract sum analysis
- Pre-tender programme

Tender sign-off
- Risk allowances
- Final tender adjustments
- Final tender figure agreed

Once the final adjudication decisions have been made, the tender summary sheet will have to be revised and any changes expressed in the tender submission. This may be a fully priced bill of quantities or simply a figure stated on a form of tender.

The form of tender will have to be signed by a responsible person and submitted in the appropriate envelope. If alternative proposals have been invited from the contractor, these should be clearly stated.

Quite frequently contractors are asked to submit two envelopes containing both a price and a quality bid. This enables the client team to judge the contractor's approach to the management of the project as well as the price. The quality bid is usually opened first and judgements are made on preselected criteria to give a quality score for the tender. If the quality bid is satisfactory then the price envelope is opened. Choice of the successful tenderer will then be made on the basis of a weighting of both price and quality (say 60/40). The CIB Code of Practice (1997) explains the tender assessment process and Morledge *et al.* (2006) gives a worked example of a tender weighting calculation.

5.18 Analysis of tender performance

The application of bidding theory is not common in construction but it is not altogether unheard of either. This is the practice of judging tender performance in a statistical and analytical form. However, many practitioners consider the technique to be of little benefit and purely the preserve of academics. Bidding theory is relevant only in stable market conditions, where the contractor is bidding against known competition in the market place. In the current volatile construction climate, this is rarely the norm. Changes in procurement procedures have led contractors to enter into partnering arrangements, with more emphasis on negotiated contracts, and hence emphasis on bidding strategies is less relevant.

Once a bid has been submitted and the contractor has committed himself to a contract price, it is no good reviewing what he should have done in the light of other competitor bid prices. 'If only we had done this or adjusted that' is crying over spilt milk and the only alternatives are to sack the estimator – because surely it is now all his fault! – or just get on with the job. Whatever the case, management rarely takes the blame for adjudication decisions.

On acceptance of the contractor's bid price, procedures will be implemented to check the priced bill, agree any bills of reduction if tenders are over the client's budget, and arrange for contracts to be signed. Letters of intent may be issued so that the contractor can prepare designs or order materials with long lead times.

Before commencing work on site, the pre-contract planning process must be started, and this is dealt with in Chapter 13.

References

Bentley, J.I.W. (1987) *Construction Tendering and Estimating.* E. & F.N. Spon.

Brook, M. (1993) *Estimating and Tendering for Construction Work.* Butterworth-Heinemann.

CIB (1997) Construction Industry Board – Working Group 3. *Code of Practice for the Selection of Main Contractors.* Thomas Telford.

CIOB (2008) *Code of Estimating Practice,* 6th edn. Chartered Institute Of Building.

CIS (The Construction Information Service) (2008) Available through IHS (formerly Technical Indexes Ltd.) at http://uk.ihs.com/products/products/htm.

Department for Business, Enterprise and Regulatory Reform (BERR) (2007) *Construction Statistics Annual.* The Stationery Office.

Hughes, W. (January 2004) Are bidding costs wasted? *Construction Manager,* Journal of the Chartered Institute of Building, pp 18.

Morledge, R., Smith, A. & Kashiwagi, D. (2006) *Building Procurement.* Blackwell Publishing.

Smith, A.J. (1995) *Estimating, Tendering and Bidding for Construction.* Macmillan.

Twort, A. C. & Rees, J. G. (2004) *Civil Engineering Project Management.* Elsevier Butterworth-Heinemann.

6 Managing the supply chain

6.1 Supply chain integration

In the main, the construction industry is concerned with one-off projects, geographically dispersed and carried out by short-term teams of designers, contractors, subcontractors and suppliers who have been assembled specifically for the project, only to be disbanded once the contract is complete.

In order to overcome this problem, some major clients have assembled their own standing lists of preferred suppliers so that their projects can be managed on a more long-term and integrated basis while retaining an appropriate degree of competition and control.

These suppliers, or supply chain partners, include architects, engineers, main contractors, specialist contractors and suppliers of a variety of goods and services appropriate to the client's projects. In the wider context of the industry at large, however, this principle tends not to penetrate the first tier of the supply chain so as to facilitate long-term relationships between contractors and *their* subcontractors and suppliers.

Consequently, one-off clients to the industry are unable to reap the benefits of supply chain integration to the same extent as repeat customers do and this leads to poor performance and client dissatisfaction.

Supply chain integration does not, however, rely entirely on successive projects for the same client to ensure collaboration and the benefits of long-term relationships and understanding.

Under the traditional system of contracting, architects and other designers rely heavily on many specialists and suppliers for design information and advice. However, unless these firms are nominated under the contract, the main contractor is likely to choose different companies to do the work once the contract has been awarded. This is not conducive to efficient working.

Informed thinking suggest that smaller and more integrated supply chains are the answer to demands from clients for increased productivity from the industry and greater regard to completion on time, budget certainty and higher standards of quality, and health and safety management. To this end, many of the top firms in the industry have changed their culture entirely by having:

- Fewer and better clients
- Fewer and better suppliers and subcontractors

Large contractors may have between 5000 and 15 000 suppliers of various sorts:

- Specialist subcontractors
- Material suppliers
- Builders merchants
- Plant hire firms
- Hire firms for small tools
- Office equipment suppliers
- Stationery suppliers
- Computer equipment and software suppliers
- Domestic firms of cleaners
- Suppliers of cleaning equipment
- Suppliers of tea, coffee and milk

The reasons for such large networks of suppliers lie in the outmoded belief that this is the way to encourage competition and thereby lower prices. The effect is quite the contrary and a great deal of time and effort is needed to obtain competitive quotations from two, three or six suppliers every time supplies or services are required.

By reducing the number of supply chain members and by entering into long-term relationships with them, many time-consuming and costly formalities can be avoided, thus adding value to the supply chain.

Therefore, instead of following the procurement procedures outlined in this chapter for, say, 6 subcontractors for each of 30 or more trades in a typical contract, the contractor can follow the same procedures with fewer subcontractors. The time saved can then be spent discussing long-term price frameworks, delivery schedules, construction programmes and health and safety and quality issues.

One of the touchy aspects of supply integration is the need to assure profits. These clearly need to be ring-fenced but many suppliers and subcontractors might be wary of revealing too much information to the contractors with whom they do business. However, in order to follow the Toyota model referred to in the Egan Report, the process of team working and integration must penetrate the entire supply chain in order to reap the available benefits.

Many examples of supply chain integration may be taken from practice, such as the major client who is constructing 20 motel complexes each year at an annual cost of £30 million. The client is able to work out a deal with a preferred contractor and both the client and contractor can integrate major suppliers and specialists into the project programme at an early stage.

Consequently, construction lead times are vastly reduced and just-in-time delivery can be assured for items such as the timber frame package, plumbing and electrical fit-out packages, the supply of fitted out bathroom and bedroom pods, etc. The whole supply chain is assured of continuity of work and the contractor assured of his ring-fenced profit margin.

6.2 Lean construction

One of the great benefits of an integrated supply chain is the opportunity to apply lean thinking to the construction process. The Egan Report 'Rethinking Construction' (Construction Task Force 1998) advocated this principle which Cartlidge (2002) defines as *the elimination of waste from the production cycle'. Cartlidge explains that 'every time waste is removed from the supply chain, value is added to the process, leading to lower costs, shorter construction periods and greater profits'.*

Part of the lean thinking approach is the use of just-in-time production where materials and components are manufactured, transported and delivered to site as and when required without the need for long lead times and stockpiling on site.

'Rethinking Construction' describes lean production as a generic version of the Toyota Production System, which is recognised as the most efficient in the world. The system is based on lean thinking principles including:

- Elimination of non-value activities which can represent up to 95% of time and effort
- Removal of waste from all activities involved in delivering the end product
- Establishment of relationships with all members of the supply chain
- Removal of delays in the design and production process using just-in-time management

Lean thinking can be successfully applied to the construction process through innovative design and assembly, including the use of off-site manufacture, prefabrication, pre-assembly and supply chain integration.

6.3 Fast-track construction

Many clients to the construction industry now demand early delivery of their project require-ments. This is especially the case with commercial clients who are looking for shorter and shorter construction periods so as to ensure that the completed facility is on line and earn-ing revenue at the earliest possible moment. As a consequence of this demand, procurement methods which overlap design, tendering and construction have emerged, but alongside these developments other approaches to the management of construction projects have developed including fast-track construction.

Kwakye (1997) describes the fast-track system as a management approach aimed at the early completion of the construction phase using a combination of innovative procurement methods, industrialisation of the construction process and the use of work package contractors in order to benefit from their expertise, especially as regards their design input. The benefits of the system include:

- Overlapping of work packages both during design and construction
- Less duplication of effort and waste
- Less uncertainty and inefficiency at work package interfaces
- The use of innovative construction methods
- Incorporation of cutting-edge technologies
- More emphasis on the standardisation, pre-assembly and modularisation of the construc-tion process

As a consequence, the fast-track system has the propensity to deliver completed projects in remarkable timescales. The result is much faster on-site construction periods such as a new superstore in 11 weeks and a new fast food outlet in 2 weeks.

The intensive construction programme which results from this approach requires high standards of planning, organisation and control. It is essential to ensure that everything is right first time and that there are zero accidents so as to prevent unnecessary delays and disruption to the programme.

Fast tracking of projects does not happen by accident and a great deal of planning and preparation is necessary to create the right client–supplier relationships. These are characterised by open and honest dealing, encouragement of innovation and the use of performance specifications in order to:

- Improve buildability by involving contractors in the design phase
- Involve first tier suppliers and works package contractors at an earlier stage
- Reduce construction lead times through pre-ordering of key materials and components (e.g. structural steel, timber frame, concrete and cladding, partitions, ductwork, air handling units, modular components, etc.)
- Arrange the direct supply of materials and components where repeat clients can benefit from volume discounts and higher quality by dealing direct with key suppliers

6.4 Subcontractors

There are very few general contractors around in the construction business these days as most projects rely on a managing contractor to organise and supervise the work and a series of specialist subcontractors who carry out the site operations.

Subcontracting is the vicarious performance of a contractual obligation where:

- The main contractor normally enters into a direct contract with each subcontractor.
- Subcontractors may sublet their work to sub-subcontractors.
- The main contractor retains contractual responsibility for the subcontractor's work.
- Subcontractors are commonly referred to as:
  - Work package contractor
  - Trade contractor
  - Specialist contractor
- Attendances are normally provided by the main contractor, which may include the provision of water and power, use of standing scaffolding, use of tower crane and removal of rubbish.
- Special attendances may be supplied, such as offloading of materials, provision of hard-standings, task lighting, etc.

6.5 Types of subcontractor – traditional procurement

Domestic subcontractor

- A subcontractor chosen and engaged directly by the main contractor to carry out a particular trade
- May be labour-only or supply and fix

- Examples – groundwork, brickwork, roof tiling, plastering. Usually appointed under a standard form of contract or using the main contractor's bespoke conditions. Under JCT 05 Standard Building Contracts, the contractor requires the architect's approval to sublet any part of the works but approval of particular subcontractors is not required
- The ICE Conditions require the contractor to notify the engineer of the extent of work to be subcontracted and the names and addresses of each subcontractor. The engineer may object to the employment of any subcontractor

Nominated subcontractor

- A subcontractor chosen by the architect or engineer (ICE7) to carry out specialist work
- Appointed on the expenditure of a provisional sum or prime cost sum included in the contract bills
- The nominated subcontractor enters into a contract with the main contractor but there may be a collateral warranty between the nominated subcontractor and the client. Nomination is not possible under the JCT 2005 Standard Building Contracts or the Engineering and Construction Contract

Named or listed subcontractor

- Under the JCT 05 Standard or Intermediate Building Contracts, the architect may provide the contractor with a list of at least three subcontractors to choose from for carrying out particular parts of the project
- The contractor can add to the list but, in any event, the subcontractor is a domestic subcontractor under the contract.

Work package contractor

- Often used as a synonym for domestic subcontractor but might also imply responsibility for a specific section or package of the works, possibly including design work
- May provide a complete service with little or no reliance on main contractor attendances

Labour-only subcontractor

- Self-employed individuals, partnerships or small firms
- Higher earnings causing disparities with other workers
- Lower tax and national insurance contributions
- Lower prices but higher overheads for the main contractor
- Lower employment costs for the main contractor (e.g. no holidays with pay, redundancy and sick pay)
- Higher output and faster work leading to quality problems
- Higher materials wastage
- Lack of training and competence testing

- Supplement skills shortages in the industry
- Lower standards of safety

6.6 Types of subcontractor – management procurement

Trade contractors

- The contractors who carry out the work under the JCT 05 Construction Management Trade Contract
- Engaged and paid directly by the client

Works contractors

- The subcontractors responsible for carrying out the works under the JCT 05 Management Works Contract
- Engaged and paid directly by the management contractor

6.7 Work package procurement

The procurement of subcontract work packages is somewhat different under management contracting and construction management contracts compared with the arrangements and negotiations on a traditional JCT 05 contract. For instance, the management contractor or construction manager has to work in conjunction with the design team's quantity surveyor in order to produce a workable budget. This may take the form of an overall budget developed from an assessment of the budget figures for individual work packages.

Deciding how to package the project is another principal difference compared to conventional contracts. Sidwell (1983) suggests a flow chart approach to the process of establishing work packages, and this concept is illustrated in Figure 6.1. However, the packages cannot be thought of as subcontract enquiries in the normal sense and careful thought is needed so as to avoid too many interface problems on site. These interfaces could lead to the situation where none of the package contractors has included for a particular item of work, or duplication may occur where several firms are responsible for the same item. Either way, extra cost or disputes could result. Also, procurement of the work packages requires special negotiating skills and package contracts have to be carefully set up so as to reduce the risk of disputes.

The process involves close liaison between the design team and the appointed management contractor or construction manager in order to establish procedures for securing the best buy.

Figure 6.2 illustrates a typical work package budget for a science park project involving the construction of an office block of some 3000 m^2 in floor area. The work packages are established at the design stage and budget figures have been produced from the scheme drawings on the basis of approximate quantities and current rates.

Provision has been made on the work package budget form for entering actual work package tender prices in order to identify variances in the budget, and to update the budget as the project develops. The individual budgets may not always be achieved in practice, of course, and a swings

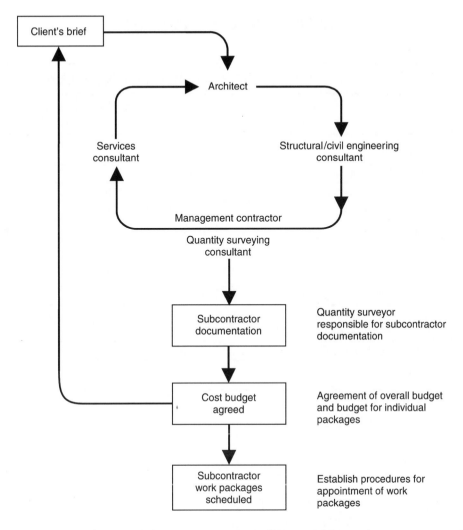

DETERMINATION OF WORK PACKAGES

Figure 6.1

and roundabouts approach may have to be taken while at the same time making sure the overall budget is maintained.

Many of the industry's experienced management contractors and construction managers adopt standard procedures for the selection and appointment of work package contractors, and examples of these are outlined below.

Establishing work package information requirements

When the scope of the various packages has been established, it is necessary to assess lead-in times for each of them prior to commencing the procurement process. In addition, information

WORK PACKAGE BUDGET

PROJECT — UNIT ONE — WARRINGTON SCIENCE PARK				
PACKAGE REF.	WORK PACKAGE	BUDGET	ACTUAL	SAVING/ INCREASE
SUBSTRUCTURE				
100	FOUNDATIONS	110 000	100 000	−10 000
101	DRAINAGE	25 000	18 000	−7 000
102	CAR PARKS	90 000	95 000	+5 000
103	LANDSCAPING	30 000		
STRUCTURE				
200	STEEL FRAME	250 000	245 000	−5 000
201	FLOOR/ROOF DECKS	40 000		
202	ROOF FINISHES	20 000		
203	EXTERNAL ENVELOPE	220 000		
204	EXT. WINDOWS/DOORS	60 000		
FINISHES				
300	INTERNAL PARTITIONS	40 000		
301	FLOOR FINISHES	35 000		
302	CEILING FINISHES	25 000		
SERVICES				
400	HEATING	150 000		
401	ELECTRICAL	100 000		
402	AIR CONDITIONING	70 000		
403	LIFTS	80 000		
404	FIRE AND ALARM	20 000		
CONTINGENCY		50 000		
TOTALS		1 415 000		
INITIAL BUDGET ASSESSMENT		1 500 000		
FINAL BUDGET ASSESSMENT				

Figure 6.2

requirements relative to each of the work packages have to be assessed and this will require the design team to work closely with each other and the management contractor/construction manager as they will have to produce the information by agreed dates prior to sending out work package enquiries.

A typical set of key dates is given below, which could be used for each of the work packages in a procurement schedule:

- Agree work package list
- Design information complete – architect/engineer/services
- Design packages to the quantity surveyor
- Schedules and quantities complete
- Tender package issued
- Enquiries sent to work package subcontractors
- Tenders returned
- Technical checks complete
- Work package negotiations complete
- Appoint work package subcontractor – letter of intent
- Arrange work package contract to be signed
- Working drawings issued
- Lead time required
- Commencement date on site planned

The establishment of key dates for each of the above items places extensive responsibility on the management contractor/construction manager. First, he has to motivate the design team to produce the information on time, clearly pointing out the consequences of failing to meet the programmed dates. Just one uncooperative member of the design team can cause the best-planned project to fall apart.

The key date requirements relative to each of the work packages may be linked to the project planning software being used for the project. Many of these packages produce key date schedules relative to pre-contract activities developed from the project network diagram.

Figure 6.3 shows a work package procurement schedule for a major project. This enables planned, current and actual dates for each work package to be monitored as the packages are awarded.

Arrangements for procuring the work packages

Selection of the individual work packages often follows a standard procedure which may also be part of a quality system. It should be noted that the extent to which selection procedures are strictly adhered to may depend on the degree of risk allocated to the work package contractor.

For each of the work packages the following selection routine may be followed:

- Select eight potential work package contractors
- Shortlist six work package contractors
- Send questionnaire to each
- Financial checks
- Visit current contracts

PACKAGE		Issue tender documents	Collate tender documents	Out to tender	Return tender	Tender appraisal	Client review	Place order	Lead-in time (weeks)	Start date on site
100/1 Brickwork and blockwork	Planned	28/3/08	1/4/08	9/4/08	2/5/08	7/5/08	28/5/08	2/6/08	13	1/9/08
	Current	28/3/08	1/4/08	9/4/08	2/5/08	7/5/08	28/5/08	2/6/08	13	1/9/08
	Actual	28/3/08	1/4/08	9/4/08	7/5/08	9/5/08	28/5/08	2/6/08	15	15/9/08
	Change	-	-	-	-	-	-	-	-	-
200/1 Wall cladding excl glazing	Planned	11/4/08	15/4/08	18/4/08	16/5/08	20/5/08	3/6/08	6/6/08	17	6/10/08
	Current	11/4/08	15/4/08	18/4/08	16/5/08	20/5/08	3/6/08	6/6/08	17	6/10/08
	Actual	11/4/08	15/4/08	18/4/08	16/5/08	20/5/08	3/6/08	Discuss rates		
	Change	-			-		-			
200/2 Rooflights	Planned	15/2/08	19/2/08	29/2/08	28/3/08	1/4/08	15/4/08	18/4/08	16	11/8/08
	Current	In abeyance								
	Actual									
	Change									
200/3 External glazing panels	Planned	25/1/08	29/1/08	8/2/08	21/3/08	25/3/08	15/4/08	18/4/08	24	6/10/08
	Current	25/1/08	29/1/08	8/2/08	21/3/08	25/3/08	15/4/08	18/4/08	24	6/10/08
	Actual	25/1/08	29/1/08	8/2/08	21/3/08	25/3/08	Postponed			
	Change									
300/1 Roofing and northlights	Planned	4/4/08	9/4/08	11/4/08	9/5/08	13/5/08	23/5/08	30/5/08	15	15/9/08
	Current	4/4/08	9/4/08	11/4/08	9/5/08	13/5/08	23/5/08	30/5/08	15	15/9/08
	Actual	2/5/08	5/5/08	7/5/08	23/5/08	27/5/08	10/6/08	13/6/08	19	27/10/08
	Change	Design delay						Import delay	Estimated	Mid Nov
400/1 Plaster/screeds	Planned	15/8/08	19/8/08	29/8/08	26/9/08	30/9/08	14/10/08	17/10/08	5	24/11/08
	Current	15/8/08	19/8/08	29/8/08						
	Actual	15/8/08	19/8/08	29/8/08						
	Change	-		-						

WORK PACKAGE PROCUREMENT SCHEDULE

Figure 6.3

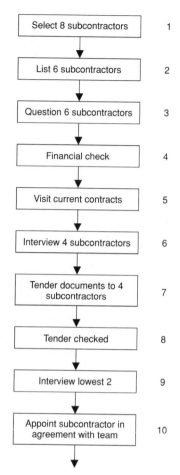

Select 8 subcontractors	1
List 6 subcontractors	2
Question 6 subcontractors	3
Financial check	4
Visit current contracts	5
Interview 4 subcontractors	6
Tender documents to 4 subcontractors	7
Tender checked	8
Interview lowest 2	9
Appoint subcontractor in agreement with team	10

SUBCONTRACTOR/WORK PACKAGE SELECTION

Figure 6.4

- Reduce shortlist and interview four contractors
- Tender documents to four contractors
- Tenders received and checked
- Interview lowest two tenderers
- Appoint work package contractor in agreement with the design team

This procedure is illustrated in Figure 6.4.

6.8 Forms of subcontract

Table 6.1 lists the main standard forms of subcontract available but there are, of course, many variations on a theme and subcontractors are frequently faced with contract conditions which they have never seen before.

Table 6.1 Standard forms of subcontract

Name of subcontract	Main provisions
JCT 05 Standard Building Subcontract	Used in conjunction with JCT 05 Standard Building Contract Provision for including procurement periods for materials and fabrication and for notice period for commencement Includes provisions for sectional completion Gives earliest and latest starting dates for the subcontract works Contractual obligation to provide information for the health and safety plan and file
JCT 05 Construction Management Trade Contract	Used in conjunction with JCT 05 Construction Management Contract Commencement date and completion period stated in appendix Interim payments apply Stage payments optional
JCT 05 Management Works Contract	Used in conjunction with JCT 05 Management Building Contract
Civil Engineering	Used in conjunction with ICE Conditions
Contractors	Period for completion stated in third schedule to contract
Association Form	Subcontract payment conditional on main contract payment (pay when paid)
The Engineering and Construction Contract Subcontract	Used in conjunction with ECC The Accepted Programme is prepared by the subcontractor and submitted to the contractor for acceptance A method statement with resources to be provided for each operation Accepted programme to be updated by subcontractor Early warning provisions apply
ACA Standard Form of Specialist Contract for Project Partnering (SPC2000)	Used in conjunction with PPC2000 Early Warning System and Problem Solving Hierarchy to operate between constructor and specialist Specialist works to be carried out in accordance with the specialist timetable Specialist to update timetable at specified periods Acceleration provisions included in contract

6.9 Construction Industry Scheme

In order to remove opportunities for the avoidance of income tax and national insurance contributions in the construction industry, since April 2007 payments made by 'contractors' to 'subcontractors' working in construction are subject to HM Revenue and Customs (HMRC) rules known as the New Construction Industry Scheme (New CIS; http://www.hmrc.gov.uk/newcis/). These rules are backed by statute.

Most 'contractors' and 'subcontractors' in the construction industry are affected by the scheme. However, under the scheme, the terms 'contractor' and 'subcontractor' have a much wider meaning than normal. Consequently, government departments and agencies, local authorities, hospital trusts and other public bodies, airport operators and large private businesses, normally known as clients, are 'contractors' under the New CIS and large 'contractors' are 'subcontractors' under the scheme.

Under the New CIS, all payments to subcontractors must take account of the subcontractor's tax status as determined by HMRC. Subcontractors who meet certain business, turnover and compliance tests may be paid without deduction of tax provided that the subcontractor's payment status has been 'verified' with HMRC by the 'contractor' making the payment.

In cases where the subcontractor is registered with HMRC, payments will be made gross (i.e. with no deductions) unless the subcontractor has chosen otherwise. In such cases the subcontractor's tax and national insurance liability must be deducted from the labour element of any payments to the subcontractor.

Under the scheme, contractors must make monthly returns to HMRC of all payments made to subcontractors and they must also make sure that subcontractors who are not limited companies are genuinely self-employed. Therefore, anyone who works regularly for, say, a painting and decorating contractor must be able to prove that they are in business in their own right or else, even though they may be self-employed, they will be construed as employees and therefore subject to PAYE deductions.

New CIS is a major change to the tax laws as they apply to construction and they are aimed at ensuring that the privilege of 'subcontractor' tax status is not abused by people who are really employees. These changes should not affect the 'normal' contractor–subcontractor relationships provided that subcontractors are properly registered with HMRC.

Private householders and non-construction businesses, etc., which spend less than £1 million a year on construction work are not regarded as 'contractors' under the scheme.

References

Cartlidge, D. (2002) *New Aspects of Quantity Surveying Practice.* Elsevier Butterworth-Heinemann.
Construction Task Force (1998) *Rethinking Construction.* Department of Trade and Industry.
Kwakye, A.A. (1997) *Construction Project Administration in Practice.* Pearson Education.
Sidwell, A.C. (1983) An evaluation of management contracting. *Construction Management and Economics,* Vol. 1. E. & F. Spon.

7 Managing risk

7.1 Risk in construction

Construction is undeniably a risky business for many reasons, including:

- Poor record of cost and time certainty for clients
- Adversarial attitudes and high levels of disputes and litigation
- The intense competition for work
- Low margins and profit risk
- The industry's poor safety and occupational health record
- Pressure from management and shareholders to produce a high return on funds invested
- Pressure on construction teams, especially site management and operatives, to save time and money
- Pressure on health and safety provision

Raftery (1994) argues that construction is nothing special compared with other industries because we simply carry out projects to a specified timescale using teams of specialists for design and construction by marshalling appropriate resources to overcome the physical and technical problems involved. He also argues that construction projects exhibit much lower levels of technical complexity than aerospace, defence or computer software projects.

Smith (1999) distinguishes between risk and uncertainty in decision-making such that a risk is a decision having a range of possible outcomes to which a probability can be attached, whereas uncertainty exists if the probability of possible outcomes is not known. He suggests that risk falls into three categories:

1. Known risks – risks that are an everyday feature of construction
2. Known unknowns – risks which can be predicted or foreseen
3. Unknown unknowns – risks due to events whose cause and effect cannot be predicted

This point is also raised by Thompson and Perry (1992) and by Edwards (1995) who say that it is the unforeseen events which can have the most significant impact. Raftery goes on to judge that risk has to be recognised, assessed and managed and that overemphasis on risk avoidance leads to overcaution and negative attitudes.

7.2 The Turnbull Report

There are many risks in construction (in common with other industries) and this has been recognised by the Institute of Chartered Accountants who commissioned a report into the management of risk as an aspect of corporate governance. This is the Turnbull Report (ICA 1999) and the recommendations from this took effect from the beginning of the year 2001.

Effectively, company directors have to show that they have proper and ongoing procedures to manage the risks to which their organisations are exposed and they have to demonstrate this in their annual report and accounts. This is not a mandatory requirement but is equivalent to other accounting standards and practices which accountants recognise. These standards are important to the extent that auditors, when they are not satisfied that the appropriate standards have been applied, will qualify the accounts accordingly. This is effectively a vote of no confidence in the accounts and can have a dramatic effect on the standing of the company and its ability to attract investors and customers. Compliance with the recommendations of the Turnbull Report has been a condition of London Stock Exchange listing since January 2001.

Turnbull suggests that risk management is an 'ongoing process for identifying, evaluating and managing significant risks' and that this must be evident in the way the affairs of the business are conducted. The report identifies risk assessment as central to this process to the extent that 'the significant internal and external operational, financial, compliance and other risks' should be identified and assessed on an ongoing basis. These significant risks include:

- Market
- Credit
- Liquidity
- Technological
- Legal
- Health, safety and environmental
- Reputation
- Business probity

7.3 Project risk

Edwards (1995) suggests that identifying hazards is an essential part of a structured approach to risk management but that it is often unidentified hazards, for which no provision has been made, that have the most significant impact. Thompson and Perry (1992) observe that all too often risk is either ignored or dealt with in an arbitrary way on construction projects, and that the practice of adding a 10% contingency is typical industry practice. Alternatively, the contractor may 'take a view' to cover for any shortfall in the estimate or inaccuracy in the client's documentation.

Figure 7.1 illustrates that risk assessment will not remove all risks on a project and that both the client and contractor must recognise that residual risks will always remain no matter what provisions are made. The aim of risk management is to ensure that such risks are managed effectively by the party best able to do so.

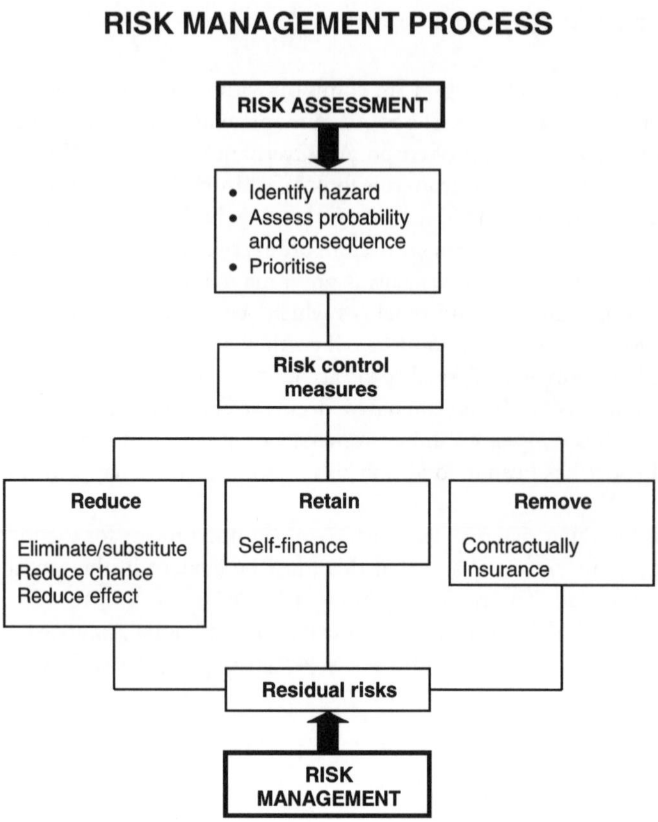

RISK MANAGEMENT PROCESS

```
              ┌─────────────────────┐
              │  RISK ASSESSMENT    │
              └─────────────────────┘
                         │
                         ▼
              ┌─────────────────────┐
              │ • Identify hazard   │
              │ • Assess probability│
              │   and consequence   │
              │ • Prioritise        │
              └─────────────────────┘
                         │
              ┌─────────────────────┐
              │   Risk control      │
              │     measures        │
              └─────────────────────┘
```

Reduce	Retain	Remove
Eliminate/substitute Reduce chance Reduce effect	Self-finance	Contractually Insurance

```
              ┌─────────────────────┐
              │   Residual risks    │
              └─────────────────────┘
                         ▲
              ┌─────────────────────┐
              │        RISK         │
              │    MANAGEMENT       │
              └─────────────────────┘
```

Figure 7.1

7.4 Risk management

The extent to which the risk can be reduced or controlled is the responsibility of management, whose function is to plan, organise, control, monitor and review the measures needed to prevent exposure to risk. To do this requires identification of the hazard, an assessment of the extent of the risk, the provision of measures to control the risk and the management of any residual risk remaining.

Unfortunately, despite all the steps that designers and managers can take, humans and human systems are imperfect and accidents and losses can and do happen.

Therefore, a roofer working at height has a chance of falling and suffering serious or fatal injury unless:

- The work can be avoided (i.e. designed out), or
- Suitable measures can be installed to prevent him falling, and
- Suitable safeguards can be provided should he happen to fall in any event

One common method of managing risks is to devise a risk register for the project. An example is shown in Figure 7.2. The CIOB Code of Practice for Project Management (CIOB

RISK REGISTER

Item	Description	Risk			Without controls			Controls	Residual risk	Action
		H	M	L	Cost impact	Time impact	Other			

Figure 7.2

2002) defines this as a 'formal record for risk identification, assessment and control actions' which may be divided into three parts:

- Generic risks – risks which are present irrespective of the project type or nature
- Specific risks – risks which are particular to the project in hand
- Residual risks – remaining risks despite the control measures proposed

Thompson and Perry (1992) suggest that risks that have not been allowed for lead to the concept of risk exposure.

7.5 Risk assessment

Management clearly needs to be concerned about these issues, both in the client organisation and within the contractor. The bottom line is the prevention of losses which are caused by exposure to risk. These losses include:

- Accidents to workers
- Accidents to the general public
- Loss of skill and experience
- Damage to property
- Loss of time and production
- Loss of money on contracts
- Loss of reputation and future business

Where there is no risk there is no chance of a loss occurring, but in the real world risk is all around. Risk is caused by hazards. BS 4778 defines a hazard as 'a situation that could occur

Table 7.1 Risk assessment.

Step	Example
1 Identify the hazard	Deep drainage excavation in bad ground
2 Identify who or what might be harmed	Pipe layers in trench
3 Evaluate the risks arising from the hazard	High risk of collapse
4 Determine the control measures required	Use steel trench drag box
5 Evaluate remaining risks	Risk of crushing/injury from excavator bucket
	Risk of falling materials
6 Record the findings of the risk assessment	Fill in risk assessment sheet
7 Make contingency plans for the residual risks	Prepare safety method statement based on risk assessment
	Supervisor to give task talk
	Permit to work required
	Banksman working with excavator
8 Review and revision	Monitor site operations and modify risk assessment where necessary
	Hold further task talk if method statement is changed

that has the potential to cause human injury, damage to property, damage to the environment or economic loss'.

In order to control risk, a risk assessment is carried out. This is a statutory requirement under the Management of Health and Safety at Work Regulations 1999 with regard to risks to health and safety. The risk assessment might follow the steps listed in Table 7.1.

Risk is measured by evaluating the chance of something happening (the hazard) and the severity or consequences if it does. This can be based on a hunch or by making an intuitive assessment of whether the risk is high, medium or low or by using a matrix to do a simple calculation (Table 7.2).

Therefore, on a scale of 1–3, if the likelihood (chance) is 2 and the severity is 3, the risk is 6 out of 9. This would be in the high-risk category. However, it must be remembered that the prediction of likelihood and severity is difficult to do with any degree of certainty.

The matrix could be used to assess commercial risk at tender stage or the risk due to a variety of possible ground conditions or the chance of a fatal or non-fatal accident to people. The effect of exposure to this risk might be the loss of money, loss or damage to property, or business losses including loss of reputation or loss of individual skill and experience due to an accident to a key worker.

Table 7.2 Risk matrix.

Severity	Likelihood		
	1	2	3
1	Low	Low	Medium
2	Low	Medium	High
3	Medium	High	High

7.6 Client risk

Thompson and Perry (1992) suggest that uncertainty is greatest at the earliest stages of a project and that time and cost overruns can invalidate the client's business case for a project by turning a potentially profitable venture into a loss-maker. They also emphasise that the risks with the most serious effects for clients are:

- Failure to keep within the cost estimate
- Failure to achieve the required completion date
- Failure to achieve the desired quality and functional requirements

Thompson and Perry's work emphasises that appropriate strategies are necessary for the control and allocation of risk and that, while risk cannot be eliminated through procurement, contractual arrangements can greatly influence how risks are managed. This is illustrated in Figure 7.3 which provides a simplified view of the balance of risk according to the procurement strategy chosen. This diagram is based on the work of Walker and Greenwood (2002) and Edwards (1995).

Some of the risk management considerations made by a client at project level are considered here.

PROCUREMENT RISK

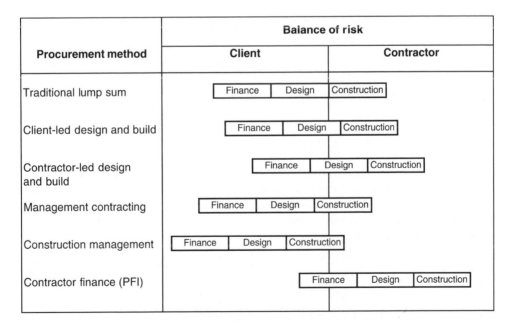

Figure 7.3

Feasibility risk

At the early stages of projects, clients must confirm the business case, identify options and develop the preferred solution. Once a scheme is sanctioned by the client, major commitments are made in terms of design, procurement and construction.

Design risk

Business decisions are all about risk and reward and the client must decide how much control is required over the design of the project. Retaining control over the quality of the design may be important for the client but the downside is that the design risk remains with the client as well. Partial or complete contractor design may remove the risk but then the client risks getting the wrong end result.

Funding risk

The contractor's income is the client's negative cash flow and arrangements must be made for available funds to draw down in order for the client to make regular monthly or stage payments for work in progress. The client's quantity surveyor must be careful not to expose the client to the risk of the contractor's insolvency by overvaluing interim payments.

Tender documentation risk

The traditional bills of quantities contract came about in order to give tendering contractors a level playing field. If traditional procurement is used and provided the design is well developed before going out to tender, quantities will be firm and all other documentation complete. Clients risk contract variations and budget overspend if they push for a fast start on site, and they may be well advised to consider more appropriate procurement arrangements in such cases.

Time risk

The obligation to complete the project on time is the contractor's responsibility and the client has redress in standard contracts through the liquidated and ascertained damages (LADs) provisions. Despite being a genuine estimate of the client's loss, LADs may not be adequate compensation if the contractor is late finishing, especially if a sell-on deal is lost or a prestigious opening deadline is missed.

Commercial risk

For most clients, buildings represent assets which are used to generate income and profits. The commercial success of a project may well be undermined if the job is delivered late or over budget or if the quality of design and construction is below the necessary standard. Clients should engage competent professional representatives to make sure this does not happen irrespective of the procurement method used for the project.

7.7 Contractor risk

Some of the biggest risks taken by contractors are at tender stage when they commit to a price and programme. Many companies now consider risk management to be an essential part of the tendering process. A contractor's risk assessment at the estimating stage may include consideration of the following risk areas.

Tender risk

At tender stage, the contractor needs to consider many factors before submitting a bid. Among these are:

- Previous experience (good/bad) working with the client team
- The financial stability of the client
- Market conditions and the level of competition for the contract
- Inflation – is a firm or fluctuating price required?
- Ground conditions and the balance of risk in the contract and method of measurement

Risks are involved with checking each section of the estimate build-up to ensure that the estimator's allowances are realistic. Final adjustments, the tactical movement of monies in the bills of quantities and the distribution of overheads and profit among the trade sections can expose the contractor to risk once the contract is awarded.

Quantity risk

The contractor must assess the accuracy of the quantities in the bills at tender stage because margins can be lost if the quantity work is subsequently reduced on remeasure. Provisional quantities in the bills pose a risk in rating up and allocation of margin.

Where quantities have been taken off by the contractor, failure to include associated labour items may add risk to the overall bid. On design and build projects, the contractor is responsible for taking off his own quantities but many serious quantity errors are often not apparent until the contract has been awarded. The contractor must then simply accept the error and grin and bear the loss (or find another way of making up the money).

Subcontractor risk

On many contracts, the contractor may simply be responsible for managing subcontractors, with very little work directly under his control. The ultimate success of the project may lie in the performance and organisation of subcontract operations. The contractor takes all the tendering risk and is especially vulnerable to subcontractors who are unwilling to stand by their quotes at tender stage.

The following check procedures may be implemented and when undertaken in conjunction with a well-managed workload should reduce the risk to the company:

- Subcontract totals should be abstracted from priced bills and checked against individual trade sheets.
- The correct selection of subcontractors and suppliers to ensure suitability for the work must be considered.
- Ensure liaison with construction, surveying and the buying department to ensure that the best subcontract packages have been selected.
- Subcontract price comparisons should highlight potential high/low rates so that assessment may be made in tender settlement meetings.
- The design input of work packages should be compared to ensure that the quotations meet design requirements. Likely cost savings proposed by subcontractors in relation to time and quality should be highlighted. Variations from the norm should be investigated.

Design risk

The contractor may be responsible for temporary works design only or may be involved in partial or complete design of the permanent works. On design and build contracts, perhaps under a JCT 05 Design and Build Contract, it is likely that most of the design work will be let to subcontractors on a 'design and build' basis. Consequently, the risk in each design package must be carefully assessed because the contractor remains wholly responsible for the quality of the work of his subcontractors and their design. The contractor's risk will be less where he is responsible for only part of the design – say under a JCT 05 Standard Building Contract with Contractor's Designed Portion because part of the design risk is carried by the client and his architect. As further risk issue is that lead-in times for design packages must be carefully considered at tender stage so that the contractor is not committed to an unrealistic programme.

Programme–time risk

Where the time for completion is stated in the tender documents, the contractor may be at risk if the client/project manager has got it wrong. On the other hand, where the tender documents require the contractor to insert his own assessment of the contract period, the contractor will be gambling on his own judgement.

It is essential for the success of the project that a realistic assessment of the construction period is backed up by a comprehensive pre-tender programme. The build-up of the contract preliminaries is based on the programme, together with major items of plant. Including contingencies for liquidated damages for non-completion (where a tight programme period has been stated) only makes the tender less competitive.

Method risk

The contractor's choice of construction method at the tender stage is crucial to winning the contract, but also fraught with risk. The ground conditions on site may be different to those expected and the type of earthwork support required may be more expensive than that allowed for in the tender. Relief may be obtained through the method of measurement but this is not always the case, especially on civil engineering projects.

It may be that the estimator's choice of plant or cranage is not feasible once the contract has been awarded, but the contractor is stuck with the allowances in the tender. This is where the contracts manager earns his money because cost-effective ideas will be needed to prevent contract losses.

Health and safety risk

Health and safety risk arises from the impact of hazards. Where there is no hazard there is no risk, but in construction there are hazards everywhere on a site. The best that can be done is to eliminate hazards in the design of the building and reduce the possible effects of residual risks through good management. The effective planning, organisation and control of construction work is central to that process.

People at risk from construction work include:

- Persons at work
- Visitors to site (including client representatives)
- General public
- Children

Documentation risk

Clarity of tender documentation is important. Bills of quantities containing extensive provisional quantities need careful pricing. Prices based on drawings and specification, or schedules of work containing extensive spot items, may prove difficult to price accurately.

Where there are defined provisional sums in the bills of quantities, the contractor has to allow for this work in his programme, even though there may be little information as to the precise nature of the work required.

The contractor needs to scrutinise the tender documentation very carefully in order to assess the implications of:

- Onerous contract terms
- Clauses deleted from standard contracts
- High levels of liquidated damages
- Unrealistic contract period
- Possible novation of the design (design and build contracts)
- Contract bonds and guarantees required

After consideration of all the risks to be assessed at the tender stage, it may be better for the contractor to withdraw from the competitive tender altogether, and spend his time negotiating work (or even consider partnering).

7.8 Tendering risk

When competition for work is fierce, contractors have to find ways of winning work *and* making a profit at the end of the job. The starting point is the estimate produced by the estimator.

Many contractors will carry out a risk assessment at tender stage and this may identify areas for making savings in the tender figure. The contractor may spot undermeasure or overmeasure in the contract bills or there may be scope for variations in the documents. Alternatively, the contractor may take into account buying 'muscle' on suppliers' and subcontractors' prices as a means of reducing the tender figure. This is frequently referred to as commercial opportunity or scope.

In the example in Table 7.3, it can be seen that the contractor's adjustments bring the tender figure (£1 780 500) below the estimator's net cost (£1 800 000). Effectively, the contractor is tendering at below net cost or, in other words, tendering at a negative margin. This is achieved simply by transferring the risk to others, principally the domestic subcontractors.

The contractor is taking a gamble in that he might not be able to squeeze down subcontractors' prices once the contract has been awarded, or the anticipated returns from variations and

Table 7.3 Tender summary.

			£	£	£
Preliminaries				130 000	
Measured work	Labour		100 000		
	Plant		60 000		
	Materials		240 000		
	Subcontract		820 000	1 220 000	
PC sums				300 000	
Provisional sums				150 000	
Net total					1 800 000
Overheads and profit 7%					126 000
Total					1 926 000
Commercial opportunity					
Materials	7.5%		240 000	18 000	
Subcontract	10%		820 000	82 000	
PC sums	2.5%		300 000	7500	
Provisional sums[a]	8%		150 000	12 000	
Preliminaries[b]	4 weeks		6500	26 000	(145 500)
Tender total					**1 780 500**

[a] Assuming rates for this work will include 7% markup and allowing for commercial opportunity of, say, 5% on materials and subcontractors.
[b] Assuming a 4-week reduction on the 20-week contract period and that all preliminaries costs are time-related.

claims may not be forthcoming. This does not mean that the contractor will necessarily lose money on the contract; all will be well, provided that allowances in the tender can be achieved in reality.

The tender documentation offers the contractor many opportunities to take commercial advantage, which can be used as a means of winning contracts and making money during the contract period. For instance, documents are frequently littered with mistakes or inaccuracies and contractors can take advantage of these. In such circumstances, the contractor might spot that the concrete work is undermeasured and that the architect will have to issue variation instructions for additional work.

At tender stage, the contractor can set the scene to make money out of this situation during the contract. By increasing his bill rates for these items of work, which will subsequently be remeasured and valued during the contract, the contractor can 'load' the rates and thus make more profit. To do this without increasing the tender sum, the contractor will take money away from some bill items and reallocate it to the undermeasured items. This technique is sometimes called rate loading.

Tender loading is a similar technique and is another way for the contractor to move risk on to the employer and make money at the same time.

Front-end loading reduces the early negative cash flow effect by increasing the margin. This is done by pricing the bill of quantities so that the margin is allocated to those items which will be carried out during the early stages of the project.

Back-end unloading is a similar technique but involves increasing both margin and net cost on early items of work. This is a dangerous practice and involves moving part of the net cost allowance for later items of work in the bill of quantities to those which are to be carried out earlier in the contract.

Harris and McCaffer (2006) explain these techniques further.

7.9 Health and safety risk

Despite having a much superior safety 'record' than our European counterparts, fatalities in the UK construction industry continue to give rise for concern and we have yet to make the 'step change' in construction health and safety expected in the 1990s. On average someone dies every week as a result of construction work and the large, well-organised contractor is no less prone to suffer a fatality than a smaller company.

Construction health and safety risk is managed through legislation and in particular:

- The Health and Safety at Work Act 1974
- The Management of Health and Safety at Work Regulations 1999
- The Construction (Design and Management) Regulations 2007

A raft of other legislation, such as the Work at Height Regulations 2005, which is not construction industry specific, also applies to construction projects.

The Health and Safety at Work Act 1974 is enabling legislation which sets out the penalties for failure to comply, and confers powers on inspectors who may require site practices to be changed if they are considered unsafe or in breach of legislation. A contractor can be served with an improvement notice or may even be prevented from continuing an unsafe operation through a prohibition notice.

The Management of Health and Safety at Work Regulations 1999 deal with assessment of risk and arrangements for and competence in the measures needed to protect individuals and prevent accidents at work. The impact of this legislation is to ensure that issues identified by risk assessments are dealt with by effective planning, organisation and control and that procedures to monitor and review such arrangements are put in place.

The Construction (Design and Management) Regulations 2007 require a construction phase health and safety plan for all but the smallest of projects and this acts as an important means of communication which is intended as a continuous theme throughout the construction phase of the project.

In practice, the construction phase plan relies on pre-construction information prepared by the design team in conjunction with the construction design and management coordinator. This is developed by the principal contractor into a safety management system for organising, implementing and monitoring health and safety arrangements on site.

The Approved Code of Practice (HSE 2007) gives comprehensive guidance concerning the contents of the construction phase plan and explains the responsibilities of those who have duties under the Construction (Design and Management) Regulations:

- Client
- Designers
- CDM coordinator (statutory appointment)
- Principal contractor (statutory appointment)
- Contractors and the self-employed

The principal contractor role is normally undertaken by the main contractor under the contract, although the statutory appointment and the appointment of the contractor under the civil contract are quite separate and distinct appointments. The duties of the principal contractor include responsibility for the health and safety management system for the project. This normally involves conducting site inductions, establishing site rules, the provision of training and information and the prevention of unauthorised access to the site.

7.10 Fire risk

Fire is an ever-present risk on construction sites, especially with respect to:

- Hot work such as welding, blowlamps, cutting and grinding
- Heating appliances, especially gas bottles in welfare facilities
- Litter, especially in rest rooms and drying areas
- Arson
- Smoking
- Burning of waste on site
- Stored materials, including adhesives and solvents

Construction (Design and Management) Regulation 2007 provides for such eventualities and requires contractors to:

- Take measures to prevent risk of injury from fire
- Provide and maintain fire-fighting equipment, fire detectors and alarm systems
- Give instructions to people in the use of fire-fighting equipment
- Give instructions to people where their work activities involve a fire risk
- Indicate fire-fighting equipment with suitable signs

Standard forms of contract provide for insurance of the works during construction and this includes the risk of fire. However, the high cost of site fires has prompted the publication of a Joint Code of Practice, 'Fire Prevention on Construction Sites' (Construction Confederation, 2006). This provides that potential fire risks are considered at design stage and also that a site fire safety plan is developed by the principal contractor. This plan must include provisions for a comprehensive fire prevention regime including:

- Organisation and responsibilities for fire safety
- Hot work provisions
- Fire escape and communications
- Fire drills and training
- Emergency procedures

The fire safety plan may be included in the construction phase health and safety plan. It should be noted that adherence to the Joint Fire Code is a contract condition in some standard forms (e.g. JCT 05).

References

BEC (2006) *Fire Prevention on Construction Sites*, 6th edn. Joint Code of Practice. Construction Confederation.

CIOB (2002) *Chartered Institute of Building Code of Practice for Project Management*, 3rd edn. Blackwell Publishing.

Edwards, L. (1995) *Practical Risk Management in the Construction Industry*. Thomas Telford.

Harris, F. & McCaffer, R. (2006) *Modern Construction Management*, 6th edn. Blackwell Publishing.

HSE (2007) *Managing Construction for Health and Safety in Construction*, Approved Code of Practice and Guidance. HSE Books.

ICA (1999) *The Turnbull Working Party Report*. Institute of Chartered Accountants.

Raftery, J. (1994) *Risk Analysis in Project Management*. E. & F.N. Spon.

Smith, N.J. (1999) *Managing Risk in Construction Projects*. Blackwell Science.

Thompson, P. & Perry, J. (eds) (1992) *Engineering Construction Risks*. Thomas Telford.

Walker, P. & Greenwood, D. (2002) *Risk and Value Management*. RIBA Enterprises.

8 The planning process

8.1 Introduction

Without planning it is difficult to envisage the successful conclusion of any project or the effective control of time, money or resources. Planning is also essential in order to deal with construction risks and devise safe working methods. An often-quoted phrase is 'fail to plan – plan to fail' and this is true throughout all stages of the process from inception through the design, tendering, construction and commissioning stages of a project.

The reasons for planning may be summarised as:

- To set a realistic time framework for the project
- To establish realistic standards and avoid 'wishful thinking'
- To aid control during the project
- To monitor performance in terms of output, time and money
- To review progress and take action when necessary to correct the situation

Chapter 16 discusses planning as an essential requirement of a good control strategy.

Programmes are commonly prepared at various stages of projects for different purposes and different audiences and these are summarised in Table 8.1.

At the very early stages of a project, it is essential for the client team to have a programme in order to aid control. Figure 8.1 shows a client's project programme which indicates the key activities during both the design and tender stages as well as highlighting the anticipated period for the construction phase. While little firm information will be available at the very early stages, the client's project manager or lead consultant should nevertheless draw up this programme based on a realistic estimate of how long the design, tendering and construction phases of the project should take. The CIOB Code of Practice for Project Management (2002) calls this the **project master schedule**, presumably to avoid confusion with the contractor's master programme.

The project master schedule shows a number of features which should be noted:

- The timescale in years and months rather than weeks indicating the strategic nature of this programme
- The use of 'negative' time to indicate the pre-construction period
- The inclusion of holiday periods without which the overall duration would be overoptimistic

Table 8.1 Planning programmes

Planning stage	Type of programme
Design	Project master schedule
Tender	Pre-tender programme
Pre-contract	Master programme
	Target programme
	Subcontractors' programme
	Procurement programme
Contract	Stage programme
	Short-term programme
	As-built programme

- The use of milestones to indicate key events in the programme (e.g. handover)
- The emphasis on the client team activities necessary to plan the design and tender stages, including obtaining statutory approvals to build
- The construction phase denoted by a single bar line
- The 'start on site' and 'handover' dates would be included in the contract appendix
- Intermittent activities shown with a dotted line

The 5-month construction period indicated on the client's master schedule in Figure 8.1 would be developed into a pre-tender programme by the contractor at the tender stage. The contract start and finish dates or contract duration would be entered in the contract appendix/contract particulars and the contractor would need to check that the construction period allowed was realistic before tendering for the contract. At the pre-contract stage, the contractor would develop the pre-tender programme into his master programme which would then be used to establish subcontract and procurement programmes etc.

8.2 Lead times

Before physical construction work or off-site fabrication or assembly works are undertaken on a project, project managers and contractors need to consider the impact of the design, procurement and manufacturing stages required for the various materials and components that make up the project. It cannot simply be assumed that all such materials are readily available 'off-the-shelf' and available for immediate delivery to site. Whilst 'just-in-time' delivery is common practice in construction, this has to be carefully planned and thought out well in advance.

Consequently, due allowance needs to be made in the project master schedule for the 'lead times' of all major components and associated trades packages as well as in the contractor's master programme for the works.

Examples of the items that may have to be considered on a project include piling, structural steelwork, cladding, roofing, suspended ceilings and lifts as well as such 'normal' elements as brickwork, blockwork and general joinery components, etc. It may come as a surprise that concrete works require a lead time of around 10 weeks at the time of writing and curtain walling systems may take some 48 weeks from order to start on site. Such constraints may

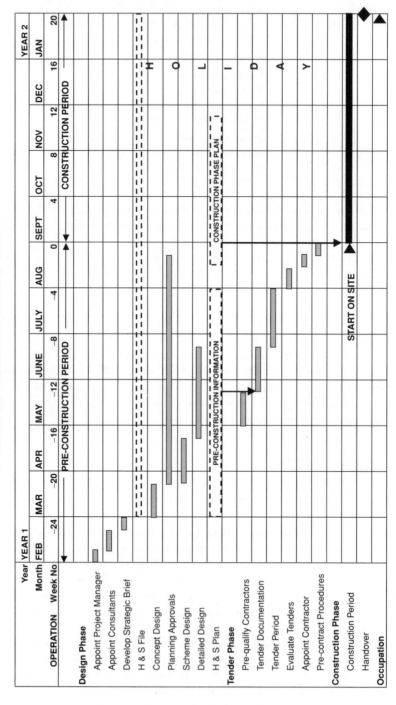

Figure 8.1

well exert considerable influence on a client's procurement strategy. The early appointment of the main contractor and the key trade packages as well as the placing of orders for important materials, components and off-site fabrication are crucial factors to be thought about at the early stages of a project.

A useful, comprehensive and regularly updated source of lead time information is available on the website of management and construction engineers, Mace (http://www.mace.co.uk/suppliers/leadtimes.aspx). This information is also published quarterly in *Building Magazine*.

8.3 Work breakdown structure

The Association of Project Management (APM) Body of Knowledge (BoK) (2006) suggests that the 'scope' of a project must be identified and defined in order to *describe what the project will include and what it will not include (i.e. what is in and out of scope)*. The BoK also suggests that the project scope needs to be managed effectively so as to keep track of the many changes that will inevitably occur and to retain control of the project.

Burke (2003) maintains that the work breakdown structure (WBS) is a key tool in 'scope management' which is used to subdivide the project into manageable packages *that can be estimated, planned, assigned and controlled*. Burke also suggests that the discipline of preparing a WBS ensures that the complete scope of the project is included in the planning and control process.

There are several methods of presenting a WBS, but they all share the same common features, that is a hierarchy or cascade arrangement together with a multi-level numbering system. Burke suggests that Level 0 is the project level and Levels 1 and 2, etc., are subdivisions of the project which enable unique reference numbers to be given to the elements or work packages that make up the project scope.

Figure 8.2 illustrates the principles of a WBS which could be adapted for a variety of circumstances.

Most linked bar chart project management software packages facilitate the use of WBS and enable activities on the programme to be 'rolled up' or expanded to an appropriate level of detail depending on the audience (e.g. client, contractor, subcontractor).

8.4 Planning stages

There is more to planning a project than meets the eye and a great deal more involved than simply producing a programme. Both the client's representative or project manager and the contractor will have many issues to think about if the project is to be successfully completed and a number of key stages will have to be negotiated in the process.

Figure 8.3 summarises some of the many complex considerations made during the client's planning process, and Figure 8.4 shows a similar overview from the contractor's viewpoint. Figure 8.4 establishes the relationship between the three planning stages undertaken by the contractor and a checklist approach is suggested for each stage to illustrate what needs to be considered. The considerations suggested are those generally undertaken by both medium-sized and large contractors but are equally valid for small contractors as well.

Clearly, there will be an interface between the client's planning process and that of the contractor at the tender stage of a traditional project, and during the administration of the

WORK BREAKDOWN STRUCTURE

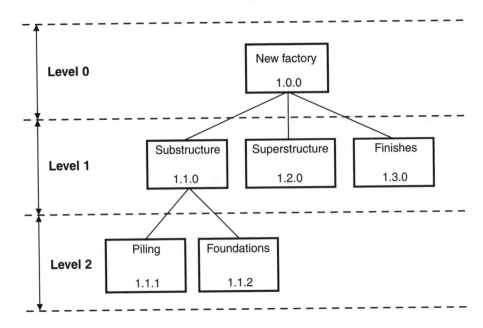

Level 2 elements may be further subdivided as required e.g.:

1.1.2   Foundations

1.1.2.1 Excavate pile cap

1.1.2.2 Cut off and trim piles

1.1.2.3 Blind base of pile cap

1.1.2.4 Fix formwork to base

1.1.2.5 Fix steel reinforcement

1.1.2.6 Set bolts in base

1.1.2.7 Place concrete

1.1.2.8 Strip formwork

1.1.2.9 Backfill working space

Each element of work is given a unique code number. This may be linked to the master programme or to items in the bills of quantities. The referencing system may be used for preparing work packages for subcontracts, for preparing budgets or for costing and linking to the contractor's CVR process.

Figure 8.2

contract on site. Where non-traditional procurement methods are used, there may be further interfaces to consider.

Planning is one of Henri Fayol's six functions of management and it starts right at the outset of a project. There are several levels of planning:

PURPOSE

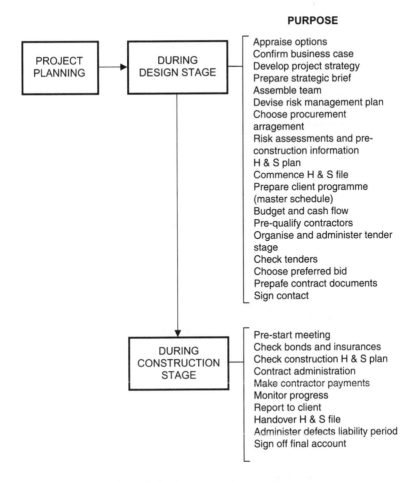

OVERVIEW OF CLIENT PLANNING PROCESS

Figure 8.3

- Project planning carried out by the client/project manager
- Pre-tender planning carried out by the tendering contractors
- Pre-contract planning carried out by the main contractor
- Contract planning carried out by the main contractor and subcontractors

8.5 Project planning

Project planning starts with the client team or organisation when a programme (or master schedule) will be prepared by the client's agent, representative or project manager. This sets out the broad framework for the project, including:

- Key dates for commencement and completion of project
- Key dates for design, tender and construction

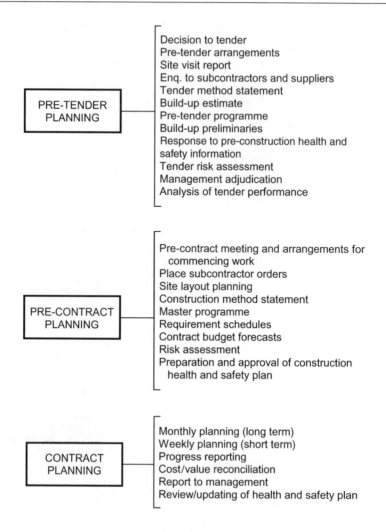

OVERVIEW OF CONTRACTOR PLANNING PROCESS

Figure 8.4

- Overall programme and phasing
- Design and tendering periods
- Key dates for commencement and completion of construction
- Stage or phased handover dates

Reasons for project planning

- To establish a realistic project master schedule or programme on which to base the appointments of consultants and contractors and commission the occupancy phase of the scheme

- To identify key dates or gateways at critical stages of the project
- To facilitate control of the design and tendering process
- To identify potential risks to progress and avoid possible delays to project completion and revenue generation
- To facilitate the arrangement and draw-down of client cash funding in order to pay for design and other professional services and make interim payments to contractors
- To establish a realistic time period for the construction stage
- To monitor actual progress and take corrective action

8.6 Pre-tender planning

At the tender stage, a pre-tender programme is usually prepared by the contractor as an aid to the tendering process. The pre-tender programme will assist the estimator to price key method-related items in the bills of quantities and also the contract preliminaries which comprise largely time-related costs.

Pre-tender planning may be defined as the contractor's planning considerations during the preparation of an estimate and its conversion into a commercial bid. The role of pre-tender planning is shown in Figure 8.4 which provides an overview of the planning process from the contractor's point of view. This clearly illustrates that pre-tender planning involves much more than simply producing a programme for tendering purposes, but concerns all aspects of a project from the initial enquiry from the client to the submission of the contractor's tender bid.

Reasons for pre-tender planning

- To establish a realistic contract period on which the tender may be based
- To identify construction methods
- To assess method-related items which affect the bid price
- To aid the build-up of contract preliminaries and plant expenditure
- To aid the tendering process

8.7 Pre-contract planning

Pre-contract planning generally takes place during the period between contract award and commencement of work on site. This is the case for a project based on a traditional competitive tender but there may be differences in procedures where other procurement arrangements are used.

Before work starts on site, the contractor will develop the pre-tender programme into the contract master programme showing the main construction operations to be carried out. Copies of this programme will be presented to the client's representative who will use it as a tool to monitor the contractor's overall progress during construction. The master programme will often show when information is required by the contractor and act as a prompt for the architect.

The master programme is the one that the client team sees, but many contractors produce an internal programme for their own use in order to save time and money. This is called a **target**

programme and is effectively a compressed version of the master programme with time taken out of the critical path. This is a commercial decision which may not work out as planned – that's contracting!

In order to help the contractor organise and manage site activities at an operational level, the target programme will need to be developed in more detail. Most of the time bars on the target programme will represent the main work packages to be carried out by various subcontractors and consequently each bar will be developed into a subcontractor programme showing the detailed activities to be carried out.

However, to ensure that work packages start and finish on time, it is now usual for the contractor to produce a procurement programme for each subcontractor. This programme will show both negative time and positive time. Negative time is the 'lead-in' time needed before work starts on site. This is required to organise the design and fabrication aspects of the package or to pre-order key materials with long lead times for delivery. Positive time is the time needed to carry out the subcontractor's work on site.

Reasons for pre-contract planning

- To provide a broad outline plan or strategy for the project
- To comply with contract conditions
- To establish a construction sequence on which the master programme may be based
- To identify key project dates
- To highlight key information requirements
- To enable the assessment of contract budgets and cumulative value forecasts
- To schedule key dates with respect to key material and subcontractor requirements

8.8 Contract planning

During the contract stage, the master programme will be further developed. For instance, a stage programme might be prepared showing part of the master programme in more detail. Alternatively, the contractor might produce a series of short-term programmes at weekly or fortnightly intervals so as to plan day-to-day work in detail.

Contract planning is done by the main contractor in order to maintain control and ensure that the project is completed on time and within the cost limits established at the tender stage. Subcontractors contribute to the process either by submitting their work programme for approval or through discussion with the main contractor.

As the contract progresses, invariably the programme changes from its original form. Delays occur, work is disrupted due to design changes and unforeseen events take place such as the discovery of bad ground or contamination. This causes delay and/or disruption to the programme which the contractor has to accommodate. These changes should be recorded on a revised programme which should be constantly updated throughout the project as work proceeds and as other problems arise. These programmes are often referred to as the as-built programme or, alternatively, the programme of the day and they are a vital tool to enable the contractor to justify his entitlement to extensions of time and/or additional payment for loss and expense.

Reasons for contract planning

- To monitor the master programme – monthly, weekly and daily
- To plan site operations in detail in the short term
- To optimise and review resources
- To keep the project under review and report on variances

8.9 Planning a project

When preparing any programme for a project, it is essential to follow a logical thought process in order to develop a realistic and workable programme. A working knowledge of the development and construction processes is essential. The level of detail shown in the programme should be commensurate with the project stage under consideration and, wherever possible, activity durations should be based on empirical data or calculation. Computer software should be used for speed, for considering 'what if' options and for high-quality professional presentation.

The planning of a project requires a logical approach involving various steps or thought processes:

- Getting a feel for the project
- Establishing key project dates
- Establishing key activities or events
- Assessing how long the activities will take
- Establishing the sequence
- Deciding which programming technique to use

These are now considered in turn.

Getting a feel for the project

It is important for the planner or project manager to get a feel for the project because a clear appreciation of the scale and complexity of the scheme helps to trigger the natural human instincts of when things look right or look wrong. These instincts are not simply based on a hunch but also on experience and familiarity with the construction process.

It is just as important to appreciate the financial scale of the project as well. Experience will often identify the project value or rate of expenditure as being inconsistent with the time allowed. To get a feel for the project, consider the following:

- Study the drawings and project documentation
- Visit the site
- Assess the scale and scope of the project
- Assess the approximate value of the project
- Consider the rate of expenditure (i.e. the relationship between value and time)

Establishing key project dates

The overall parameters of a project will be determined by establishing the key dates. Some of these will be established by the client and his advisors and conveyed to the contractor in the tender documentation, and others will be common sense. They will include:

- Project start and finish dates
- Sectional or phased completion dates
- Holiday periods
- Commissioning or handover

Establishing key activities or events

The next step is to determine the key activities or tasks to be carried out, together with any important events which should be included in the programme.

Activities are tasks or jobs to be done which have a time value. Obtaining planning permission is an activity because it can take several months, or even years, to complete. Events are points in time by which things must happen and have no time value. For example, the start of construction work is an event which triggers a series of activities which do have a time value.

The activities on the programme will vary according to the stage of the project. Some suggestions are listed here.

Checklist of key activities/events during the design stage

- Brief the design team
- Make professional appointments
- Notify the project to the Health and Safety Executive (Form F10)
- Start the health and safety file
- Obtain planning permission
- Apply for building regulation approval
- Pre-qualify contractors
- Appoint contractor(s)

Checklist of key activities/events during the tender stage

- Prepare tender documents
- Complete the pre-construction health and safety information
- Period for tendering
- Evaluate tenders
- Compile contract documentation

Checklist of key activities/events during the pre-contract stage

- Appoint project staff
- Develop the construction phase health and safety plan

- Prepare requirement schedules
- Pre-start meeting
- Check the construction phase health and safety plan
- Permit start of construction work (the client has a statutory duty under Construction (Design and Management) 2007 Regulation 16)

Checklist of key activities/events during the construction stage

- Set up site establishment
- Groundworks or substructure
- Frame/external envelope
- Floors
- Roof structure and cladding
- Building watertight
- Mechanical, electrical and ventilation installations
- Finishes
- External works and drainage
- Practical or substantial completion
- Clear site

Assessing how long the activities will take

Assessing the duration of activities is not an exact science. At the early stages of a project, when little detailed information is available, a great deal of reliance must be placed on judgement and experience. Later on, calculations can be performed by considering the relationship between the quantity of work to be done and the output or rate of production anticipated. For example:

$$\frac{\text{Quantity}}{\text{Output per hour}} = \textbf{Hours}$$

$$\frac{\text{Hours}}{\text{Number of hours per day (8)}} = \textbf{Days}$$

$$\frac{\text{Days}}{\text{Number of days per week (5)}} = \textbf{Weeks}$$

Checklist

- The RIBA Plan of Work gives detailed information on the tasks undertaken by members of the design team from which time assessments can be made
- Many local authorities provide information and statistics relating to planning applications and the time taken to deal with them
- The Joint Consultative Committee for the Building Industry publishes codes of procedure which give indicative time allowances for tendering under various procurement arrangements

- EC tendering rules (see Chapter 5) establish times for the notification, tender periods and evaluation of public sector tenders
- The RICS Building Cost Information Service (BCIS) provides detailed tender analyses and contract periods for a wide variety of construction projects
- Any number of estimating books provide average outputs for a wide variety of building and civil engineering work. The quantity of work divided by the output per day will give the required duration in days
- Contractors are the best source of information but this unfortunately is rarely published
- Inevitably, a degree of common sense and experience is invaluable in assessing activity durations and sometimes there is no substitute for an educated guess

Deciding which programming technique to use

There is no strict rule as to which programming technique should be employed. This needs to be considered in the light of the size and complexity of the project in hand, any personal preferences and whether there are any stipulations in the contract documentation.

Bar charts are the easiest to use but they can give misleading results because there is no strict logic imposed on the programme. It may be better to use linked bar charts or arrow or precedence diagrams to overcome this problem. However, for repetitive work such as housing projects, line of balance may be preferred, or for roadworks, tunnelling or repetitive civil engineering work, time-chainage diagrams could be the best application to use.

When using project management software packages, it is usually advisable to draw out the programme on paper first. This helps to establish the correct logic and avoid the possibility of getting in a tangle when working at the computer screen – which is particularly likely to happen when making the logical links between activities, as it is not always possible to see the whole picture on screen.

References

Association for Project Management (2006) *APM Body of Knowledge.* APM Publishing.

Burke, R. (2003) *Project Management,* 4th edn. John Wiley & Sons.

CIOB (2002) *Chartered Institute of Building Code of Practice for Project Management,* 3rd edn. Blackwell Publishing.

9 Programming techniques

9.1 Introduction

A variety of programming techniques are available to the client's project manager or the contractor's planner and these can be used according to the type and complexity of the project concerned. The following are considered in this chapter:

- Bar charts
- Linked bar charts
- Arrow diagrams
- Precedence diagrams
- Line-of-balance diagrams
- Time-chainage diagrams

The technique adopted in any particular case is largely a matter of personal preference but the recipient of the information needs to be considered because programmes based on overly complex techniques may be counterproductive to effective communication.

9.2 Bar charts and linked bar charts

History and development

Kempner (1980) and other management writers recognise that Henry Gantt first introduced bar charts for ship-building projects in the early 1900s by popularising the graphical presentation of work versus time. Gantt belonged to the Scientific Management school of thinking of the late nineteenth and early twentieth centuries which included Taylor, Fayol and Gilbreth, among others. Gantt charts were the first scientific attempt to consider work scheduling against time. They have now become the basis of the modern bar chart, which has several variants.

Before the advent of powerful modern computers, bar charts were prepared by hand, often using pre-prepared blank sheets containing a column for the project activities and squares for drawing the bar lines. It was quite common to see the bar chart on the site agent's wall in the site cabin, with progress coloured in with crayon/coloured pencil and a vertical string pinned in place to denote the current date.

Limitations in the power and flexibility of bar charts resulted in the adoption of more sophisticated techniques in the 1950s and 1960s, in particular network analysis using arrow and later precedence diagrams. However, in more recent times, the development of the linked bar chart technique has led to a resurgence in the use of bar charts, mainly as a consequence of developments in project management computer software. There are a number of project management software packages available which employ linked bar charts as the preferred display. These include:

- C S Project Professional – Crest Software
- Easyplan – Asta Development
- Microsoft Project – Microsoft Ltd.
- Powerproject – Asta Development
- Project Commander – PMSC Software

Principles of bar charts and linked bar charts

The bar chart is laid out with the timescale in days/weeks/months/years along the top axis and a list of tasks or activities down the left-hand side. The time required for each activity is represented by a horizontal line (or bar), with the length of the line indicating the duration of the activity.

The software packages available for producing bar charts require the user to enter a logic by creating links between the activities on the programme, which overcomes the major disadvantage of Gantt charts. Users are able to present bar charts in a professional manner at any level of the planning process and programmes can be readily updated, even on site, with the aid of laptop computers.

One of the problems with the traditional Gantt charts is the tendency for the planner/manager to work backwards from the completion date stated in the contract documents in establishing the detailed programme. The resulting bar chart is often no more than wishful thinking. It has been said that the bar chart often suffers from a 'morning glory complex' – it blooms early in the project and is nowhere to be seen later on! This problem is overcome to some extent with the linked bar chart as this form of display forces the programmer to think about the logic of the programme as each link is developed.

Bar charts are well suited to depicting construction sequences and are readily understood at all levels of management. They can be used to develop the programme prepared at the tender stage into the master programme and likewise into the short-term planning throughout the contract period. Bar charts are easily and readily updated at weekly and monthly intervals. A colour coding system may be introduced for progress recording, but most software packages have facilities for project tracking and updating which provides an accurate record of progress on the contract for future reference.

Most architects and site managers tend to have problems in understanding anything other than bar chart displays, even where network analysis has been stipulated in the contract documents as the project planning technique to be employed. However, bar charts have limitations and a key disadvantage is that they do not show dependency. Consequently, it is not easy to see the interrelationship between activities and how dependent they might be on one another. This is not so bad on a simple project but, where there is a large number of activities on the programme, real problems can arise for the manager. The problem can be overcome by using the linked bar chart.

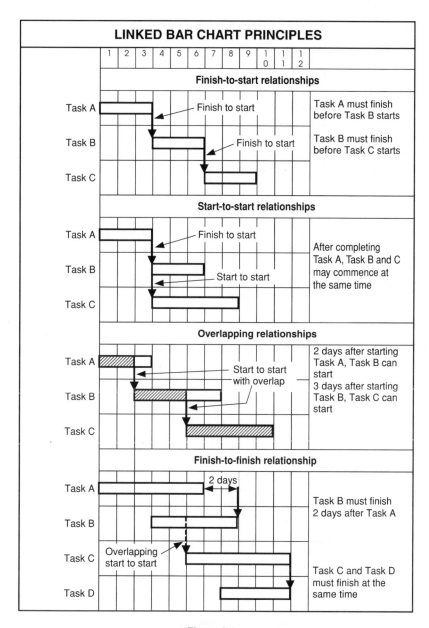

Figure 9.1

Figure 9.1 illustrates the relationships used on linked bar charts. These include:

- Finish-to-start relationships
- Start-to-start relationships (overlaps)
- Finish-to-finish relationships

Advantages of bar charts and linked bar charts

- Simple format readily understood at all levels of management
- Applicable at all stages of the planning process: project planning, pre-tender, and pre-contract and contract planning
- Particularly useful in showing the relationship between the pre-tender programme, master programme and short-term programmes
- Clearly mimics the construction sequence – the use of linking between bars facilitates the overlapping of related operations
- Easily updated at weekly and monthly intervals for review purposes and progress reports
- Key milestone symbols may be introduced to highlight critical dates with regard to key contract stages, information requirements and as an aid to subcontractor and materials procurement
- Resources may be shown on the bar chart, which clearly relates labour, plant and subcontractors to the rate of working and helps the manager to see problems of continuity and waste
- Facilitates the production of labour histograms, value–time forecasts, cumulative labour and plant forecasts and other project budgets so that actual progress may be charted against planned progress
- Enables the contractor to quickly and simply fulfil contractual requirements to submit a programme prior to commencement of work on site
- Readily updated with information such as key site deliveries, progress to date and delays, making it simple to produce an 'as built' programme (or 'programme of the day') which may prove an asset to the contractor in forming contractual claims. The effect of the receipt of late information on programmed operations can be monitored and reported. Copies of the contract position at the date of a specific occurrence may provide evidence of the resulting delay
- The bar chart programme can be used to form the basis of financial forecasting for both the client and the contractor
- Computer printouts in full colour can be scaled down to A4 size, which is invaluable for management reports and ease of handling

Disadvantages of bar charts and linked bar charts

- Gantt charts do not show dependency and therefore do not clearly indicate which operations directly relate to the successful completion of the project. This makes it difficult to apply management by exception
- Logical links are used to overcome this problem but these can become confusing and difficult to interpret on complex projects
- Consequently, complex interrelationships cannot be clearly shown

Developing a linked bar chart

Creating a useful and professional-looking bar chart using modern software packages is a relatively straightforward task requiring a couple of hours to learn the package and then it is just a question of adopting a logical and systematic approach. Of course, it does help to know how buildings are put together as well!

One of the common packages which is simple to use, relatively inexpensive and gives good results is Microsoft Project. Figure 9.2 illustrates the following three basic steps in developing a simple programme using this package.

- *Step 1*

  Decide on the most appropriate timescale for the programme. Months on the major scale and weeks on the minor scale are usually best for most programmes. For short-term planning and short-duration projects, where a finer level of detail is necessary, weeks and days would be more appropriate. Next, list the various tasks/activities that make up the project and sort these into order using the cut and paste facility. Choose the activities carefully. Too few activities will make the programme of little use, whereas too many will be cumbersome and difficult to print out and read. Think about the possibility of using rolled-up tasks (called summary tasks in Microsoft Project) which can be expanded into subtasks at a later stage. Note that the computer defaults to a duration of 1 day for each activity.
- *Step 2*

  Insert summary tasks and add durations (weeks are generally best) to the activities by calculation or from experience. Make any changes to the number or order of the activities before going on to Step 3.
- *Step 3*

  Add logic to the programme by linking relevant activities to one another. Think about which activities must come first, which must follow and which may happen at the same time. Overlaps or delayed starts can be introduced by choosing the appropriate relationship from the menu (e.g. start-to-start) and by adding the necessary lag time.

Further features are easily added such as holiday periods, key milestones (events), resources and cost information.

9.3 Network analysis

History and development

In 1956, E. I. duPont de Nemours and Company established a team to study new management techniques for the company's engineering functions. One of the first areas to be considered was the planning and scheduling of construction work. Data were input into a UNIVAC 1 computer in the form of construction sequences and activity durations in order to generate a schedule of work.

In early 1957, J.W. Mauchley, J.E. Kelley Jr. and M. Walker developed the basic principles of the Critical Path Method (CPM). A test group was set up to apply the new technique to a chemical plant project in Kentucky. In 1958, Mauchley Associates developed a series of training programmes to spread knowledge and use of the method throughout industry. The development of PERT (*program evaluation and review technique*) was originated by the Special Projects Office for the Polaris missile programme.

DEVELOPING A BAR CHART (using Microsoft Project)

Step 1

ID	Task Name	Duration	Nov 1-5	Dec 6-9	Jan 10-14	Feb 15-18
1	SET UP SITE	1d				
2	BASEMENT PILING	1d				
3	RC SLAB	1d				
4	RC FRAME	1d				
5	GROUND SLAB	1d				
6	RC COLUMNS & BEAMS	1d				
7	FLOORS	1d				
8	PCC CLADDING	1d				
9	WINDOWS	1d				
10	ROOFING	1d				
11	M&E	1d				
12	FINISHES	1d				
13	DRAINAGE	1d				
14	ROADS & PAVINGS	1d				
15	LANDSCAPING	1d				
16	CLEAR SITE	1d				

Step 2

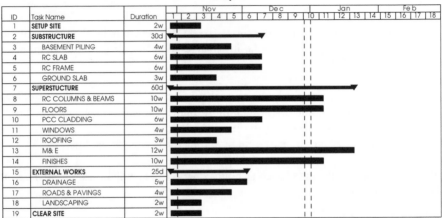

ID	Task Name	Duration
1	**SETUP SITE**	2w
2	**SUBSTRUCTURE**	30d
3	BASEMENT PILING	4w
4	RC SLAB	6w
5	RC FRAME	6w
6	GROUND SLAB	3w
7	**SUPERSTUCTURE**	60d
8	RC COLUMNS & BEAMS	10w
9	FLOORS	10w
10	PCC CLADDING	6w
11	WINDOWS	4w
12	ROOFING	3w
13	M&E	12w
14	FINISHES	10w
15	**EXTERNAL WORKS**	25d
16	DRAINAGE	5w
17	ROADS & PAVINGS	4w
18	LANDSCAPING	2w
19	**CLEAR SITE**	2w

Step 3

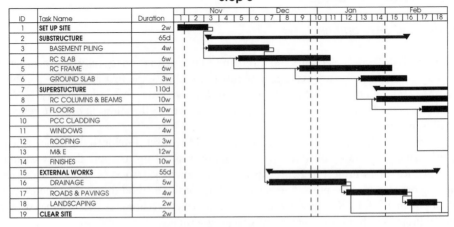

ID	Task Name	Duration
1	**SET UP SITE**	2w
2	**SUBSTRUCTURE**	65d
3	BASEMENT PILING	4w
4	RC SLAB	6w
5	RC FRAME	6w
6	GROUND SLAB	3w
7	**SUPERSTUCTURE**	110d
8	RC COLUMNS & BEAMS	10w
9	FLOORS	10w
10	PCC CLADDING	6w
11	WINDOWS	4w
12	ROOFING	3w
13	M&E	12w
14	FINISHES	10w
15	**EXTERNAL WORKS**	55d
16	DRAINAGE	5w
17	ROADS & PAVINGS	4w
18	LANDSCAPING	2w
19	**CLEAR SITE**	2w

Figure 9.2

In the UK, development work on CPM was undertaken by the Building Research Establishment and a number of papers were published by Nuttall and Jeans (1960a,b).

The original method for presenting networks was the arrow diagram which consists of a series of arrows linked together with circles or nodes. Each arrow represents an activity and the critical path can be calculated by performing a forward and backward pass through the network. This method of presentation has practical limitations, however, and so precedence diagrams were developed in the early 1970s by the Cementation Company as an alternative approach to network analysis. Precedence diagrams consist of a series of boxes linked with lines or arrows where the boxes represent the activities and the lines represent the logic. The precedence format can more readily be applied to works of a civil engineering nature. In practice, precedence diagrams are far more widely used than arrow networks because they are more flexible and more easily reflect the ways things happen.

The initial CPM applications developed were suitable only for large, complex projects and required the assistance of mainframe computers to analyse the data. Data input and output data retrieval were slow and difficult to interpret – in fact the progress of the project had often changed by the time the data were analysed and reported on. Reece (1989), in his article on the future of project planning, reviewed the rise and fall of CPM. The resurgence in the use of CPM in the late 1980s was mainly due to the introduction of the personal computer and developments in user-friendly project planning software. Now, linked bar charts are more popular for most projects but, for the larger schemes, CPM still has an important part to play, especially with the advent of cheaper software and hardware.

Project management-based software is available in CPM format which at the time of writing includes:

- Micro Planner Expert – Micro Planning International
- Plantrac – Computerline Ltd.
- Primavera SureTrak– Primavera Systems Inc.

Planning managers have now moved away from using arrow diagrams and precedence formats due to the wider availability of more sophisticated and powerful linked bar chart software.

Perhaps the days of network analysis are now over?

Arrow diagrams

The distinguishing feature of these diagrams is that the arrow represents the activity and the circle or node between the arrows is the event. By numbering the events, the arrow activities can be identified. This is usually done in numerical order starting at the beginning of the network and progressing to the end, ensuring that the number at the tail of the arrow is smaller than that at the head.

Each activity is given a duration, and the earliest and latest event times of the activity can be calculated by making forward and backward passes through the network. These times are recorded in the node or event circles. From this information, a schedule can be produced which will facilitate calculation of the total float or float time (spare time) for each activity. Dummy activities, which usually have no duration or value, can be introduced to indicate dependencies not shown by the arrow activities.

The project management software carries out the time analysis for the programme sequence developed by the planner. This allows the operations to be sorted into priority, thereby enabling the critical path to be calculated and highlighted. The critical path is defined as the longest route through the sequence of operations which must be undertaken in order to complete the project.

Precedence diagrams

Precedence diagrams follow the same logical procedures as arrow networks except that the activities and their dependencies are drawn differently. The precedence diagram consists of a series of boxes interlinked with lines. The box or node represents the activity and the linking arrow indicates the relationships of the activities to one another. The box contains an activity label or name and duration. There is space for the earliest and latest start and finish times of the activity and a reference number may also be included if required.

Both the boxes and the lines may be given a time value. The time given in the box represents the duration of the activity, while any time on the line or arrow adds a dependency which might be a lead or a lag as required. Precedence diagrams do not require dummies to preserve the logic of the relationships and each node is ascribed a unique activity number.

One of the key reasons for the growth of precedence diagrams is the limitation of arrow diagrams when, for instance, one activity is required to start before the preceding activity is completed. This means either dividing the preceding activity into smaller parts or introducing a dummy with a time value.

The precedence approach introduced the idea of activity boxes, rather than activity arrows, which permits a number of different relationships to be expressed between activities. This approach relates more closely to the real situation on a construction project and this practicality makes the technique more popular. The relationships which can be included are:

- Finish-to-start
- Start-to-start
- Finish-to-finish
- Start-to-finish

This makes the precedence display easier to follow and permits the introduction of time constraints on the logical links without the need to include dummies or ladders.

9.4 Relationships between arrow diagrams, precedence diagrams and linked bar charts

It is important to understand the relationship between arrow diagrams, precedence diagrams and linked bar charts.

The common feature of all three methods is that there is a logic which links the activities and this logic creates dependencies in the programme. These dependencies are used to determine the order and sequence of the work and, together with the activity durations, may be used to calculate the overall duration of the project. These calculations are called 'critical path analysis' because they reveal those activities with no free time (i.e. zero float), and it is these activities which determine the longest route through the programme. This is called the 'critical path'.

Arrow diagrams and precedence diagrams are known as 'networks' (because that is what they look like) and the analysis of the diagrams is referred to as 'critical path or network analysis'.

There are two ways of presenting networks:

- using arrow diagrams where the arrow represents the activity and the circles join the activities together
- precedence diagrams where the activities are shown as boxes and the boxes are linked together with lines or arrows

Linked bar charts are simply Gantt charts (or bar charts) where the activities are linked with arrows to show dependencies. The various links are illustrated in Figure 9.1 – these may be at the beginning or end of the activities or somewhere in between (this is sometimes called 'mid-bar linking').

The different activity relationships for the arrow, precedence and linked bar chart formats are shown in Figures 9.3–9.7:

- Figure 9.3 indicates finish-to-start relationships in all three formats
- Figure 9.4 and 9.5 indicate start-to-start relationships or methods of introducing overlaps into the sequence of work
- Figures 9.6 and 9.7 indicate finish-to-finish relationships and start-to-finish relationships, again in all three formats.

On a construction project, operations on site mainly overlap one with another. In the example shown in Figure 9.4, the setting up site activity starts at the commencement of the project on day 1 (time 0) and takes 6 days to complete. Reduced level excavation work starts on day 4 (time 3) and takes 4 days to complete. On the bar chart display a 3-day overlap has been introduced between the starting dates of the two operations.

The overlapping of operations is easier to express on either a linked bar chart or in a precedence format, whereas overlapping on an arrow diagram is much more difficult (see Figure 9.4). This cannot be done by starting an overlapping activity part-way along an arrow but involves breaking activities down into more detailed operations (see 'set up site' activity in Figure 9.4 which has been split in two). Alternatively, overlaps can be created by using a 'ladder' diagram which necessitates complex lead and lag terminology (see Figure 9.8 for an example of this).

9.5 Arrow diagrams

Basic principles

The stages involved in preparing a network based on arrow diagram principles are as follows:

(1)   Develop a basic logic diagram (doodle diagram) which is based on the sequence of work

FINISH-TO-START RELATIONSHIPS

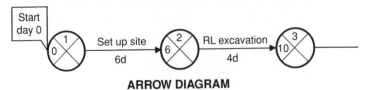

ARROW DIAGRAM

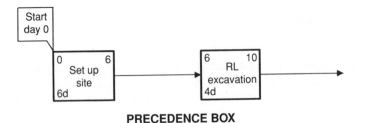

PRECEDENCE BOX

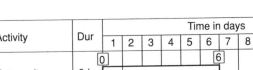

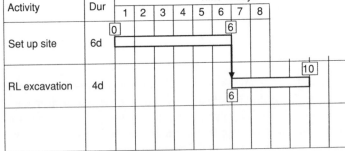

LINKED BAR CHART

Figure 9.3

(2) Enter the operation descriptions, durations and event numbers

(3) Apply the *forward pass* working from left to right through the diagram and calculate the earliest event times for each activity. At an intersection, always carry the highest number forward. Establish the overall project period and enter on the finish flag.

(4) Apply the *backward pass*, working from right to left back through the diagram and establish the latest event times for each activity. At intersections always carry the lowest number back

(5) Establish the *float times* for each activity and highlight the *critical path* (the operations which have zero float)

(6) Convert the arrow diagram to a bar chart format as required, for example an earliest or latest start bar chart display.

START–TO–START RELATIONSHIPS
(Overlapping Activities)

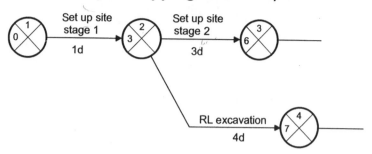

ARROW DIAGRAM

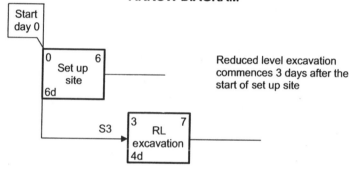

Reduced level excavation
commences 3 days after the
start of set up site

PRECEDENCE BOX

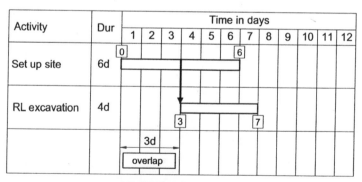

LINKED BAR CHART

Figure 9.4

START-TO-START RELATIONSHIPS

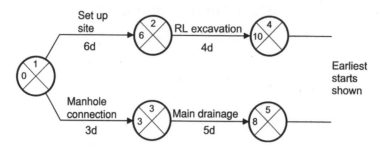

Earliest starts shown

ARROW DIAGRAM

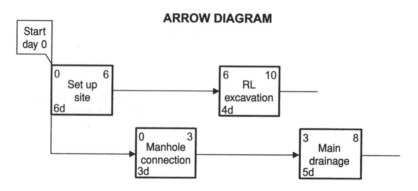

PRECEDENCE BOX

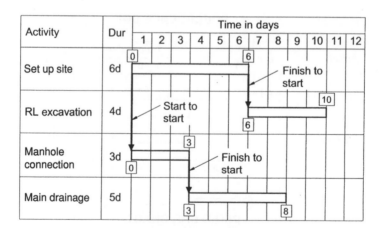

LINKED BAR CHART

Figure 9.5

FINISH-TO-FINISH RELATIONSHIPS

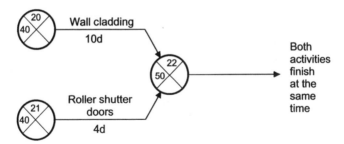

Both activities finish at the same time

ARROW DIAGRAM

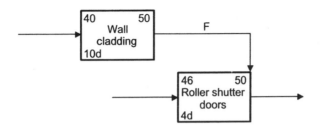

PRECEDENCE BOX

Activity	Dur	Time in days											
		40	41	42	43	44	45	46	47	48	49	50	51
Wall cladding	10d	40										50	
Roller shutter doors	4d							46					

LINKED BAR CHART

Figure 9.6

START-TO-FINISH RELATIONSHIPS

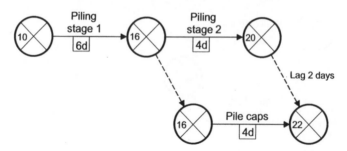

ARROW DIAGRAM

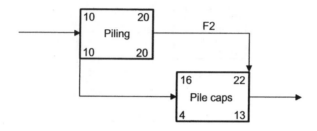

PRECEDENCE BOX

Ref	Operation	Dur	Time in days
			8 9 10 11 12 13 14 15 16 17 18 19 20 21 22 23 24
12	Piling	10d	2 day lag
13	Pile caps	4d	6 day lead

LINKED BAR CHART

Figure 9.7

Arrow diagram example

Example – Retaining wall project

Figure 9.8 indicates the plan and section of three bays of a concrete retaining wall. A schedule of the operations relating to the sequence of work has been developed as follows:

Activity/operation	Duration (days)
Excavate base	2
Concrete blinding	1
Fix formwork	2
Fix steel reinforcement	2
Concrete base	1
Strike formwork	1
Erect precast wall units	3

The basic logic diagram for developing an arrow or precedence diagram for Section A of the retaining wall foundation is shown in Figure 9.9. The figure also indicates the basic arrow diagram for the three sections of the foundations after entering activity descriptions, durations and event numbers.

Figure 9.10 illustrates the analysed diagram after applying the forward and backward pass.

Floats have been indicated and the critical path highlighted passing through the activities with zero float. The overall duration for the sequence has been established as 21 days.

Figure 9.11 indicates an earliest start bar chart based on the arrow diagram. The critical operations have been highlighted in the upper part of the display.

9.6 Precedence diagrams

Basic principles

The stages involved in presenting and analysing a sequence of work in a precedence format are shown in Figure 9.12. Seven operations are to be undertaken in the sequence illustrated. The standard notation used within each precedence box is shown. Analysis follows similar principles to analysing an arrow diagram as follows:

(1) Develop a basic logic diagram in relation to the construction sequence
(2) Enter data – operation description, duration and reference box number in the precedence boxes
(3) Enter relationships, that is finish-to-start, start-to-start, and start-to-finish relationships.
(4) Analyse the diagram by applying the *forward* and *backward* pass. Enter start and finish flags
(5) Calculate the *floats* and establish the operations with zero float – highlight the *critical* operations. The precedence boxes are much easier to analyse, as the floats are simply the difference between the numbers in the corner of the precedence boxes.
(6) Convert the precedence sequence to a bar chart, if required

PRECAST CONCRETE RETAINING WALL AND FOUNDATION

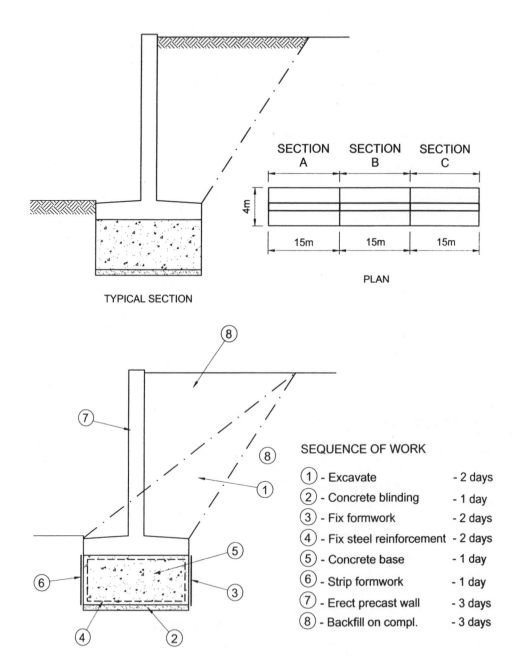

TYPICAL SECTION

PLAN

SECTION A | SECTION B | SECTION C

4m

15m | 15m | 15m

SEQUENCE OF WORK

(1) - Excavate   - 2 days
(2) - Concrete blinding   - 1 day
(3) - Fix formwork   - 2 days
(4) - Fix steel reinforcement   - 2 days
(5) - Concrete base   - 1 day
(6) - Strip formwork   - 1 day
(7) - Erect precast wall   - 3 days
(8) - Backfill on compl.   - 3 days

Figure 9.8

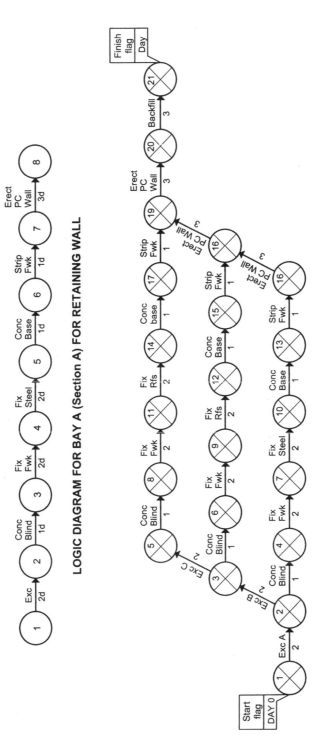

LOGIC DIAGRAM FOR BAY A (Section A) FOR RETAINING WALL

INITIAL NETWORK DIAGRAM

Figure 9.9

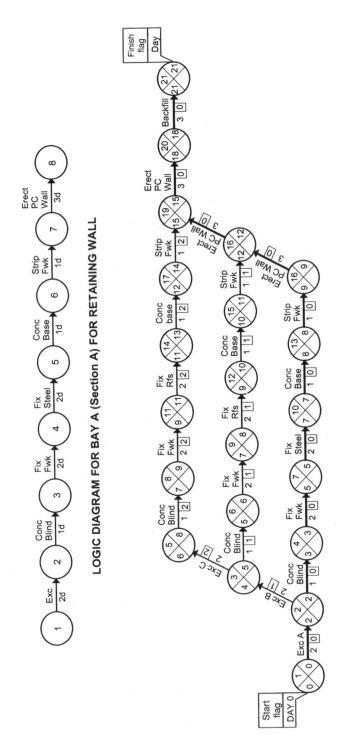

LOGIC DIAGRAM FOR BAY A (Section A) FOR RETAINING WALL

INITIAL NETWORK DIAGRAM

Figure 9.10

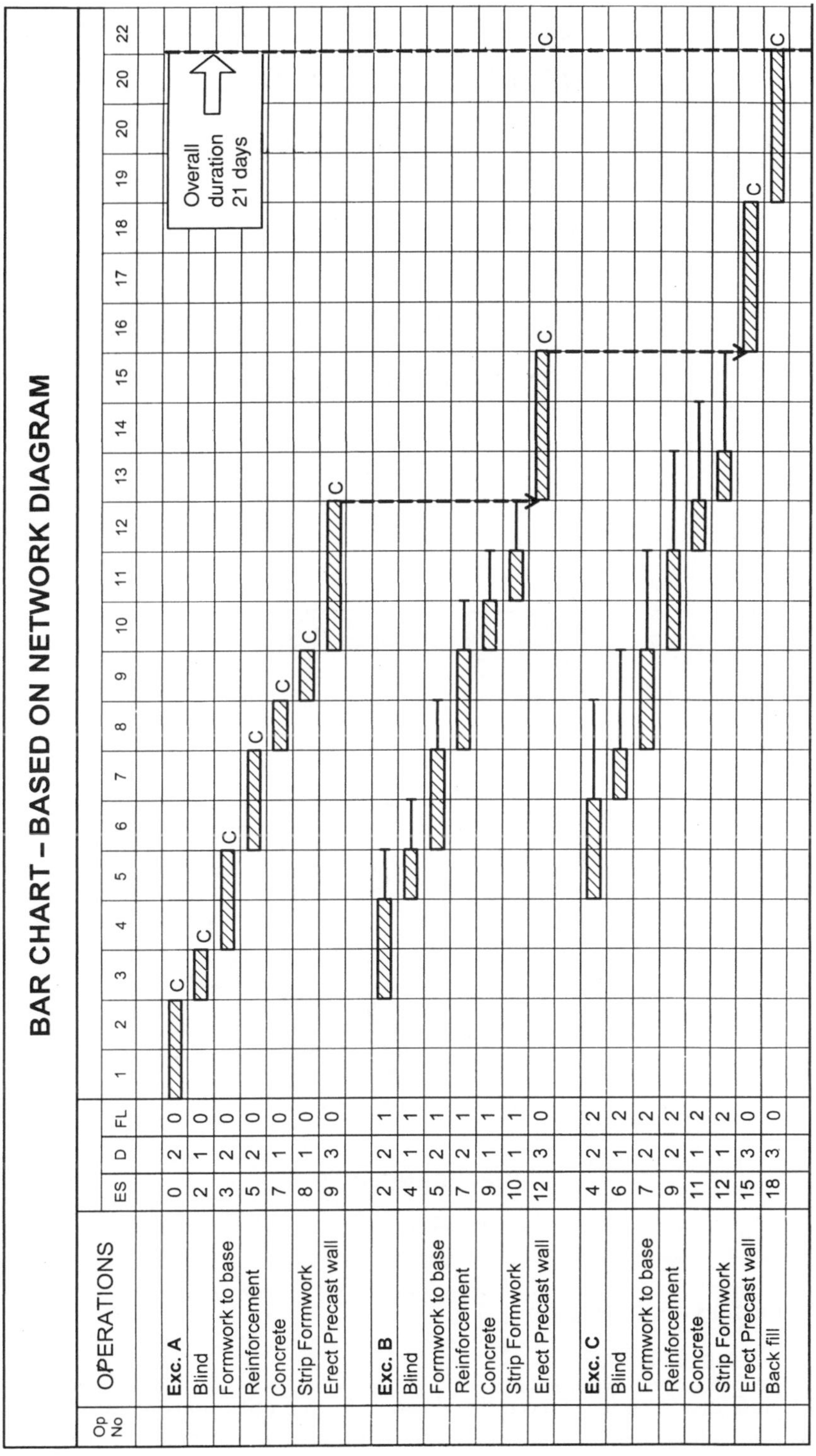

BAR CHART – BASED ON NETWORK DIAGRAM

Op No	OPERATIONS	ES	D	FL	1	2	3	4	5	6	7	8	9	10	11	12	13	14	15	16	17	18	19	20	21	22
	Exc. A	0	2	0																						
	Blind	2	1	0																						
	Formwork to base	3	2	0																						
	Reinforcement	5	2	0																						
	Concrete	7	1	0																						
	Strip Formwork	8	1	0																						
	Erect Precast wall	9	3	0																						
	Exc. B	2	2	1																						
	Blind	4	1	1																						
	Formwork to base	5	2	1																						
	Reinforcement	7	2	1																						
	Concrete	9	1	1																						
	Strip Formwork	10	1	1																						
	Erect Precast wall	12	3	0																						
	Exc. C	4	2	2																						
	Blind	6	1	2																						
	Formwork to base	7	2	2																						
	Reinforcement	9	2	2																						
	Concrete	11	1	2																						
	Strip Formwork	12	1	2																						
	Erect Precast wall	15	3	0																						
	Back fill	18	3	0																						

Overall duration 21 days

Figure 9.11

PRINCIPLES OF PRECEDENCE DIAGRAMS
PRECEDENCE NOTATION

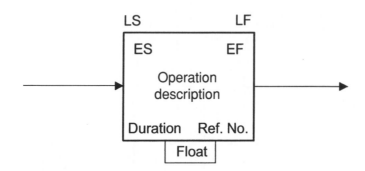

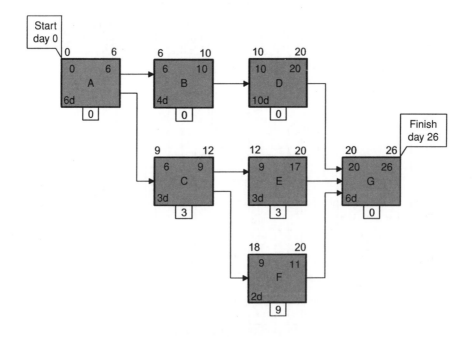

PRECEDENCE BOX

Figure 9.12

Precedence diagram examples

Example 1 – Establishing site

Figure 9.13 indicates an initial precedence diagram for ten operations required for the establishment of a site. Figure 9.14 shows the analysed precedence diagram for the operations

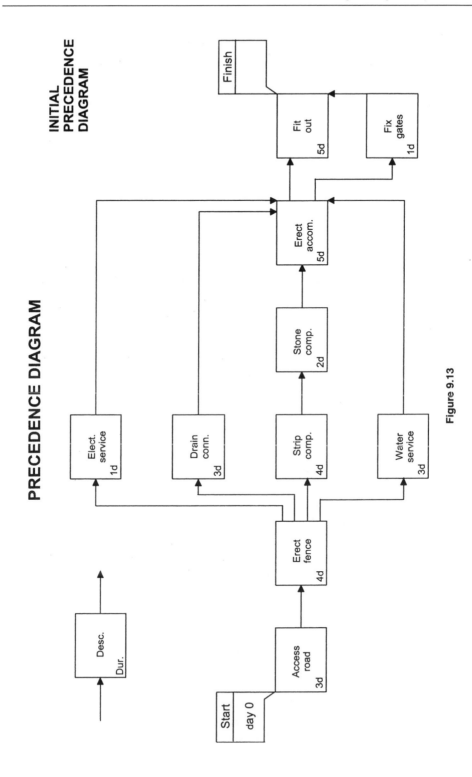

PRECEDENCE DIAGRAM

INITIAL PRECEDENCE DIAGRAM

Figure 9.13

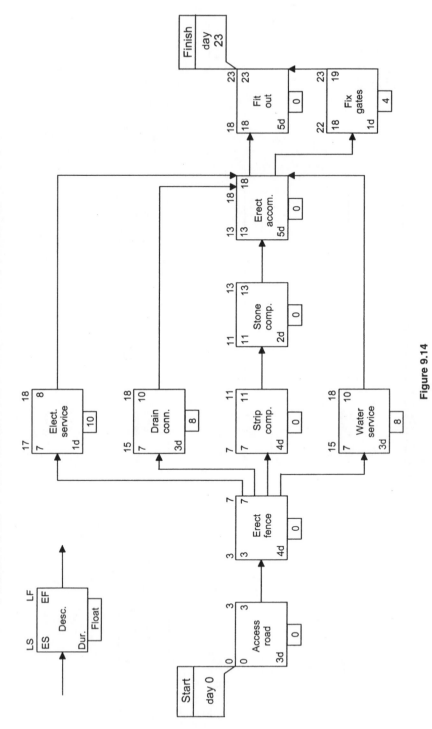

ANALYSED PRECEDENCE DIAGRAM – ESTABLISHING SITE

Figure 9.14

involved in the sequence. All relationships on the diagram are finish to starts. The overall time for the complete sequence is 23 days. The critical operations have been highlighted (operations with zero float). The float is simply the difference between the numbers in the corner of the boxes, that is (LS minus ES) or (LF minus EF). LS, latest start; ES, earliest start; LF, latest finish; EF, earliest finish.

Figure 9.15 illustrates an earliest start bar chart with the critical operations shown first.

Example 2 – Factory project

Figure 9.16 indicates the initial precedence diagram for nine operations during the construction of the foundations and ground floor slab of a small factory building. Finish-to-start, start-to-start and finish-to-finish relationships have all been introduced and the critical path established. Figure 9.17 shows the analysed diagram indicating an overall project period of 34 days.

The analysis indicates the simplicity in analysing the sequence and the establishment of the critical path. The precedence diagram has been further presented as earliest start bar chart in Figures 9.18.

In Figure 9.19, monetary values have been allocated to each operation and the monthly cumulative value assessed. The cumulative value forecast is based on the earliest start situation and is presented in both graphical and tabular format in Figure 9.20.

Authors' note: The use of network diagrams to establish cumulative value forecasts is a useful management tool for the financial control of projects.

9.7 Line of balance (elemental trend analysis)

History and development

A line-of-balance diagram comprises a series of inclined lines which represent the rate of working between repetitive operations in a construction sequence. This approach was developed by the National Building Agency.

The application of line of balance to construction was pioneered by Lumsden (1965) and became recognised as the best planning method for repetitive work such as housing.

The technique has been widely used for the planning of refurbishment works, new build housing and flats and has also been applied to civil engineering works as illustrated by Harris and McCaffer (2006) and Cormican (1985). It is not unusual to see bar chart displays incorporating a line-of-balance diagram to illustrate the programming of any repetitive sections of the works.

Line of balance is a visual display of the rate of working of different activities on a programme. The ideal line-of-balance display shows all balance lines running parallel to each other, but in practice this is often difficult to achieve.

Principles of line of balance

Figure 9.21 shows the logic diagram for three operations, A, B and C, in a construction sequence. Figure 9.22 indicates the line-of-balance diagram for each of these activities showing different rates of working from one operation to the other. As can be observed from the diagram,

ESTABLISHING SITE – Bar chart based on precedence diagram

Time in weeks

OPERATION	ES	Dur	FL	1	2	3	4	5	6	7	8	9	10	11	12	13	14	15	16	17	18	19	20	21	22	23
Access road	0	3	0	CR																						
Erect fence	3	4	0	CR																						
Strip compound	7	4	0	CR																						
Stone compound	11	2	0	CR																						
Erect accom.	13	5	0	CR																						
Fit out	18	5	0	CR																						
Electrical service	7	1	10																							
Drain connection	7	3	8																							
Water service	7	3	8																							
Fix gates	18	1	4																							

CRITICAL OPERATIONS

TARGET COMPLETION DATE

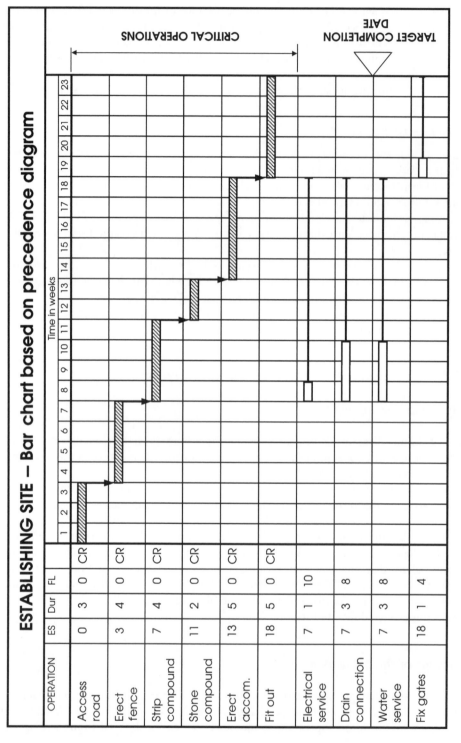

Figure 9.15

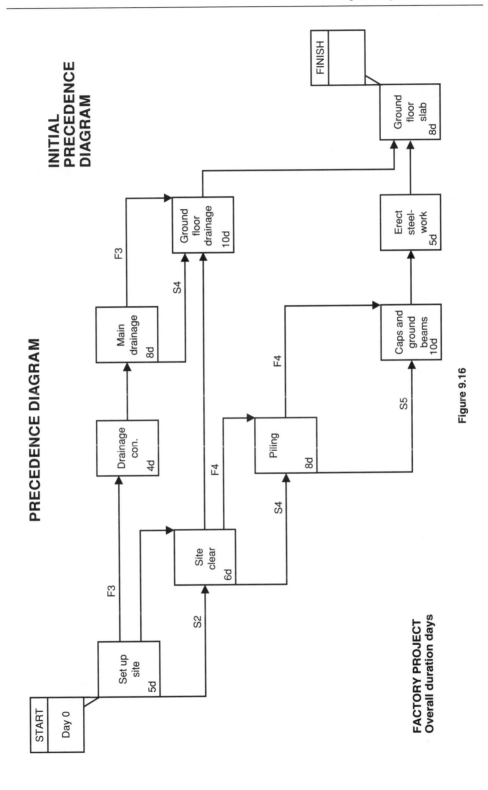

PRECEDENCE DIAGRAM

INITIAL
PRECEDENCE
DIAGRAM

Figure 9.16

FACTORY PROJECT
Overall duration days

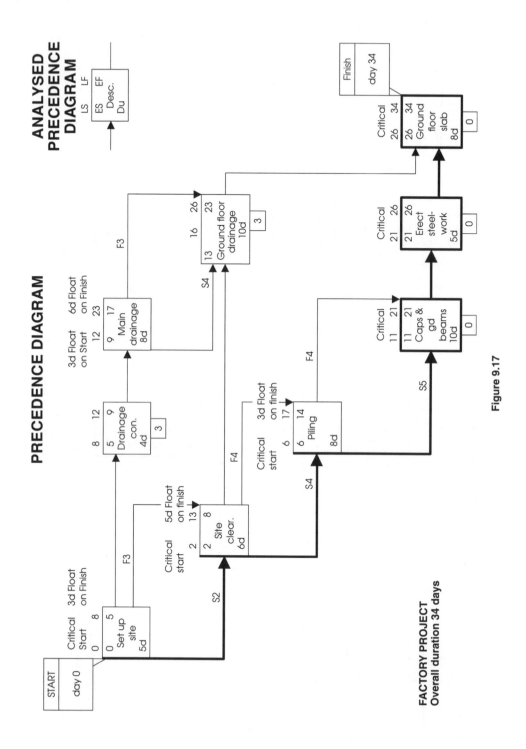

PRECEDENCE DIAGRAM

ANALYSED PRECEDENCE DIAGRAM

FACTORY PROJECT
Overall duration 34 days

Figure 9.17

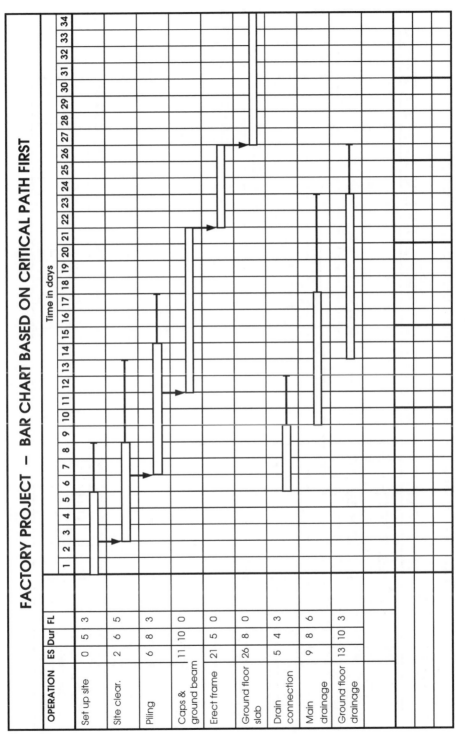

Figure 9.18

FACTORY PROJECT – BAR CHART (from precedence diagram)

Time in days

OPERATION	ES	DV	FL	VALUE	1	2	3	4	5	6	7	8	9	10	11	12	13	14	15	16	17	18	19	20	21	22	23	24	25	26	27	28	29	30	31	32	33	34
Set up site	0	5	3	5000	1	1	1	1	1																													
Site clear.	2	6	8	12000			2	2	2	2	2	2																										
Piling	6	8	7	24000							3	3	3	3	3	3	3	3																				
Caps & ground beam	11	10	0	20000												2	2	2	2	2	2	2	2	2	2													
Erect frame	21	5	0	50000																						10	10	10	10	10								
Ground floor slab	28	8	0	16000																											2	2	2	2	2	2	2	2
Drain connection	5	4	3	4000						1	1	1	1																									
Main drainage	9	8	9	8000										1	1	1	1	1	1	1	1																	
Ground floor drainage	13	10	3	10000														1	1	1	1	1	1	1	1	1	1											
Prelims	£1000 per day			34000	1	1	1	1	1	1	1	1	1	1	1	1	1	1	1	1	1	1	1	1	1	1	1	1	1	1	1	1	1	1	1	1	1	1
Daily value					2	2	4	4	4	4	7	7	5	5	7	7	8	8	5	5	5	4	4	4	4	12	12	11	11	11	3	3	3	3	3	3	3	3
Cumulative value					2	4	8	12	16	20	27	34	39	44	49	56	63	71	76	81	86	90	94	98	102	114	126	137	148	159	162	165	168	171	174	177	180	183
Weekly value									16					28					32					22				50					23				12

Figure 9.19

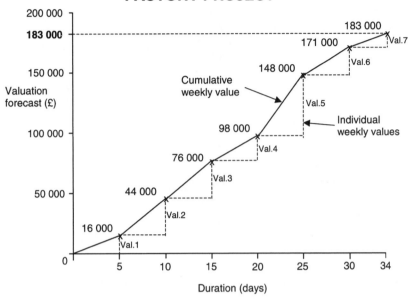

VALUE–TIME FORECAST
FACTORY PROJECT

TABULAR CUMULATIVE WEEKLY VALUE FORECAST			
Week	Day	Cumulative value	Weekly value
1	5	16 000	16 000
2	10	44 000	28 000
3	25	76 000	32 000
4	20	98 000	22 000
5	25	148 000	50 000
6	30	171 000	23 000
7	34	183 000	12 000

Figure 9.20

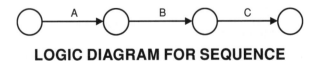

LOGIC DIAGRAM FOR SEQUENCE

Figure 9.21

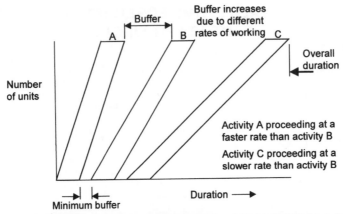

LINE-OF-BALANCE DIAGRAM – NON-PARALLEL WORKING

Figure 9.22

operations B and C are out of balance with operation A. However, by increasing the number of gangs employed on operations B and C, the line-of-balance diagram as shown in Figure 9.23 can be achieved. This results in a considerable saving in the overall project period.

Figure 9.22 introduces the idea of 'buffers' at the start or finish of an operation in order to build some degree of flexibility into the programme. This also allows the preceding operation to move clear of the work area or house unit before the next operation commences.

Figure 9.24 shows a logic diagram for a construction sequence where operations B and C may start or finish together prior to the commencement of operation D. Figures 9.25 and 9.26 illustrate the balance diagram for these two situations depending on whether operation B is working at a slower or faster rate than operation C.

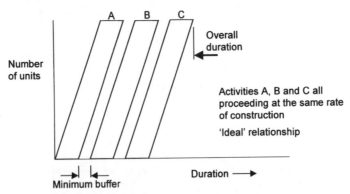

LINE-OF-BALANCE DIAGRAM – IDEAL RELATIONSHIP

Figure 9.23

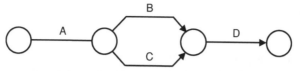

LOGIC DIAGRAM FOR SEQUENCE

Figure 9.24

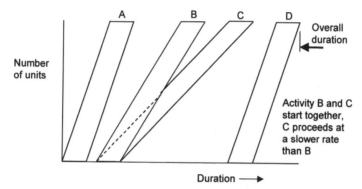

LINE-OF-BALANCE DIAGRAM – OPERATIONS B AND C COMMENCING TOGETHER

Figure 9.25

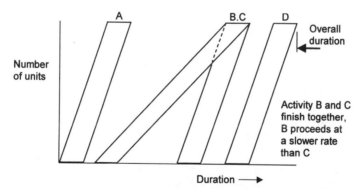

LINE-OF-BALANCE DIAGRAM – OPERATIONS B AND C FINISHING TOGETHER

Figure 9.26

Developing a line-of-balance diagram

Example – Access deck project

Figure 9.27 illustrates the plan and section of a reinforced concrete-elevated access platform. The work involves the construction of six bays of elevated deck together with the associated pile supports and capping beams. The sequence of the activities is shown in the logic diagram in Figure 9.28:

- Operation 1: Install piles – 2 days
- Operation 2: In situ beams – 4 days
- Operation 3: Precast concrete deck – 1 day

ACCESS DECK PROJECT

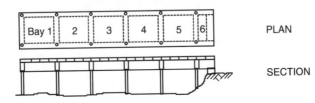

PLAN AND SECTION — REINFORCED CONCRETE PLATFORM

Figure 9.27

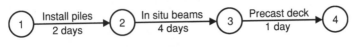

LOGIC DIAGRAM

Figure 9.28

On the basis of utilising one gang of men on each operation, the line-of-balance diagram as shown in Figure 9.29 has been developed. The planner/project manager is required to consider the most economical proposal for speeding up the project.

From the line-of-balance diagram, it can be observed that operation 2 is out of balance with operations 1 and 3. By increasing the number of gangs on operation 2 (from one gang to two gangs), the balance lines will move into a better overall parallel position.

Figure 9.30 indicates the procedure for producing a histogram of labour resources from a line-of-balance diagram. This is achieved by considering the start and finish date of each of the labour gangs engaged on each operation.

Figure 9.31 shows the redrawn line-of-balance diagram using two labour gangs on operation 2. This results in an overall time reduction of 10 days. The revised labour resources are shown in histogram form below the line-of-balance diagram.

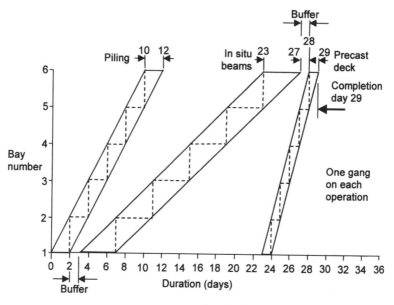

LINE-OF-BALANCE CHART

LINE-OF-BALANCE SCHEDULE – REINFORCED CONCRETE PLATFORM

Figure 9.29

As can be seen from the above example, line-of-balance planning provides a visual display of the rate of working across the whole project and enables decisions to be made in relation to the use of labour. It enables the planner to start with a forecast handover rate per week and then to produce a forecast of the labour resources needed to achieve it.

The line-of-balance programming technique is rarely understood by construction managers and planners and is still very much underused in the construction industry. Line of balance tends to be favoured only by those who have a thorough grasp of the principles and application based on experience. Where it is company policy to use line of balance, it will be used, but this is in only a small number of companies. Line of balance is only really applicable to repetitive operations on refurbishment and housing projects and therefore does not have the same wide application as bar charts or networks.

Practical applications

Example – Repetitive housing project

Line of balance has gained a foothold in a number of the large house-building firms where the repetitive part of a construction sequence can be readily integrated into a bar chart programme.

The following example illustrates the application of line of balance to a housing project where a contractor requires a programme for the construction of ten house units. The five operations that occur in the construction sequence are shown in Table 9.1.

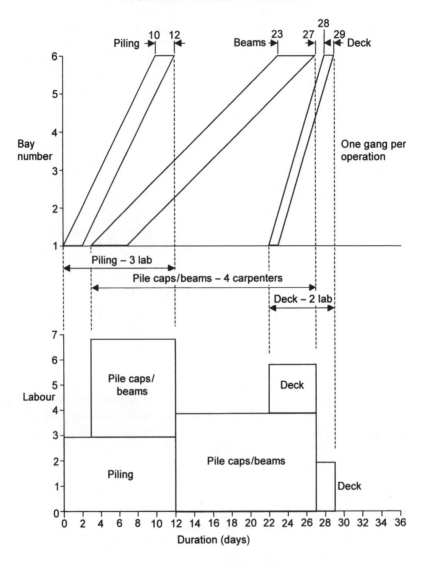

ACCESS DECK PROJECT

LABOUR RESOURCES – REINFORCED CONCRETE
PLATFORM

Figure 9.30

ACCESS DECK PROJECT

LINE-OF-BALANCE CHART BASED ON TWO GANGS
WORKING ON BEAMS AND CAPS

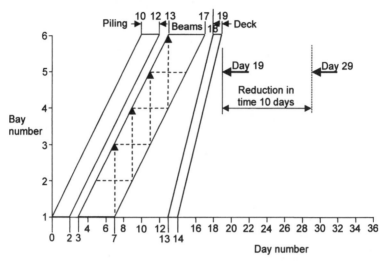

REVISED LINE-OF-BALANCE SCHEDULE

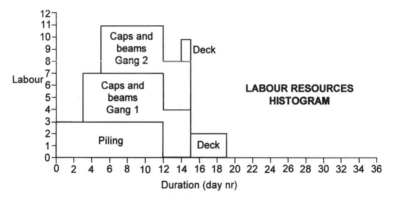

**LABOUR RESOURCES
HISTOGRAM**

REVISED LABOUR RESOURCES

Figure 9.31

- *Step 1*
  Draw the sequence logic for the five operations as shown in Figure 9.32.
- *Step 2*
  Assess the start and finish date of each operation in the construction sequence for the first and last units. This enables the balance lines to be plotted. Use the following formula in

Table 9.1 Five construction operations.

Operation	Duration per unit in weeks	Number of gangs
Foundations	2	2
External walls	4	3
Roof construction	1	1
Internal finishes	4	3
External works	2	2

Note: No of house units = 10.

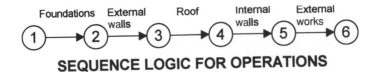

SEQUENCE LOGIC FOR OPERATIONS

Figure 9.32

each case:

$$\frac{\text{number of units less } 1 \times \text{duration of operation}}{\text{number of gangs used}}$$

Operation 1–2 (foundations)

Start of unit 1 = **week 0**

Finish of unit 1 = **week 2** (i.e. the duration of the operation)

Start of unit $10 = 0 + \dfrac{(10 \text{ units} - 1) \times 2}{2 \text{ gangs}} = \dfrac{18}{2} = \textbf{week 9}$

Finish of unit 10 = week 9 + 2 weeks = **week 11**

- *Step 3*
  Plot the balance lines on squared paper. Figure 9.33 shows the balance line drawn for the foundations operation.
- *Step 4*
  Assess the start and finish date for the next operation:

Operation 2–3 (external walls)

NB: Allow a minimum buffer of 1 week between one operation and the next. This will be either at the start or finish of the next operation depending on the rate of working.

Start of unit 1 = week 2 + 1 week buffer = **week 3**

Finish of unit 1 = week 3 + duration of 4 weeks = **week 7**

Start of unit $10 = \text{week } 3 + \dfrac{(10 \text{ units} - 1) \times 4}{3 \text{ gangs}} = 3 + \dfrac{36}{3} = \textbf{week 15}$

Finish of unit 10 = week 15 + 4 weeks = **week 19**

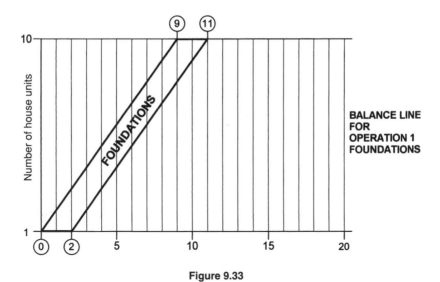

Figure 9.33

Before plotting the balance line for this operation, consider the rate of construction between the foundations and external walls operations.

Foundations

$$\text{Slope of balance line} = \frac{\text{duration}}{\text{no. of gangs}} = \frac{2}{2} = 1.00$$

External walls

$$\text{Slope of balance line} = \frac{\text{duration}}{\text{no. of gangs}} = \frac{4}{3} = 1.33$$

Strictly speaking, the numbers do not really mean anything except that the lower number for the foundations operation gives an approximate indication that this operation is progressing at a faster rate than the external walls activity. This quick calculation indicates from where to plot the balance line for the external works operation. If the number is higher than that of the preceding operation, plot from the bottom of the diagram, otherwise the operations will bump into each other. If the number is lower than that of the preceding operation, plot from the top. In either case, be sure to include the minimum buffer allowance.

- *Step 5*
  Plot the balance line for external walls as shown in Figure 9.34.
- *Step 6*
  Assess the start and finish date for the next operation:

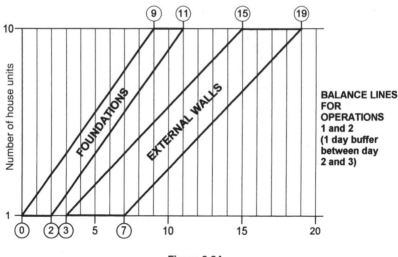

Figure 9.34

Operation 3–4 (roof construction)

Start of unit 10 = week 19 + 1 week buffer = **week 20**

Finish of unit 10 = week 20 + 1 week duration = **week 21**

Start of unit 1 = week 20 − $\frac{(10-1) \times 1}{1 \text{ gang}}$ = 20 − 9 = **week 11**

Finish of unit 1 = week 11 + 1 week duration = **week 12**

- *Step 7*
  Consider the rate of construction between the external walls and roof operations.

External walls

Slope of balance line = **1.33**

Roof construction

Slope of balance line = $\frac{1}{1}$ = **1.00**

As the roof operation is progressing at a faster rate than the external walls, the relationship between the balance lines will commence at the finish of the external walls operation for house unit 10.

- *Step 8*
  Plot the roof construction balance line starting from the top of the diagram, as shown in Figure 9.35.
- *Step 9*
  Complete the calculations for the remaining operations. These have been added to Figure 9.35 to indicate an overall completion at week 32.

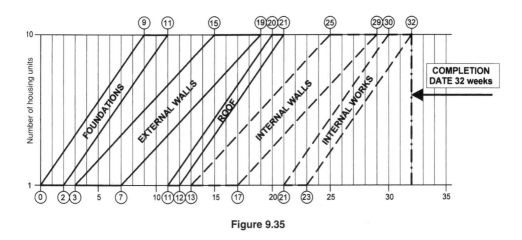

Figure 9.35

9.8 Time-chainage diagrams

History and development

In simple terms, the time-chainage diagram or location–time chart referred to by Cormican (1985) is a combination of the bar chart and line-of-balance scheduling formats, and it is from these programming techniques that time-chainage principles have been developed.

Time-chainage diagrams have been widely applied on major roadworks projects and in the development of the motorway system in the UK for many years. The technique was also used for the planning of tunnelling and fixed equipment installation on the Channel Tunnel project and its application is discussed in *Proceedings of the Institution of Civil Engineers: The Channel Tunnel* (ICE 1992). Time-chainage diagrams, like their close cousin the line of balance, are applicable only for limited types of project and therefore are not as widely appreciated in the industry as bar charts and network techniques. Nevertheless, the technique has distinct attributes and advantages on projects where it is important to depict:

- The order of activities or operations
- Where activities are happening locationally
- How activities must progress in relation to direction and distance
- Time, key dates and holidays, etc.

Principles of time-chainage diagrams

The time-chainage form of presentation enables the time dependencies between activities to be shown, together with their order and direction of progress along the job. These diagrams are most usefully employed as a planning tool on projects such as motorways and major highway works, pipelines, railway track work, tunnelling, etc.

Projects of this nature can be viewed as mainly linear in nature. In other words, construction starts at one point and proceeds in an orderly fashion towards another location. This would be typified on a highway project by activities such as fencing, drainage, road surfacing and road markings.

To some extent, this type of work calls for a different planning technique because bar charts would not be useful in giving locational information, and precedence/arrow diagrams would not reflect the time–location relationship which clearly exists on such projects.

In this respect, most operations take place on a forward travel basis with the gang starting at one point or chainage and moving along the job. As one activity leaves a particular location, other activities can take its place. This ensures the correct construction sequence and avoids over-intensive activity in one location.

Most of the lines on the time-chainage diagram have no appreciable thickness. This is because the time spent by each gang at a particular location is relatively small and the gang moves along the site quite quickly. Examples of this are drainage, road surfacing and safety barrier erection on a motorway.

Retaining walls would also constitute a linear activity but would tend to occupy any particular location (or chainage) for a more appreciable time due to the nature and duration of the construction operations involved.

With earthworks cut and fill operations, the situation is different in that earthworks plant will occupy a particular cut or fill zone for some time before moving to another location.

Bridges, culverts and underpasses, on the other hand, are 'static' operations and can be viewed as individual 'sites' in their own right. Such activities act as restrictions and forward travel activities may have to be programmed around them. For instance, on a highway project, drainage work may be interrupted by a bridge site and consequently the contractor will have to return later to finish the drainage once the bridge nears completion.

Various types of time-chainage representations are possible, but basically the diagram comprises two axes, time and distance, with the various activities shown as lines or bars on the chart. Linear activities are represented with a line or bar which is positioned on the chart to show its commencing and completion chainages, and is inclined in the direction of progress at an angle consistent with the anticipated duration of the operation. A static activity, such as a bridge on a motorway, is represented by a line or thin bar positioned at a particular location or chainage, with the duration of the activity expressed by the length of the line or bar. Activity labels are annotated on the respective line or bar to distinguish one operation from another.

Figures 9.36 and 9.37 illustrate two methods of presentation of time-chainage diagrams. Chainages can be located on either the vertical or horizontal axes with time (usually in weeks) shown on the other axis.

Developing a time-chainage diagram

Drawing a time-chainage diagram is not as easy as it looks and a good deal of practice is required. Consider the following steps:

- *Step 1*
  Consult the project layout drawings and note the chainage positions. Main chainages on a highway project are at 1000 m intervals.

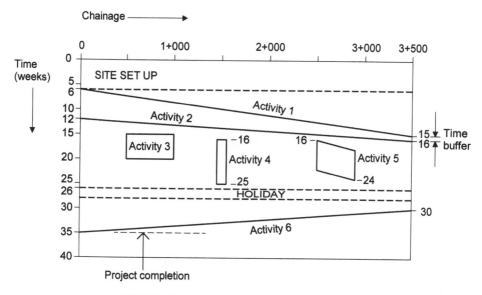

TIME-CHAINAGE DIAGRAM
Horizontal Format

Figure 9.36

- *Step 2*

  Draw an outline time-chainage diagram with time along one axis and distance or chainage along the other, using either the horizontal or vertical format. Add main holiday periods allowing 2 weeks for Christmas and 1 week for Easter.

- *Step 3*

  List main programme activities or operations in approximate construction sequence. Include activities for site set-up or mobilisation and clear site at the end. Estimate the duration of each programme activity in weeks.

- *Step 4*

  Fill in the site 'set-up' and 'clear site' activities on the time-chainage diagram. Using your preferred format, plot the appropriate number of weeks over the entire length (chainage) of the project.

- *Step 5*

  Decide in turn where and when each activity will take place. For linear activities (e.g. drainage), start at the appropriate location (chainage) and week number and draw a line for the correct distance (chainage) and time (weeks). For static activities (e.g. bridgeworks), draw a thin box at the appropriate chainage with a length representing the activity duration.

- *Step 6*

  Complete all activities on the list. It is sometimes helpful to produce an outline bar chart programme to help clarify the correct time-chainage display.

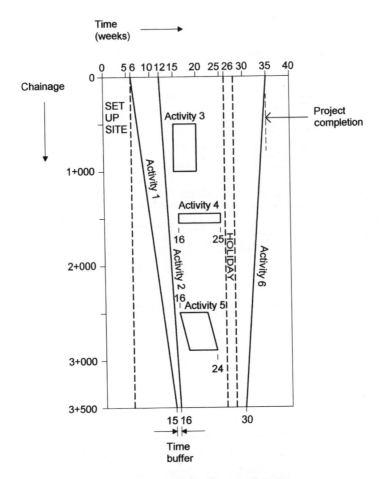

TIME-CHAINAGE DIAGRAM
Vertical Format

Figure 9.37

Practical applications

Take, for example, a project to carry out improvements to an existing highway 3.5 km long comprising the following activities:

- Activity 1 – Fencing
- Activity 2 – Drainage
- Activity 3 – Bulk earthworks
- Activity 4 – Footbridge
- Activity 5 – Retaining wall
- Activity 6 – Road surfacing

The start of the job is at chainage 0 + 000 and the project finishes at chainage 3 + 500. A footbridge is required to cross the highway at chainage 1 + 500 which would be 1500 m up chainage from the beginning of the job. Distances or locations can be conveniently determined by these chainages.

OPERATION	TIME/LOCATION	EXAMPLE
Set up site	Starts at time 0 Duration 6 weeks	Offices for contractor and engineer
Activity 1	Starts at week 6 Duration 9 weeks Finishes week 15 Forward travel up chainage	Fencing
Activity 2	Starts week 12 Duration 4 weeks Finishes week 16 Forward travel up chainage	Drainage

Activity 3	Starts week 15
	Duration 5 weeks
	Finishes week 20
	Occupies zone between
	    chainage 0+500 and 1+000
	    for 5 weeks
		Bulk earthworks

Activity 4	Starts week 16 Duration 9 weeks Finishes week 25 Takes place at chainage 1+500	Footbridge

Activity 5	Starts week 16
	Duration 8 weeks
	Finishes week 24
	Takes place between chainage
	    2+500 and 2+900
	Forward travel up chainage at the rate
	    of 50 metres per week
		Retaining wall

Activity 6	Starts week 30 Duration 5 weeks Finishes week 35 Forward travel down chainage	Road surfacing
Holiday	Starts week 26 Duration 2 weeks Finishes week 28	Christmas shut down

TIME-CHAINAGE DIAGRAMS
Explanation

Figure 9.38

This chainage referencing system is also useful on site where the contractor will usually position chainage boards along the job to enable anyone to quickly identify exactly where they are on the site.

The advantage of showing chainage along the horizontal axis is that this more readily resembles the way the drawings are laid out and the programme can therefore more easily be related to what has to be constructed. On the other hand, time–chainage diagrams showing time on the horizontal axis may be easier to read by those familiar with bar charts. The choice is a matter of personal preference.

Figure 9.38 explains how the different activities are plotted on the horizontal and vertical time–chainage diagrams in Figures 9.36 and 9.37. Close inspection of Figures 9.36 and 9.37 will reveal time buffers between activities similar to those used in line-of-balance/elemental trend analysis.

The highway project case study in Chapter 21 illustrates the practical considerations involved in developing a time–chainage diagram for a major highway project.

9.9 Accelerating the project

Project acceleration and time–cost optimisation

There are many circumstances in which the contractor may wish to speed up work on a contract, including being behind programme and having to increase production in order to minimise extensive liquidated damages. Alternatively, the client may have requested the contractor to indicate the additional costs of completing the project earlier than the contract completion date.

In practice, a department store client, for instance, may request an earlier occupation date for the building in order to take advantage of the winter or summer sales. The additional profit created by the earlier opening may well exceed the contractor's additional costs.

In order to balance the time savings against the costs of speeding up the work, optimisation studies are undertaken in order to consider the various options available. These studies allow the client and/or contractor to assess the effect on the direct and indirect costs of reducing the overall project period, and this can then be compared with the potential profits or savings in liquidated damages due to earlier completion.

Many writers refer to this method of analysis as 'time–cost optimisation', 'least cost optimisation' or 'crash costing', but perhaps a more appropriate term would be 'project acceleration' as this is really what it is all about.

Whatever you decide to call it, the first step is to understand the terminology used and then the principles of the process, illustrated here by way of two worked examples.

Project acceleration terminology

Consider the network arrow for the activity 'electrical services':

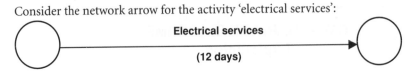

Electrical services

(12 days)

Normal time, normal cost

This is the usual time that would be needed to carry out the electrical services work under normal circumstances, estimated at 12 days with a cost of £10 000.

Crash time, crash cost

The crash time is the maximum time the operation can be compressed by increasing the resources. This reduction in time leads to an increase in the direct cost. The revised cost is called the crash cost. In this example, the accelerated time or crash time is to be 8 days at a total cost or crash cost of £18 000.

Cost slope

The cost slope represents the cost of accelerating any of the project activities by one unit of time (in the above case, 1 day).

In order to achieve a reduction in the overall project duration at the least possible cost, the activities on the critical path of the programme must be compressed as much as possible. This is done by first considering the activities with the least cost slope.

The cost slope of the electrical services activity is expressed as:

$$\frac{\text{increase in cost}}{\text{reduction in time}} \quad \text{or} \quad \frac{\text{crash cost less normal cost}}{\text{normal time less crash time}}$$

$$= \frac{£18\,000 - £10\,000}{12\,\text{days} - 8\,\text{days}} = \frac{£8000}{4\,\text{days}}$$

$$= £2000\ \text{cost slope}$$

Activity ranking

Once the arrow or precedence relationship has been analysed and the critical activities identified, each of the activities on the critical path are ranked in order of their cost slopes, starting with the least expensive.

It is obviously more economical to apply reductions in the project time to the less expensive activities first in order to achieve the required reduction in the overall project period. As the ranking is applied to the network sequence, the float times and cost slopes of non-critical activities must be considered as at some point these may become critical.

Direct costs

These are the costs associated with carrying out activities on the programme including:

- Labour
- Plant
- Materials

- Subcontractors (if applicable)
- Overheads and profit

When an activity is accelerated, the corresponding direct costs will increase. This may be due to the need to supply additional resources in the form of increased labour, plant and material requirements. Accelerating an operation may also involve overtime or weekend working. Also, incentives in the form of bonus payments may have to be made to ensure that the task is completed on time. Other direct costs may be incurred, such as additional formwork. This will reduce the number of formwork uses originally envisaged by the estimator and thus add to the direct cost of acceleration.

Acceleration may require extra direct supervision, such as foremen and gangers, to cover weekend working and the supervision of additional labour gangs.

Indirect costs

These are the time-related costs of the project which change as the project duration changes. They are normally included in the contract preliminaries and will include:

- Project supervision
- Site hutting and accomodation
- Site office telephones, heating and lighting
- Vans and site transport

When an activity is accelerated, the corresponding indirect costs will decrease. This is due to the reduction in project duration which directly affects the contract preliminaries. The contractor will, in principle, be on site for a shorter period and therefore the client will expect some reduction in the site administration costs or preliminaries.

Total project cost

This is the summation of the direct and indirect costs. The total cost is usually expressed at the normal time and at the optimum project duration.

Optimum project duration

The optimum project duration occurs at the point where the most beneficial least cost situation occurs, taking into account both direct and indirect costs. In order to establish the least cost situation, the cost increase for each unit of time reduction must be considered.

Figure 9.39 illustrates the relationship between direct cost, indirect cost and project duration. The summation of the direct and indirect costs (i.e. the total project cost) is also indicated. The optimum duration is shown on the total cost graph and is the date at which the costs rise most significantly.

RELATION BETWEEN INDIRECT COST, DIRECT COST AND TOTAL PROJECT COST

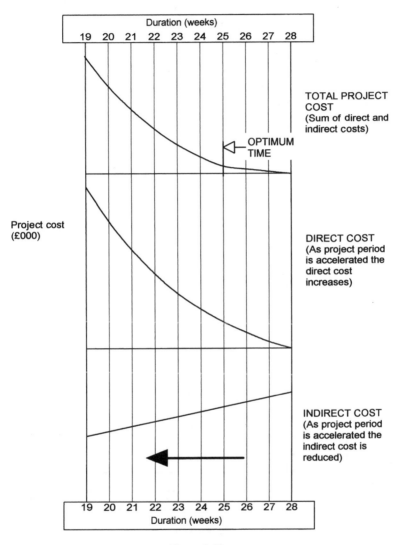

Figure 9.39

Project acceleration principles

Example – based on a network arrow diagram

Figure 9.40 shows a construction sequence involving activities A to I in arrow diagram format. The tabular data indicate the normal times, normal costs, crash times and crash costs. The indirect costs (or time-related preliminaries) of the project amount to £2000 per week.

INITIAL ARROW DIAGRAM

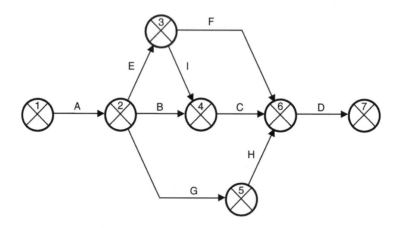

PROJECT DATA				
Activity	Normal time	Normal cost	Crash time	Crash cost
A	6	12 000	4	20 000
B	8	24 000	4	48 000
C	3	18 000	2	20 000
D	6	18 000	4	24 000
E	6	36 000	4	54 000
F	10	10 000	6	50 000
G	5	20 000	3	30 000
H	8	40 000	6	50 000
I	2	20 000	2	20 000
Summation	£198 000			
Indirect costs £2 000 per week				

Figure 9.40

Figure 9.41 illustrates the analysed arrow diagram based on the normal time situation which gives an overall project duration of 28 weeks. The total project cost, based on normal time, has been calculated at £254 000 which is made up of direct costs of £198 000, plus £56 000 (28 weeks × £2000) of indirect costs.

Two scenarios will now be considered.

Scenario 1: A reduction of 5 weeks in the project duration in order to complete the project by week 23. The effect on the direct and indirect costs will be considered in order to achieve this reduction in time.

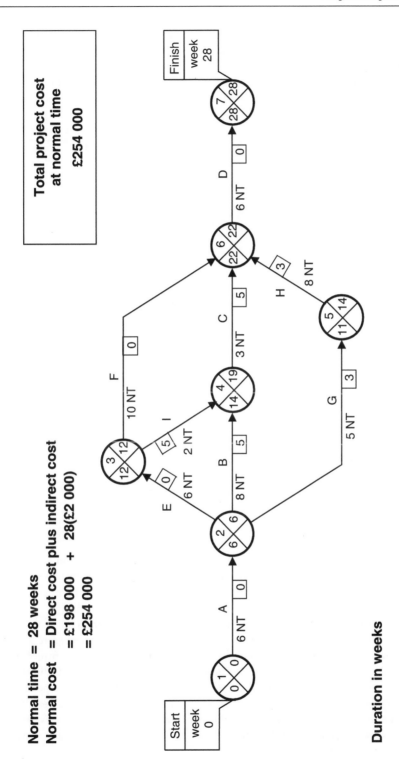

NORMAL TIME/NORMAL COST ANALYSIS

Figure 9.41

In order to reduce the project duration by 5 weeks, it will be necessary to consider the cost slopes of all activities. The assessment of the cost slopes and their appropriate ranking are indicated in Figure 9.42.

From Figures 9.41 and 9.42 it can be seen that in order to reduce the project duration by 5 weeks, it will be necessary to reduce the duration of activities on the critical path in rank order (from the least expensive to the more expensive cost slopes). In Figure 9.41, the critical path follows activities D, A and E in rank order. By using the crash durations for these activities, a 5-week reduction in the overall project period can be achieved, as shown in Table 9.2. This gives a crash cost of £267 000, as shown in Table 9.3.

Table 9.2 Reduction in project period.

Activity	Normal time (weeks)	Crash time (weeks)	Reduction in time (weeks)	Cost slope (£)	Increase in direct cost (£)
D	6	4	2	3000	6000
A	6	4	2	4000	8000
E	6	5	1	9000	9000
Reduction in overall time =			**5**		
Increase in direct cost due to this					
reduction in time =					**23 000**

Table 9.3 Total project costs at week 23.

Direct costs =	£198 000 + £23 000 =	£221 000
Indirect costs =	23 weeks @ £2000 =	£46 000
Total project cost		**£267 000**

Therefore, in order to achieve an acceleration of 5 weeks, the project cost will be increased by £13 000, that is:
Crash cost = £267 000
Less normal cost = £254 000
Acceleration cost = **£13 000**

Figure 9.43 shows the revised arrow diagram analysis using the crash times on activities A, E and D. This analysis also indicates the revised float times on the non-critical activities.

Scenario 2: The effect on the project cost for each week's reduction in time, in order to assess the least cost situation.

The change in the direct and indirect costs due to each week's reduction in the project time is indicated in Figure 9.44. This relationship is presented in graphical form in Figure 9.45.

The relationship between the indirect and direct costs for each week of the project from weeks 28 to 18 can be observed in Figure 9.45. The summation of the direct and indirect costs is displayed and the point on the graph where the cost suddenly increases is the position of the optimum time and least cost situation. This occurs at the end of week 24. At week 25, the change in cost per week alters from £2000 to £7000 per week.

The project costs have been analysed back to week 18 in order to provide an overall assessment of the project cost situation.

COST SLOPE ASSESSMENT

Activity	Normal time	Crash time	Saving in time	Normal cost (£000)	Crash cost (£000)	Increase in cost (£000)	Cost slope	Float	Order/ Ranking
A	6	4	2	12	20	8	4000	Zero	2nd
E	6	4	2	36	54	18	9000	Zero	3rd
F	10	6	4	10	50	40	10000	Zero	4th
D	6	4	2	18	24	6	3000	Zero	1st
B	8	4	4	24	48	24	6000	5	Non critical operations
G	5	3	2	20	30	10	5000	3	
H	8	6	2	40	50	10	5000	3	
I	2	2	0	20	20	0	0	5	
C	3	2	1	18	20	2	2000	5	

Figure 9.42

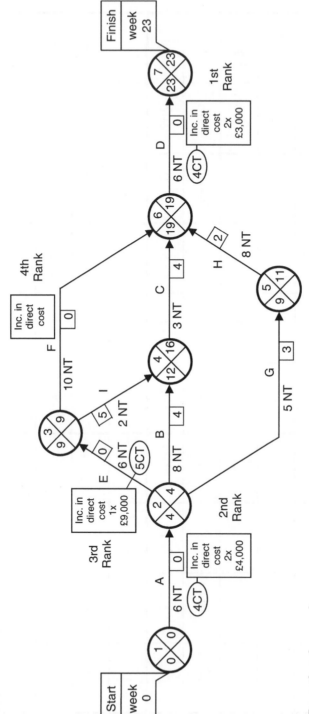

Analysis based on accelerated project period of 23 weeks

REVISED ARROW DIAGRAM

Figure 9.43

Duration in weeks

Week No. start	Direct cost	+ve	Indirect cost	−ve	Aggregate	Total project cost (£000)	Activity ranking
28	198		56			254	
27		+3		−2	+1	255	Activity D
26		+3		−2	+1	256	Activity D
25		+4		−2	+2	258	Activity A
24		+4		−2	+2	260	Activity A
23		+9		−2	+7	267	Activity E
22		+9		−2	+7	274	Activity E
21		+10		−2	+8	282	Activity F
20		+10		−2	+8	290	Activity F
19		+10		−2	+8	298	Activity F
18		+10		−2	+8	306	Activity F

CALCULATION OF CHANGE IN DIRECT AND INDIRECT COST PER WEEK AFTER APPLYING RANKING

Figure 9.44

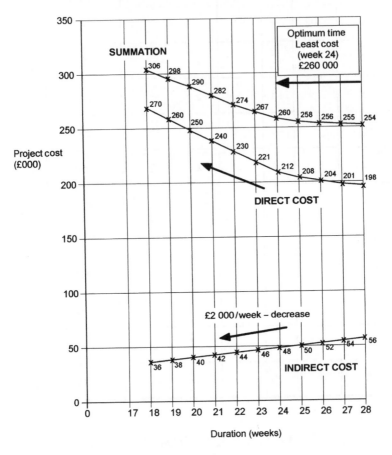

INDIRECT COST, DIRECT COST AND TOTAL PROJECT COST RELATIONSHIP

GRAPHICAL PRESENTATION

Figure 9.45

References

Cormican, D. (1985) *Construction Management: Planning and Finance.* Construction Press.

Harris, F. & McCaffer, R. (2006) *Modern Construction Management,* 6th edn. Blackwell Publishing.

ICE (1992) *Proceedings of the Institution of Civil Engineers.* The Channel Tunnel Part 1: Tunnels. Thomas Telford.

Kempner, T. (1980) *A Handbook of Management,* 3rd edn. Penguin Books.

Lumsden, P. (1965) *Programming House Building by Line of Balance.* National Building Agency.

Nuttall, J.F. & Jeans, R.E. (1960a) *The Critical Path Method.* BRE Current Papers, Construction Series 3.

Nuttall, J.F. & Jeans, R.E. (1960b) *CPM Applied to Building Site Control.* BRE Current Papers, Construction Series 12.

Reece, G. (May 1989) The future of project planning. Construction Manager, *Journal of the Chartered Institute of Building.*

Part B   **Programming**

10 Developing construction sequences

10.1 The thought process

The preparation of any construction programme begins with a structured and logical thought process and this is illustrated in Figure 10.1, which shows the keys steps when preparing a programme for a series of construction operations. Added to this, the would-be planner requires a thorough knowledge of construction technology and a practical understanding of construction sequences in order to develop a realistic programme. Without this knowledge, a realistic and viable programme is well-nigh impossible to achieve.

This is not to say that the knowledge and experience that goes into the contractor's programme implies any particular degree of accuracy. Lowsley and Linnett (2006) emphasise that the programme is merely a *reasonable view* of the project based upon the information available at the time. The planner's art is to develop an achievable construction sequence, within the contractually required timings, that fully relates to the contractor's available resources. Lowsley and Linnett further explain that *it is most unlikely that the works will be undertaken strictly in accordance with the planned programme* but that this simply *illustrates one of the many possibilities of how the work may progress.*

Durations on a master programme are usually shown in weeks, whereas greater detail is shown in short-term programmes where construction sequences are normally planned in days. The opportunity to chose an appropriate timescale is available with all the widely used commercial software packages most of which use the linked bar chart as the principal display. Whilst there are a small number of 'high-level' packages available which use precedence (activity on the node) displays, the majority of construction firms use linked bar charts except on the large and complex projects or where demanded by the client.

SEQUENCING OF CONSTRUCTION OPERATIONS

THE THOUGHT PROCESS

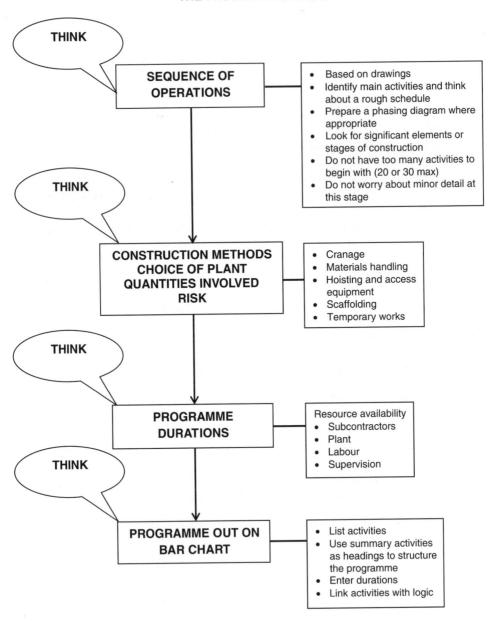

Figure 10.1

The commonly used linked bar chart packages include the following:

Name	Developer	Brief summary
Powerproject	Asta Development	Powerful market leading package with construction specific templates and full functionality for critical path analysis, detailed float calculations, resourcing and costing, baselines, progress recording and high-quality printouts.
Project Commander	PMSC Software	A user-friendly construction industry specific package ideal for construction managers and students alike which is easy to learn and produces a colourful and professional linked bar chart with facilities for baselines, progress recording, rudimentary resourcing and costing and good quality printout.
CS Project	Crest Software	Fully functional and powerful software used extensively in construction allowing task and time management, resourcing, costing, progress tracking and reporting and high-quality printouts. A CS Project 'Lite' version is available at reduced cost.
Easyplan	Asta Development	A cut-down version of Powerproject which looks the same and allows resourcing and progress recording but not full critical path analysis, float calculations, baselines or costing.
Microsoft Project	Microsoft Ltd.	Not designed for construction, but nevertheless easy to learn and with good functionality for all main baseline, progress tracking, resourcing and costing requirements as well as a professional-looking printout.

10.2 Programme considerations

It is essential that those responsible for preparing a programme fully understand the link between the sequence of operations to be undertaken and the practicalities on site. The planner often relies on his own experience to assess the sequence of work, but if he is not familiar with this he should call on the assistance of his manager or an experienced colleague.

In order to think through the likely sequence of operations, it is a good idea to use simple sketches or sequence diagrams similar to those illustrated in Figures 10.2, 10.4 and 10.6. Sequence diagrams may simply be presented on the corner of working drawings, and this helps transmit ideas to the site manager and those doing the work as well as providing an aid to the planner. This practice is widely used in Denmark and France as a means of both expressing the sequence of work and devising safe working methods.

DEVELOPING CONSTRUCTION SEQUENCES
BASEMENT EXAMPLE (Concrete Basement)

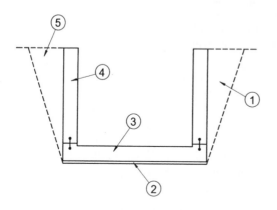

SEQUENCE OF WORK

(1) - Excavate/earth support

(2) - Blind base slab

(3) - Floor construction

(4) - Wall construction

(5) - Backfill earth

Work sequence is also dependent on the basement size as some operations will overlap

Figure 10.2

Example 1 illustrates how this thought process can be developed into a practical programme of work:

Example 1: Concrete basement	
Sequence diagram	Figure 10.2 shows a cross section through an in situ concrete basement.
Key considerations	The sequence of work requires consideration of the ground conditions as this directly influences the earth support methods to be adopted on site. This might include the use of sheet piling or a battered excavation as illustrated.
Main programme activities	1. Excavate basement/earthwork support 2. Concrete blinding to floor slab 3. Floor construction – Formwork, reinforcement and concrete 4. Wall construction – Formwork, reinforcement and concrete 5. Backfill to basement
Linked bar chart	Figure 10.3 indicates an overall period of 19 days.

PROGRAMME – BAR CHART SEQUENCE DIAGRAM

BAR CHART – CONCRETE BASEMENT

Op No	Operation	Dur Days	Days 1	2	3	4	5	6	7	8	9	10	11	12	13	14	15	16	17	18	19	20
1	EXCAVATE	3d	▨▨▨											F to S Rel.								
2	BLIND BASE	1d				▨								F to S Rel.			19 days construction period					
3	FLOOR – FWK	4d					▨▨▨▨							Ft to St Rel.								
	REINFORCEMENT	4d							▨▨▨▨					Ft to St Rel.								
	CONCRETE BASE	1d											▨									
4	WALL – FWK/RES	6d											▨▨▨▨▨▨				S to S F to St					
	CONCRETE WALL	1+2d												Pour ① ▨		Pour ② ▨▨						
5	BACKFILL	2d																		▨▨		

Relationships shown on bar chart
Durations vary according to available resources

Figure 10.3

Differing construction methods can often dictate a different programme of work even though, on the face of it, the situation is similar. For instance, Figure 10.3 shows a similar basement to that described above but this time using brickwork rather than reinforced concrete walls. The resultant programme has a much longer duration than the concrete basement mainly because of the more complex basement floor construction and the need to waterproof the basement with asphalt 'tanking' between the two skins of brickwork:

DEVELOPING CONSTRUCTION SEQUENCES
BASEMENT EXAMPLE - BRICK BASEMENT

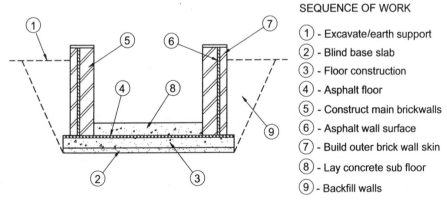

SEQUENCE OF WORK

① - Excavate/earth support
② - Blind base slab
③ - Floor construction
④ - Asphalt floor
⑤ - Construct main brickwalls
⑥ - Asphalt wall surface
⑦ - Build outer brick wall skin
⑧ - Lay concrete sub floor
⑨ - Backfill walls

Figure 10.4

Example 2: Brick basement	
Sequence diagram	Figure 10.4 shows an example of a brick basement with an in situ concrete floor.
Key considerations	The sequence involves more complex operations than the concrete basement due to the integration of the asphalt work into the programme.
Main programme activities	1. Excavate/earthwork support 2. Concrete blinding 3. Concrete slab including rebar and formwork 4. Asphalt floor 5. Construct brick outer wall 6. Asphalt wall surface 7. Build outer brick skin 8. Concrete sub-floor 9. Backfill
Linked bar chart	Figure 10.5 shows a 27-day construction period and illustrates the additional complexity of the traditional form of construction.

PROGRAMME – BAR CHART SEQUENCE DIAGRAM

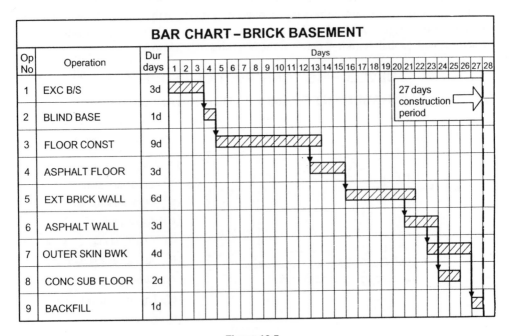

BAR CHART – BRICK BASEMENT

Op No	Operation	Dur days	Days
1	EXC B/S	3d	
2	BLIND BASE	1d	
3	FLOOR CONST	9d	
4	ASPHALT FLOOR	3d	
5	EXT BRICK WALL	6d	
6	ASPHALT WALL	3d	
7	OUTER SKIN BWK	4d	
8	CONC SUB FLOOR	2d	
9	BACKFILL	1d	

27 days construction period

Figure 10.5

10.3 Resource considerations

The planner needs to make decisions in relation to the resources available to carry out a specific work sequence because labour, plant and subcontractors are finite resources. Labour, and some trades in particular, may be difficult to find 'on demand' and good quality subcontractors invariably have more work than they can handle. Consequently, good 'subbies' can often 'pick and choose' their work and so subcontract lead times and availability are important questions for the planner to ponder over. The planner needs a reality check if he simply assumes that everyone will slot into his programme conveniently.

In order to overcome resource availability problems, it is becoming common to use multi-skilled labour gangs who are capable of multitasking. For example, a labour gang may be able to undertake excavation work, fix steel reinforcement, position bolt boxes and place concrete. This allows the gang to move from one section to another without having to wait for other trades, such as joiners, when undertaking the construction of a repetitive sequence of foundation bases.

The development of a method statement is important for underpinning the programme, and this will assist the contractor to develop his preferred approach to the work and the resources needed. However, this cannot be prepared in isolation without thinking about the safety of construction operations and the development of safe working methods. The planner must also think about the statutory requirement to reduce risk to health and safety. This is done by carrying out risk assessments and by devising suitable control measures, and these issues must be considered as part of the process of developing the method statement.

The assessment of operational durations based on approximate quantities is a skill to be developed by the planner, but an even greater skill is the ability to determine realistic times for specialist work based on experience and past records.

10.4 Achieving continuity of work

During the development of a programme sequence, consideration must be given to the continuity of work for the gangs involved. This relates to plant, labour and subcontract operations.

When constructing a number of pile cap bases, the objective should be to move plant and labour from bases A to B to C, etc. Consideration should also be given to the work balance between related operations. Excavation, formwork and concreting operations, for instance, progress at different speeds and therefore the plant, labour and other trade gangs involved need balancing in order to develop cycles of work.

In such a case, it would be uneconomical to keep each operation running at the same rate as the leading operation, and it would be unrealistic to suppose that the steel fixing and carpentry gangs could be balanced with the concreting gang simply to keep the concrete gang continuously employed. Consequently, the planner would need to look for other activities on the programme where labour could be usefully employed in periods of 'downtime' whilst waiting for other trades or gangs to catch up.

Figure 10.6 illustrates some of these issues relative to the construction of a group of pile caps, and Example 3 shows the thought process in developing a linked bar chart for this operation on the basis of a group of four pile caps. This results in a 4-day cycle for each pile cap group. However, in practice, it is likely that greater continuity of work will be achieved by excavating

DEVELOPING CONSTRUCTION SEQUENCES
PILE CAP FORMATION

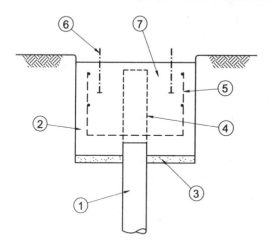

SEQUENCE OF CONSTRUCTION

(1) - Insert bored/driven pile

(2) - Excavate pile cap

(3) - Concrete blinding to base

(4) - Crop pile
      (cut off surplus pile)

(5) - Fix steel reinforcement

(6) - Fix holding down bolts

(7) - Concrete pile cap,
      i.e. no formwork assumed

Programme sequence is
based on groups of 4 pile caps

Figure 10.6

the pile caps over a continuous 4-day period and then blinding the bases in 2 continuous days starting part way through the excavation works. Example 3 refers:

Example 3: Concrete pile cap	
Sequence diagram	Figure 10.6 indicates a section through an in situ concrete pile cap.
Key considerations	The sequence of programmed operations will be: 1. Insert bored piles – subcontract operation 2. Excavate pile cap and expose pile 3. Concrete blinding to pile cap base 4. Crop pile (cut off surplus pile shaft) 5. Fix steel reinforcement 6. Fix holding down bolts or column starter bars 7. Concrete pile cap
Main programme activities	It is to be assumed that no formwork will be required to the sides of the base due to good ground conditions. The linked bar chart shown in Figure 10.7 is based on working on pile caps in groups of four and using a multitask labour gang (operatives experienced in undertaking a range of tasks). A 4-day cycle has been developed for each pile cap group.
Linked bar chart	Figure 10.7 shows the overall duration for constructing 16 pile caps as 19 days.

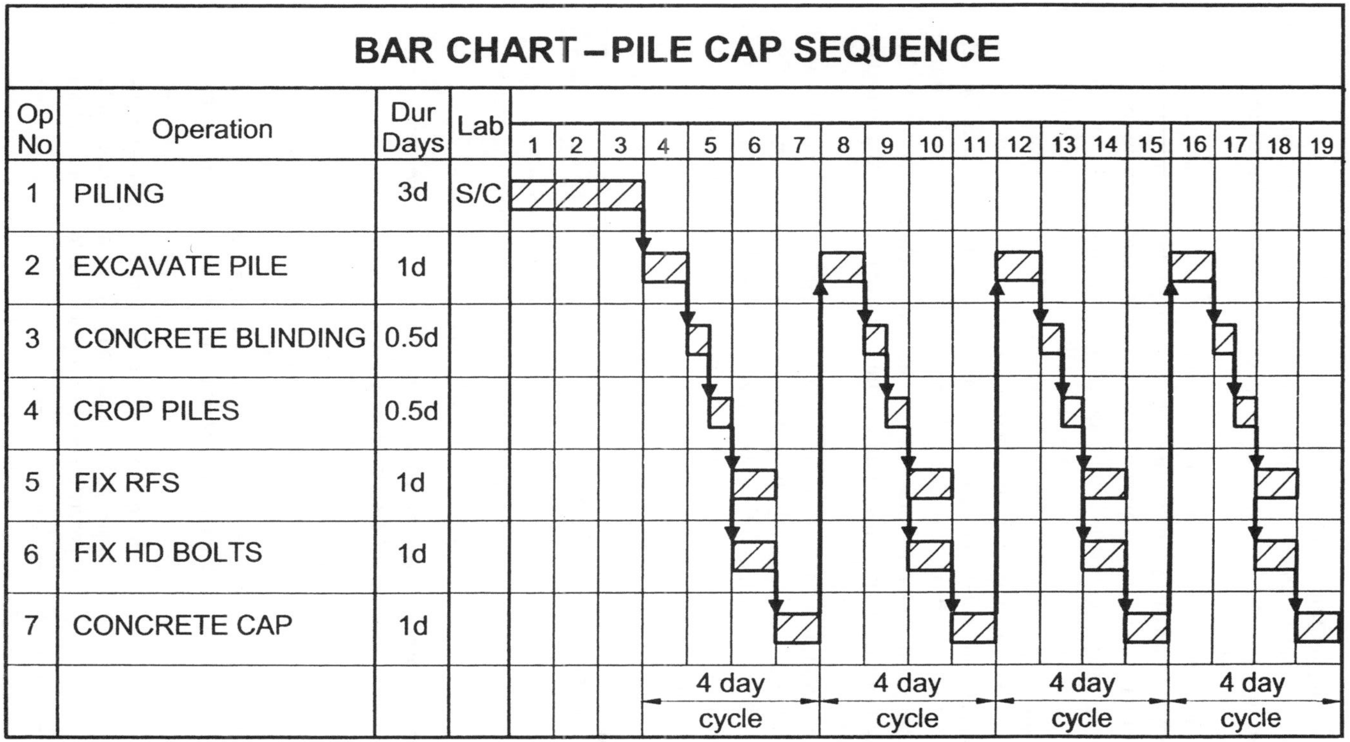

BAR CHART – PILE CAP SEQUENCE

Op No	Operation	Dur Days	Lab	1	2	3	4	5	6	7	8	9	10	11	12	13	14	15	16	17	18	19
1	PILING	3d	S/C																			
2	EXCAVATE PILE	1d																				
3	CONCRETE BLINDING	0.5d																				
4	CROP PILES	0.5d																				
5	FIX RFS	1d																				
6	FIX HD BOLTS	1d																				
7	CONCRETE CAP	1d																				

4 day cycle   4 day cycle   4 day cycle   4 day cycle

Note: By introducing 2 gangs, the work would be considerably speeded up and the excavation plant excavating the pile caps would be used more efficiently.

Figure 10.7

10.5 Sequence studies

An important aspect of planning which concerns contractors is the efficient use of resources and nothing eats away profits more quickly than plant and labour working inefficiently or standing idle. Ensuring continuity of work is essential irrespective of whether the main contractor is doing the work or it is sublet.

Subcontractors are just as keen as main contractors to ensure continuity of work, as they normally want to get in and out as quickly as they can. Good subcontractors are usually busy and want to move on to their next job quickly. Additionally, should the main contractor cause them delay or disruption or not provide attendances such as cranage and scaffolding on time, contractual claims will undoubtedly arise, and this will have a damaging effect on the main contractor's bottom line profits.

All this is especially true where repetitive work is being carried out, such as housing or multi-storey construction. Following trades or operations cannot afford to be delayed by slower preceding activities and attention has to be paid to this in the detailed planning of the project. While perfect continuity of work is not possible in practice, sequence studies enable the contractor to analyse construction activities at the resource level in order to make best use of the labour and plant available.

Figure 10.8 illustrates this point with an example of a precast concrete retaining wall. It can be seen from the linked bar chart that the programme is driven to a large extent by the need to provide perfect continuity of work for the subcontractor engaged to place the precast concrete wall units. Continuity of work has also been maintained for the excavation work, fixing base formwork and rebar operations.

Example 4: Precast concrete retaining wall	
Sequence diagram	Figure 10.8 illustrates a plan and section through a precast concrete retaining wall and in situ concrete foundation.
Key considerations	The foundation is to be poured in 15 m bay sizes.
Main programme activities	The sequence of programmed operations will be: 1. Excavate base 2. Place concrete blinding 3. Erect formwork to sides of base 4. Fix steel reinforcement in base 5. Concrete the base slab 6. Erect precast concrete wall sections Backfill to rear of retaining wall on completion of all bays.
Linked bar chart	The overall construction period shown on the linked bar chart in Figure 10.9 is 17 days. Note the finish-to-start link between the excavation work, formwork, rebar and precast concrete wall operations in order to achieve continuity.

PRECAST CONCRETE RETAINING WALL AND FOUNDATION

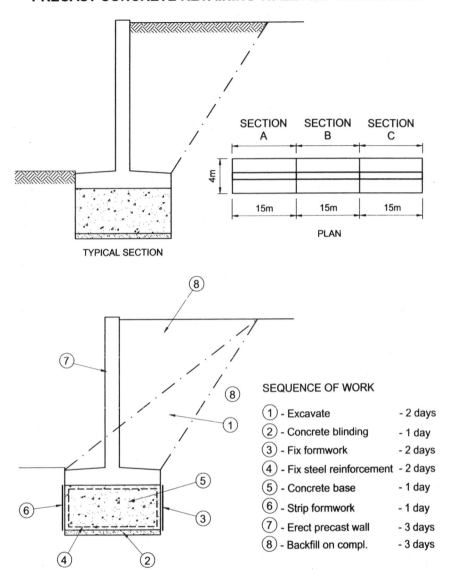

TYPICAL SECTION

PLAN

SEQUENCE OF WORK

(1) - Excavate — - 2 days
(2) - Concrete blinding — - 1 day
(3) - Fix formwork — - 2 days
(4) - Fix steel reinforcement - 2 days
(5) - Concrete base — - 1 day
(6) - Strip formwork — - 1 day
(7) - Erect precast wall — - 3 days
(8) - Backfill on compl. — - 3 days

Figure 10.8

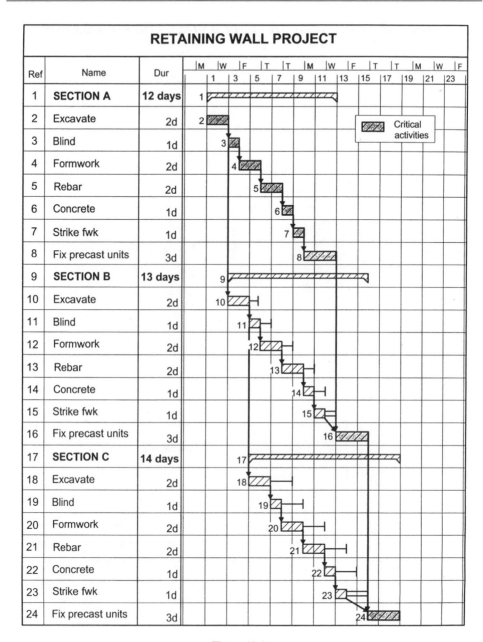

Figure 10.9

The use of sequence studies is planning at a fine level of detail. However, where alternative methods of construction are being considered for a particular operation, a sequence study provides a means of determining the time and cost implications of the choices available.

Examples 5 and 6 illustrate worked examples of sequence studies for the construction of two concrete-framed buildings using two different methods of construction.

The first example is for the construction of an unusual ten-storey precast concrete building which is circular on plan (see Figure 10.10). The second example is a rectangular eight-storey in situ reinforced concrete-framed building (see Figure 10.12).

Detailed explanations for each example are given below. NB: A further sequence study example for the construction of a reinforced concrete basement is given in the City Road project in Chapter 21.

Example 5: The construction of two floors of a ten-storey precast concrete crosswall-framed building	
Sequence diagram	Figure 10.10 indicates the plan and section through one floor of a ten-storey precast concrete building. The 24-m diameter frame is of crosswall construction. The sequence of constructing one floor of the frame is indicated on the diagram.
Key considerations	The precast concrete wall units must be bedded in mortar and temporarily propped until the structure is stable. This requires additional labour and increases the complexity of construction.
Main programme activities	Sequence of work 1. Erect external circular wall panels 2. Erect internal crosswalls including core areas 3. Erect precast floor units Form in situ perimeter strips and grout up wall joints
Linked bar chart	A linked bar chart programme for two floors of the frame is illustrated in Figure 10.11 indicating a cycle time of 6 days per floor. NB: This example is part of a more detailed case study (Eastlands Project) in Chapter 20.

SEQUENCE OF WORK ~ PRECAST CONCRETE FRAME

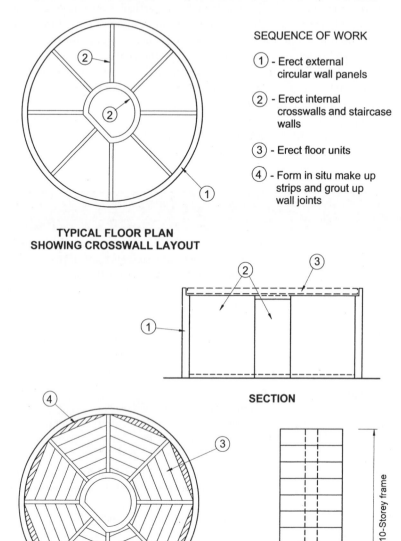

SEQUENCE OF WORK

① - Erect external circular wall panels

② - Erect internal crosswalls and staircase walls

③ - Erect floor units

④ - Form in situ make up strips and grout up wall joints

TYPICAL FLOOR PLAN SHOWING CROSSWALL LAYOUT

SECTION

FLOOR SLAB UNITS

ELEVATION

10-Storey frame

Figure 10.10

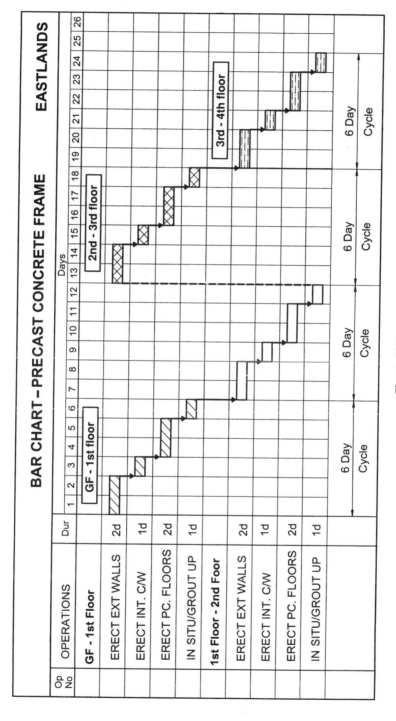

BAR CHART – PRECAST CONCRETE FRAME　　EASTLANDS

Op No	OPERATIONS	Dur
	GF - 1st Floor	
	ERECT EXT WALLS	2d
	ERECT INT. C/W	1d
	ERECT PC. FLOORS	2d
	IN SITU/GROUT UP	1d
	1st Floor - 2nd Foor	
	ERECT EXT WALLS	2d
	ERECT INT. C/W	1d
	ERECT PC. FLOORS	2d
	IN SITU/GROUT UP	1d

Days: 1 2 3 4 5 6 7 8 9 10 11 12 13 14 15 16 17 18 19 20 21 22 23 24 25 26

GF - 1st floor
2nd - 3rd floor
3rd - 4th floor

6 Day Cycle　　6 Day Cycle　　6 Day Cycle　　6 Day Cycle

Figure 10.11

Example 6: The construction of two floors of an eight-storey in situ concrete-framed building

Sequence diagram

1. Figure 10.12 shows the plan and section of an eight-storey concrete-framed building.

Figure 10.13 is a sequence diagram for the column and floor construction sequence. The movement of the column formwork and table forms for each floor is also shown.

Key considerations

1. The building is of flat slab construction and contains 24 columns per floor. Table forms are to be used for the floor slab construction. A fixed tower crane is to be used for all lifting operations, that is formwork, reinforcement and the column concrete pours. The larger floor pours are to be concrete pumped.

2. Formwork resources are critical for the success of the project. Sufficient formwork is to be available for 12 columns – these will be used twice on each floor level with some degree of remaking taking place after the fifth floor has been completed. Sufficient table forms are to be made available for two complete half-floor slabs. These will be moved up the building as each floor slab is completed.

Formwork striking times are of critical importance to the construction sequence. Vertical column sides can be stripped overnight – horizontal slab soffits, however, must be left in position for a minimum period of 7 days. Early stripping times may be achieved by proving high early cube strengths at say 4 days. A 4-day stripping time is therefore to be allowed in the programme.

Main programme activities

Construction sequence per floor

Columns per floor (durations per 12 columns)

1. Make column formwork	3 days
2. Fabricate column reinforcement	3 days
3. Fix column rebar	1 day
4. Fix formwork to columns	2 days
5. Concrete columns	1 day
6. Strike formwork to columns	1 day

Floor slab construction (durations per half a floor)

1. Fix table forms in position	4 days
2. Fabricate and fix rebar	4 days
3. Concrete floor slab (curing time = 4 days)	1 day
4. Strike floor soffit	2 days

Linked bar chart

Figure 10.14 shows a linked bar chart sequence for the construction of two floors of the building.

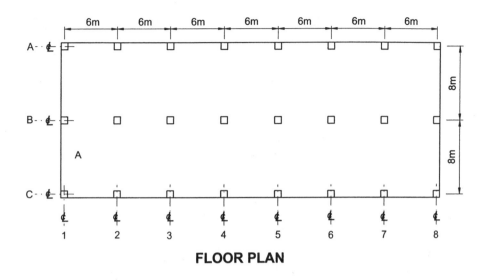

FLOOR PLAN

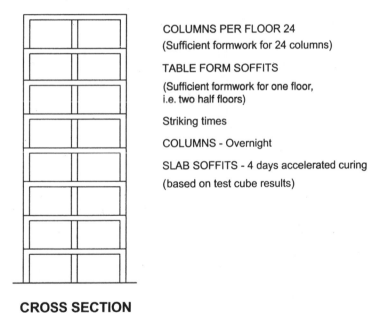

COLUMNS PER FLOOR 24
(Sufficient formwork for 24 columns)

TABLE FORM SOFFITS

(Sufficient formwork for one floor,
i.e. two half floors)

Striking times

COLUMNS - Overnight

SLAB SOFFITS - 4 days accelerated curing

(based on test cube results)

CROSS SECTION

8-STOREY REINFORCED CONCRETE FRAME BUILDING

Figure 10.12

CONSTRUCTION SEQUENCE FOR COLUMNS AND SLAB SOFFIT

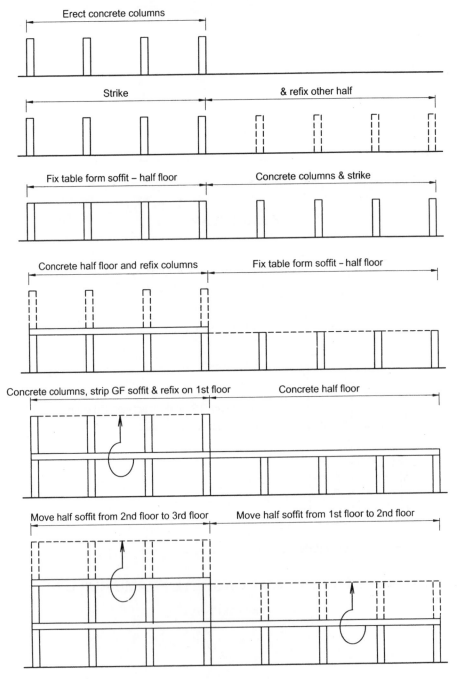

Figure 10.13

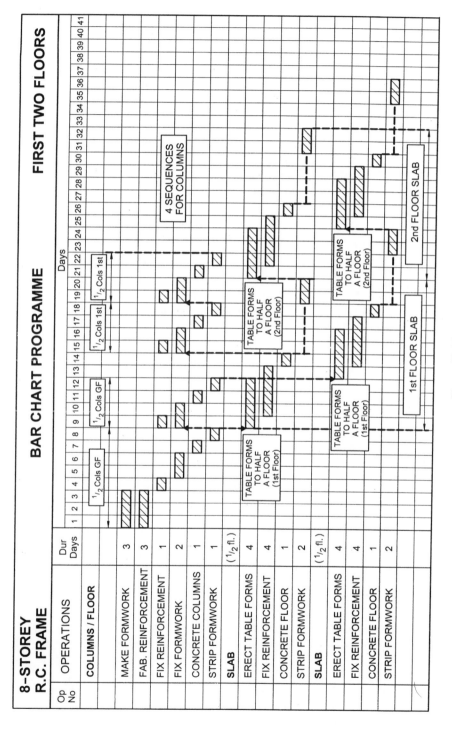

Figure 10.14

Reference

Lowsley., S. & Linnett., C. (2006) *About Time – Delay Analysis in Construction.* RICS Books.

11 Method statements

11.1 Introduction

The preparation of method statements forms an essential part of the contractor's planning process, as these underpin the programme and explain how the work is to be undertaken. Method statements may be categorised into three distinct formats:

- The tender method statement
- The construction or work method statement
- The safety method statement

Figure 11.1 summarises the purpose of each type of method statement and indicates the relationship between them.

11.2 Definition

A method statement is a written explanation of the proposed methods of construction relating to specific site operations, building elements or stages of work.

Example of **specific operations** that might be included in a method statement:

- excavation works to foundations
- demolition works
- laying precast concrete floor units

Stages of work may relate to:

- Groundworks to damp proof course
- Brickwork to first floor
- Brickwork to eaves
- Site drainage and manholes

Where the method statement is based on **elements** of the building these might include:

- Substructure work
- Superstructure frame
- External envelope etc.

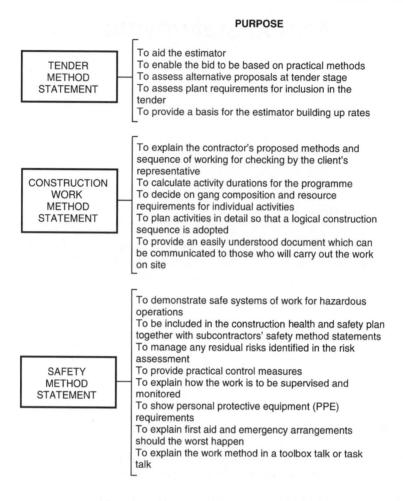

PURPOSE

TENDER METHOD STATEMENT
- To aid the estimator
- To enable the bid to be based on practical methods
- To assess alternative proposals at tender stage
- To assess plant requirements for inclusion in the tender
- To provide a basis for the estimator building up rates

CONSTRUCTION WORK METHOD STATEMENT
- To explain the contractor's proposed methods and sequence of working for checking by the client's representative
- To calculate activity durations for the programme
- To decide on gang composition and resource requirements for individual activities
- To plan activities in detail so that a logical construction sequence is adopted
- To provide an easily understood document which can be communicated to those who will carry out the work on site

SAFETY METHOD STATEMENT
- To demonstrate safe systems of work for hazardous operations
- To be included in the construction health and safety plan together with subcontractors' safety method statements
- To manage any residual risks identified in the risk assessment
- To provide practical control measures
- To explain how the work is to be supervised and monitored
- To show personal protective equipment (PPE) requirements
- To explain first aid and emergency arrangements should the worst happen
- To explain the work method in a toolbox talk or task talk

OVERVIEW OF METHOD STATEMENT FORMATS

Figure 11.1

Method statements may be presented in a written or tabular format, and they often include details of the likely hazards that might be encountered during the method statement operations which forms part of the risk assessment process. Some planners also include safety measures as specific activities on the construction programme such as the provision of access equipment or the installation of safety nets. Programme relationships may also be included in the descriptive notes on the method statement.

The tabular format is probably simpler and easier to read but, as with all standard forms, there is a space restriction on the content. Using a written or prose format (rather like a mini report) gets over this problem and, provided it is well laid out, can be both useful and comprehensive. Both types are commonly used, according to personal preference. Whichever is chosen, they should not be too long or overly complex as they are an essential aid to conducting activities on site in accordance with a well thought-out and agreed procedure, and they need to be easily communicated at site level.

It is essential that the purposes of method statements at the various stages of the planning process are understood. Method statements convey different meanings to clients, consultants, subcontractors and main contractors. A method statement is not simply a list of construction operations with notes written alongside.

It is common practice to combine the construction and safety method statement for each work activity into a single document. This makes sense because safety is an integral factor in the planning process. A disadvantage of doing this is that the combined method statement may become too complicated.

11.3 Preparation of method statements

Construction method statements are prepared by the contractor and subcontractors at various stages of a project and are now an essential component of construction best practice. There is no standard method statement format and a wide variety of different formats can be found in practice. Each individual preparing a method statement will have a personal interpretation of what is required, but the main thing is that the job is thought through carefully and a safe and efficient method of working is determined.

Some clients may request that method statements are submitted at the tender or negotiation stage of a project or, alternatively, the contractor may be required to prepare a method statement for scrutiny by the client's representative prior to commencing a major operation. Subcontractors are also routinely required to submit method statements and risk assessments prior to commencing work. It should be noted that risk assessments are a legal requirement but method statements are not.

11.4 Pre-tender method statement

The pre-tender method statement forms the basis of the pre-tender programme and, together, they outline the sequence and method of construction upon which the tender is to be based. The pre-tender method statement should indicate how the major elements of work will be carried out and may highlight areas where new or alternative methods are being considered. The estimator may consider several methods in order to arrive at the best price or the shortest duration for the project.

While there are no strict rules to follow, for most projects the pre-tender method statement will be a simple outline document conveying the contractor's general approach to the construction methodology, but with little detail. It may contain details of gang sizes and outputs, plant and supervision requirements but, normally, there will be no detailed calculations. The tendering period is often too short to afford the luxury of spending much time on this stage of the process.

For civil engineering tenders, greater reliance is placed on producing a well-developed construction method at tender stage as an aid to accurate pricing of both the permanent and temporary works.

Pre-tender method statements are particularly useful when the estimator or planner needs to assess the costs of any alternative proposals. It is a good idea to involve members of the contracts staff in its production to allow those ultimately responsible for undertaking the work to provide some practical input and ideas at the estimate stage, thereby developing a sense of ownership.

The pre-tender method statement may be prepared by the contracts manager in the medium-sized contracting business but within the larger company, input will also be available from the planning department which often works in conjunction with the estimators.

Influence on price at the tender stage

Decisions relating to the choice of construction method are an integral part of the tendering process, and these decisions directly influence the rates included in the estimator's price build-up. Consequently, methods allowed in the estimate build-up influence the final tender sum. Sometimes the contractor pricing a project may wish to spread the majority of the 'method risk' onto his subcontractors in order to keep his price competitive, but it is not, however, common practice to require a subcontractor to submit a method statement to substantiate their rates at the tender stage. Such issues may need to be thrashed out at the negotiation or pre-contract stages.

An extract from a typical tender method statement is shown below indicating a selection of operations each with alternative method considerations. Each method has its own influence on the price ultimately included in the estimate.

Reduced level excavation average 500 m deep

Quantity: 1200 m^3

Method 1
Excavate using 20 tonne hydraulic back actor excavator loading 10 m^3 wagons direct and cart to tip off site. Construct temporary haul road using reclaimed brick hardcore to provide access for wagons.

Method 2
Doze material into site spoil heaps, re-excavate using large hydraulic excavator and cart to tip off site.

Concrete slab to 5 m wide access road, average 200 mm thick

Quantity 40 m^3

Method 1
Place ready mixed concrete by direct discharge from delivery vehicles alongside bays and level by metal tamp using alternate bay construction.

Method 2
Place concrete by mobile concrete pump in one 40 m^3 pour. Permission to be obtained from consulting engineer for proposed method.

Hoist and fix lightweight cladding system to elevations of eight-storey building

Quantities as per drawing ECL/121/E

Method 1
Lightweight panels and associated support system to be fixed from double mast climbers located at 4.00 m centres.

Method 2
Lightweight panels and support system to be fixed from an external perimeter scaffold. Materials to be raised to each floor level via a scaffold hoist.

Tender decisions on plant selection

Plant is an expensive commodity and a contract award may be won or lost purely on the methods of lifting and materials handling included in the final bid. Each project requires individual consideration of appropriate plant and methods and many factors have to be accounted for prior to making a final choice:

- Nature of the site and surrounding buildings
- Site access and working space
- Overhead power lines and other services
- Crane radius restrictions on use of adjoining air space due to nearby roads or railway lines etc.

It is important to carefully develop realistic cost estimates for mobile cranes, hydraulic lifters and tower cranes, etc., for inclusion in the tender bid but also to bear in mind that these are merely 'tender allowances'. Consequently, site managers may decide to change the estimator's methods at the pre-contract or construction stage provided that such decisions can be justified on the grounds of increased efficiency and provided that the costs are within those allowed in the tender.

Tender proposals for materials handling and general cranage

EXAMPLE – Castlefield project

A £28 million project involves the construction of two eight-storey apartment blocks and a basement car park. The basement area is approximately 50 m × 38 m and covers the majority of the site. The apartment blocks are of in situ concrete flat slab construction and incorporate two staircase towers each 7 m². The towers are to be constructed by slipform construction.

Figure 11.2 is a site layout plan showing the location of the blocks and the surrounding canal basin area which are in the heart of Manchester. The site is surrounded by a series of canals and space for the delivery and storage of materials is limited. Both the basement construction and the erection of the two concrete frames pose major materials handling problems at the tender stage.

The construction of the basement and the two tower blocks present different problems. The basement involves handling quantities of reinforcement and large formwork panels for the wall construction, whereas the tower blocks require lifting facilities for formwork and reinforcement for columns and table form units for the upper floor construction. Concreting operations can readily be accommodated using either mobile or static concrete pumps.

PROPOSAL 1

In Figure 11.3, two tower cranes are proposed:

- Tower crane 1 will be located outside the basement area between the two staircase towers. The 20-m jib radius will cover all lifting operations for the staircase slipform construction and will also partly service the construction of the eight-storey concrete frames.

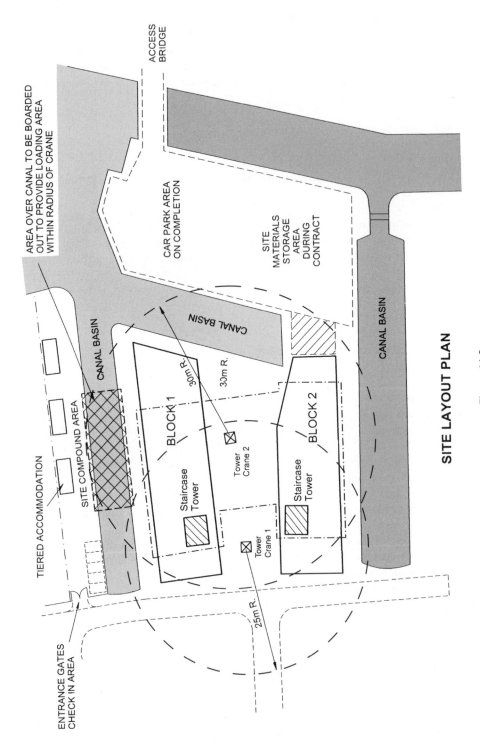

ACCESS BRIDGE

AREA OVER CANAL TO BE BOARDED OUT TO PROVIDE LOADING AREA WITHIN RADIUS OF CRANE

CAR PARK AREA ON COMPLETION

SITE MATERIALS STORAGE AREA DURING CONTRACT

CANAL BASIN

CANAL BASIN

CANAL BASIN

TIERED ACCOMMODATION

SITE COMPOUND AREA

BLOCK 1

BLOCK 2

Staircase Tower

Staircase Tower

30m R.

30m R.

Tower Crane 2

Tower Crane 1

25m R.

ENTRANCE GATES CHECK IN AREA

SITE LAYOUT PLAN

Figure 11.2

CRANAGE PROPOSALS AT TENDER

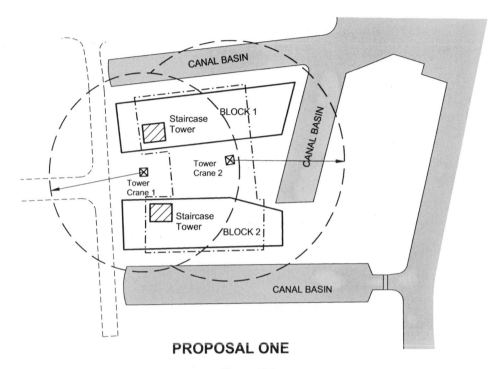

PROPOSAL ONE

Figure 11.3

- Tower crane 2 is to be located in the basement area and is to have a jib radius of 30 m. The crane is to be erected as soon as excavation works are complete and is intended to provide all lifting operations within the basement area (i.e. rebar, formwork and concrete to the smaller pours).

All major concrete pours to floor slabs and walls will be undertaken by static concrete pumps and smaller pours with crane and skips.

PROPOSAL 2

As shown in Figure 11.4, the basement materials handling operations are to be undertaken using two low pivot jib mobile cranes located along each side of the basement. On completion of the basement area, a 20-m radius tracked tower crane is to be located alongside each of the apartment blocks. Each crane will easily reach across each block in order to handle formwork and reinforcement and the concrete pours to the columns.

All major floor pours will be undertaken by static concrete pump and smaller pours with crane and skips.

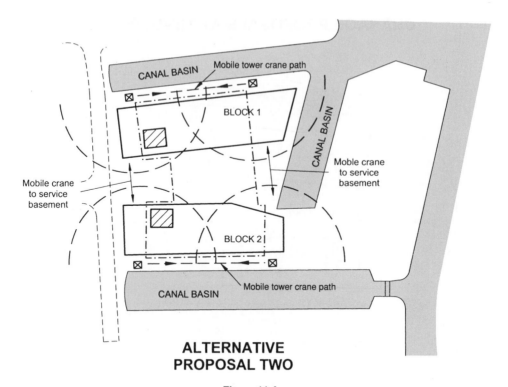

ALTERNATIVE
PROPOSAL TWO

Figure 11.4

11.5 Construction method statements

Construction method statements are prepared in order to explain the contractor's proposed working methods and demonstrate how the durations and sequence of work shown on the master programme will be achieved. At the pre-contract stage in the project, method statements will be prepared in detail in order to be sure that durations are accurate and reliable. Method statements may be presented as written statements or in tabular form.

The main uses for method statements are:

- To calculate activity durations for the programme
- To decide on gang compositions and thus the resourcing requirements for individual activities
- To plan activities in detail so that a logical construction sequence is adopted
- To provide an easily understood document which can be communicated to those who will carry out the work on site.

Preparing method statements requires discipline and logic in that the steps needed to complete an activity have to be thought out. Hand in hand with this is consideration of what plant and labour will be used, as this is fundamental to achieving a sensible working method and sequence. Temporary works and working space requirements will also have to be considered as will subcontractors' attendance requirements for packages such as piling and structural steel erection.

In order to calculate the duration of an activity, the output of the gang has to be considered. Outputs used by estimators are not suitable for planning as these relate to measured bill items where both main quantities and smaller items are included. In brickwork, for instance, the bills of quantities will include an item for facing brickwork, say, and there will also be items for laying the damp-proof course, for closing cavities and for building in lintels. All these items would make calculations too complex for planning purposes, and thus 'incidental' items should be reflected in an overall output for the facing brickwork quantity alone.

The activity duration in days would then be given by:

$$\frac{\text{Quantity of brickwork in m}^2}{\text{Output of gang/h} \times 8\,\text{h/day}}$$

where output is expressed in m^2/h.

Alternatively, the quantity of bricks in thousands can be calculated and divided by the output of the gang in bricks laid per day. This is fairly straightforward but the difficult bit is to judge what the output should be. This will depend on factors such as:

- The gang make up
- The type of work
- Location of the work
- The standard expected in the specification

Construction method statement – example

Castlefield project

A reinforced concrete basement 50 m × 38 m × 5 m deep is to be constructed within a secant piled cofferdam in order to cut off water ingress from surrounding canal basins. Figure 11.5 shows a plan and section through the basement. The base slab to the basement is to be piled to form the foundations to two eight-storey reinforced concrete framed buildings.

Method statements are to be prepared for the following stages of work:

- Excavation to basement
- Concreting of basement slab

Headings used in method statement

- Description of works
- Work sequence
- Resources
- Assessment of significant risks
- Control measures
- Personal and protective equipment

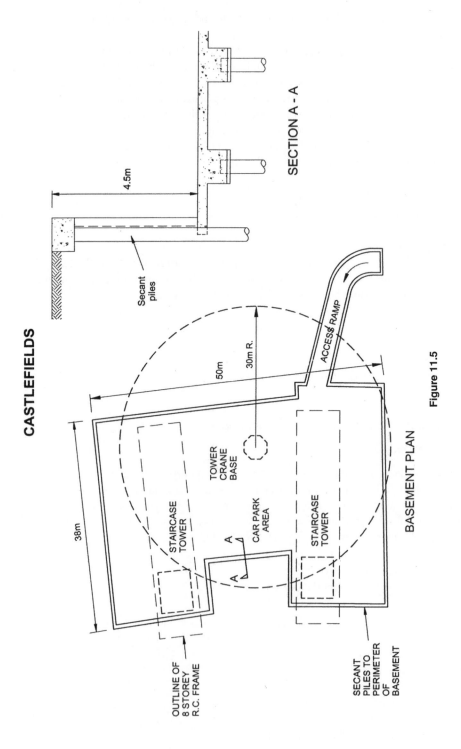

CASTLEFIELDS

SECTION A - A

4.5m

Secant piles

BASEMENT PLAN

50m

30m R.

TOWER CRANE BASE

38m

ACCESS RAMP

STAIRCASE TOWER

CAR PARK AREA

STAIRCASE TOWER

OUTLINE OF 8 STOREY R.C. FRAME

SECANT PILES TO PERIMETER OF BASEMENT

Figure 11.5

- To whom information has been submitted
- Monitoring compliance

CASTLEFIELD PROJECT CONSTRUCTION METHOD STATEMENT EXCAVATION TO BASEMENT

Description of works

Excavation works and associated removal of spoil to reduce the ground to the required level allowing construction works to the basement piling and floor to commence. Construction works include pile caps and bases within the basement area.

The basement excavation is to take place between the secant piled wall. Access to the 5-m deep basement is via a proposed access ramp. Pedestrian access is to be via scaffold access towers. The volume of excavated material is approximately 9800 m³. The programmed duration for the works is 4 weeks (20 days).

Work sequence

Prior to excavation works taking place, the following procedures must be carried out:
- The area is to be cleared and kept clear of unnecessary materials during the period of work.
- With pedestrian segregation precautions in place (minimum – scaffold tube barrier), the engineer will set out the position and line and levels for the excavation works.
- Where there is evidence of existing services in the line of the excavation, trial holes will be dug by hand to establish the nature of the service and the line and depth of each.
- Excavation will proceed between grid lines L to A and 4 to 12. The hydraulic back actor will load suitably positioned tipper wagons. Once loaded, the vehicles will leave the site via the agreed access route. This procedure will be repeated until the correct line and level has been reached. Suitable barriers and signage for deep excavations will be maintained during the whole of the works.
- Access to the basement area will be provided by two scaffold staircases, both fully boarded, and clearly signed as access points

Resources
Supervision

- Contracts manager
- Site foreman
- Groundworks foreman
- Site engineer.

Labour

- One experienced banksman
- Two groundworkers
- Two competent excavator drivers
- The tipper wagons are to be hired.

Plant

- Two 360° hydraulic excavators
- De-watering equipment or static pumps
- Scaffold tube barrier fencing
- Access scaffold platforms

- 8–10 m³ capacity tipper wagons
- Signage for pedestrian access, vehicular access and deep excavations

Significant hazards

- Moving plant and machinery
- Access to basement
- Underground services
- Excavation collapse
- Re-fuelling of plant
- Objects/persons at height
- Pedestrian access to basement

Control measures

- All operatives to have received a site induction prior to commencing work on site
- All operatives must read and understand the method statement and risk assessments before any work commences
- Plant and equipment must all have current certification and road fund licences (the use of Guinness labels as tax discs is prohibited!!)
- Permits for all work to be obtained prior to work commencing
- Groundworks foreman to inspect excavations daily prior to work commencing
- Site cleanliness and general housekeeping to be monitored to ensure safe access and egress to excavations

Personal Protective Equipment

Minimum site requirements to include:

- High-visibility jacket, steel-toe-capped boots with steel mid-sole, safety helmet, gloves and eye protection, when required

Authors' note: Further to the contractor's intended method as described above, the actual methods adopted during the project are described below:

- Excavation was undertaken using two 30-tonne hydraulic back actors with 1.5-m³ buckets loading 8–10-m³ capacity wagons. Vehicular access during excavation works was via the permanent ramped roadway into the basement area.
- Actual duration of the excavation work:

$$\text{Excavation quantity} = 9800 \text{ m}^3$$
$$\text{Machine output} = 25 \text{ loads per day}$$
$$= 250\text{--}300 \text{ m}^3 \text{ per day}$$

Activity duration is given by:

$$\frac{9800 \text{ m}^3}{250 \text{ m}^3/\text{day}} = 40 \text{ days for one excavator or 20 days for two machines}$$

- A 30-cm stone-piling mat was laid over the basement area prior to piling operations commencing.

CASTLEFIELD PROJECT CONSTRUCTION METHOD STATEMENT BASEMENT FLOOR SLAB

Description of works

The concreting of a 450-mm thick basement floor slab and associated thickenings. The floor is to be poured in 5-m wide bays using alternate bay construction. Approximate total quantity of concrete = 1000 m^3

Work sequence

Reinforcement

The reinforcement will be fixed in accordance with the engineer's drawings. The steel reinforcement will be lifted into the basement with the tower crane and fabricated in situ. Prior to concreting, the reinforcement will be checked by the site engineer and signed off before the pour takes place

Formwork

Vertical stop ends will be formed between the floor bays using Expamet or similar and incorporate a hydrophilic water bar

Concrete work

- The main contractor requires the basement slab to be constructed in the work sequence indicated in Figure 11.6
- Work is to proceed from area A to D. A detailed bay layout is shown in Figure 11.6 for the proposed area A
- Cube tests are to be taken for every 50 m^3 of concrete poured. Testing will take place at 7 and 28 days. At the end of each pour, the slab surface will be cured using an appropriate method

Resources

Supervision

- Site manager
- General foreman

Labour

- Six joiners
- Eight steel fixers
- Six labourers

Plant

- Compressor and tools
- Bar croppers
- Disc cutters
- Cranage
- Concrete pump

Significant hazards

- Moving plant and vehicles
- Collapse of excavations
- Work at height
- Objects falling from height
- Collapse of shutters

- Working with concrete
  - contact with eyes
  - vibration from pokers
  - mould oil
  - power tools

Control measures

- All operatives to undergo site induction
- Access to the basement area will be via staircase scaffolds

Personal protective equipment

- All operatives to wear hard hats, safety boots, high-visibility vests and gloves
- Goggles to be worn when using compressed air for blowing out the reinforcement area of dirt and grit.

11.6 Safety method statements

The construction method cannot be considered in isolation from safety and part of the pre-contract planning process is to think about the provision of safe systems of work. This is a statutory requirement under the Health and Safety at Work etc. Act 1974. Safe systems are commonly expressed in the form of safety method statements.

Safety method statements covering the main stages of the works are prepared at the pre-contract stage. They will be based on risk assessments which are a statutory requirement under the Management of Health and Safety at Work Regulations 1999. Whilst safety method statements are not required by statute, it is common practice for contractors to prepare them and, together with subcontractors' risk assessments and method statements, they form part of the site safety management system.

There must be no deviation from a method statement, as this can lead to confusion and possibly accidents. Where a change in working method is required, this should be discussed and the method statement changed formally before going ahead with the work. In this respect, it is crucial that the workforce is notified either by a toolbox talk or task talk and any subcontractors notified of the impact on their operations. On no account must a method statement be changed on an ad hoc basis as there have been a number of multiple fatal accidents for this reason. One of these concerned the demolition of a multi-storey reinforced concrete building. The normal site foreman went on holiday and his stand-in came to site without being briefed. He adopted a method of working which was contrary to the agreed method and an 18-tonne excavator fell on to the floor below as a result. Two workmen were crushed to death.

SEQUENCE OF CONSTRUCTING MAIN FLOOR AREAS
CASTLEFIELD PROJECT

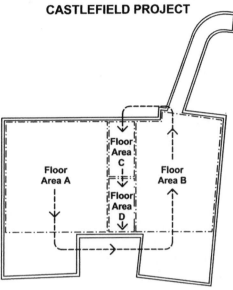

BASEMENT FLOOR – CONCRETE SEQUENCE
MAIN WORK AREAS

Concrete to basement wall
to commence on completion
of bays ① to ③

Pouring dates for
each bay to be
recorded and
marked up on
progress drawing

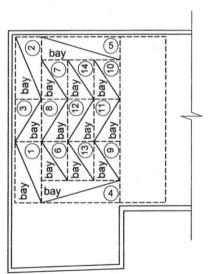

POURING SEQUENCE FOR MAIN FLOOR AREA

Figure 11.6

12 Planning for safety

12.1 Introduction

It is generally accepted that good health and safety practice is part of good management and not a separate or additional consideration. Good managers think about health and safety issues as a matter of course and integrate this into their planning and control strategies whilst good workers take the time to think about the consequences of their actions both for their own well-being and that of others.

12.2 Definitions

- **Health** relates to the physical and mental well-being of persons at work, or those affected by work activities, to the extent that they are affected by exposure to noise, dust, chemicals or other dangerous materials or processes.
- **Safety** is concerned with the protection of workers, passers-by or visitors to site from physical injury from work activities such as moving plant and machinery, work at height, falling materials or tools, collapse of excavations or scaffolding or overturning of cranes or other lifting equipment.
- **Welfare** relates to the provision of facilities for those at work so as to maintain their health and well-being and includes the provision of site accommodation with proper washing, drying and sanitary facilities, the provision of drinking water, eating and rest facilities and the provision of first-aid facilities, etc.

Health and safety considerations begin at the top of the organisation, and managers must make sure that there is legal compliance on all sites and that there is a positive and active safety culture within the organisation.

At tender stage, the contractor must make sure that his price reflects the cost associated with developing safe systems of working on site within a suitable organisation structure and that the pre-tender programme allows sufficient time for the planning, resourcing and control of site safety.

At the contract planning stage, detailed measures must be put in place to ensure that safety, health, welfare and environmental issues are properly managed and do not pose a risk to those working on or affected by the project. It is suggested that site managers will be concerned with five main issues:

(1) ensuring **compliance** with the law
(2) making sure that there is an appropriate **safety management system** in place for the particular site in question
(3) the identification of **hazards** that might give rise to accidents on site
(4) **assessing risk** arising from the hazards in question and putting in place appropriate **control measures** to deal with the risks
(5) devising **safe systems of work** for the project in hand

12.3 Legal framework

The Health and Safety at Work etc. Act 1974 (HSWA) is the primary health and safety legislation in the UK which is known as 'enabling legislation' because it gives Ministers of the Crown the power to enact other legislation under it. Some of the subsidiary or subordinate legislation has developed from European Union directives, for example the Management of Health and Safety at Work Regulations 1999 and Construction (Design and Management) Regulations (CDM) 1994 and 2007.

The main thread running through all UK health and safety legislation is the **management of risk**. This is achieved, in the main, through a 'goal setting' or 'objective' style of legislation which emphasises the setting of standards to be achieved rather than a rigid set of rules to follow.

Legislation of particular relevance to construction is summarised in Figure 12.1, although the list is by no means exhaustive.

12.4 Health and safety policy

Section 2(3) of the HSWA requires firms with more than four employees to have a written **health and safety policy** comprising:

- General statement
  - Aims (not measurable)
  - Objectives (measurable)
- Organisation
  - Structure (organisation chart – see Figure 12.2)
  - Health and safety roles
  - Responsibilities
- Arrangements (i.e. systems and procedures):
  - A policy and procedures manual
  - Standard forms
  - Legislation register
  - Construction phase plan (as required by CDM 2007)
  - Other manuals including control of substances hazardous to health (COSHH)
  - Company risk control guidelines
  - Risk assessments and method statements

HEALTH AND SAFETY LEGISLATION

Statutory Instrument	Brief summary of key requirements
Health and Safety at Work etc Act 1974 - HSWA	• General duties of care • Safe systems of work • Health and safety policy • Inspection and enforcement
Management of Health and Safety at Work Regulations 1999	• Risk assessment • Principles of prevention • Formal arrangements for planning, organisation, control, monitoring and review of safety measures • Communication, co-operation and co-ordination
Construction (Design and Management) Regulations 2007 - CDM	• Creation of duty holders • Preventative design • Pre-construction information • Construction phase plan • Health and safety file • Notification of project to HSE • Detailed requirements for construction work
Work at Height Regulations 2005	• Avoidance of risk • Organisation, planning, competence • Selection and inspection of work equipment • Scaffolding plans & specific requirements for guard rails
Lifting Operations and Lifting Equipment Regulations 1998 - LOLER	• Strength and stability of lifting equipment • Lifting and lowering of persons • Safe operation of lifting equipment • Planning and supervision of lifts • Thorough examination and regular inspection
Provision and Use of Work Equipment Regulations 1998 - PUWER	• Duty holders responsibilities • Suitability, maintenance and inspection • Information, instruction and training • Conformity of equipment; Roll-over protection
Control of Substances Hazardous to Health Regulations 2002 - COSHH	• Assessment of health risks • Prevention/control of exposure • Exposure limits, monitoring and surveillance • Provision of information
Construction (Head Protection) Regulations 1989	• Provision of suitable head protection • Duty to ensure it is worn • Power to make site rules
Personal Protective Equipment at Work Regulations 1992	• Provision of PPE by employers • Assessment of risk • Maintenance, cleaning and replacement
Manual Handling Operations Regulations 1992	• Duty of care • Risk assessment
Noise at Work Regulations 1989	• Duty of care • Assessment of exposure • Action levels
Electricity at Work Regulations 1989	• Duties on employers and workers • Prevention of danger • Competent persons
Control of Asbestos at Work Regulations 2006	• Duty to manage • Plans of work • Licensing • Control measures: air monitoring and testing
Confined Spaces Regulations 1997	• Safe systems of work • Competence and training • Emergency arrangements

Reporting of Injuries, Diseases and Dangerous Occurrences Regulations 1995 - RIDDOR	• Reporting of specified events to HSE:
	  o Major injuries
	  o Over 3-day accidents
	  o Work-related disease
	  o Dangerous occurrences

Figure 12.1

HEALTH AND SAFETY ORGANISATION

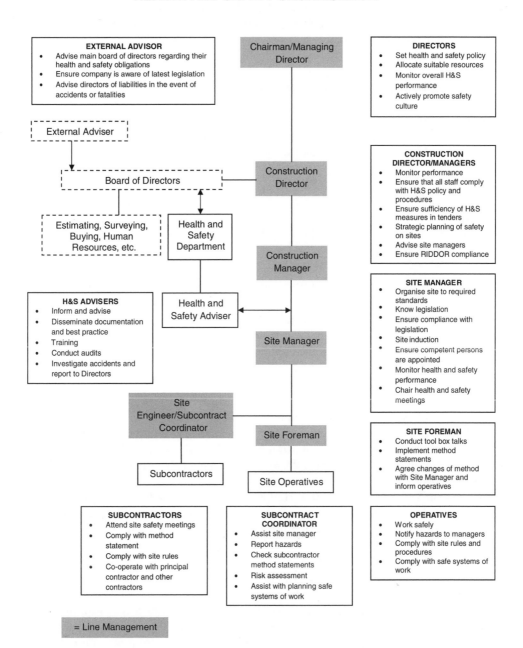

EXTERNAL ADVISOR
- Advise main board of directors regarding their health and safety obligations
- Ensure company is aware of latest legislation
- Advise directors of liabilities in the event of accidents or fatalities

Chairman/Managing Director

DIRECTORS
- Set health and safety policy
- Allocate suitable resources
- Monitor overall H&S performance
- Actively promote safety culture

External Adviser

Board of Directors

Construction Director

CONSTRUCTION DIRECTOR/MANAGERS
- Monitor performance
- Ensure that all staff comply with H&S policy and procedures
- Ensure sufficiency of H&S measures in tenders
- Strategic planning of safety on sites
- Advise site managers
- Ensure RIDDOR compliance

Estimating, Surveying, Buying, Human Resources, etc.

Health and Safety Department

Construction Manager

H&S ADVISERS
- Inform and advise
- Disseminate documentation and best practice
- Training
- Conduct audits
- Investigate accidents and report to Directors

Health and Safety Adviser

Site Manager

SITE MANAGER
- Organise site to required standards
- Know legislation
- Ensure compliance with legislation
- Site induction
- Ensure competent persons are appointed
- Monitor health and safety performance
- Chair health and safety meetings

Site Engineer/Subcontract Coordinator

Site Foreman

SITE FOREMAN
- Conduct tool box talks
- Implement method statements
- Agree changes of method with Site Manager and inform operatives

Subcontractors

Site Operatives

SUBCONTRACTORS
- Attend site safety meetings
- Comply with method statement
- Comply with site rules
- Co-operate with principal contractor and other contractors

SUBCONTRACT COORDINATOR
- Assist site manager
- Report hazards
- Check subcontractor method statements
- Risk assessment
- Assist with planning safe systems of work

OPERATIVES
- Work safely
- Notify hazards to managers
- Comply with site rules and procedures
- Comply with safe systems of work

= Line Management

Figure 12.2

Hughes and Ferrett (2005) say that aims (e.g. for a safe and healthy workplace) will probably remain in place following policy revisions whilst objectives will change each year (e.g. specific targets or campaigns). Company policy should reflect the genuine aims and objectives of the firm and should not be a generic document.

12.5 Health and safety management

Successful health and safety management (HSE 1997) is widely recognised as the 'bible' of good health and safety principles and practice and it sets out clear guidance on the key elements of a sound health and safety management system:

- Policy
- Organisation
- Planning
- Measuring performance
- Audit and review

The 4 Cs are fundamental to a positive health and safety culture and these are:

- Control – top-down commitment and collective endeavour
- Cooperation – participation and ownership, sharing knowledge and experiences
- Communication – sending out the right signals, talking and listening to each other
- Competence – awareness, application, skill, knowledge and training

HSG65 (HSE 1997) explains that organisations need to manage health and safety with the same degree of expertise and to the same standards as other core business activities if they are effectively to control risks and prevent harm to people. Responsibility for this lies with line management and Figure 12.2 illustrates how this might work within a medium-sized company. Also illustrated are some of the main health and safety responsibilities at each level of management.

Regulation 5 of the Management of Health and Safety at Work Regulations 1999 requires employers to have appropriate arrangements in place for the effective planning, organisation, control, monitoring and review of the **preventative and protective measures** required by regulation 4, schedule 1:

- Avoid risks and evaluate those that cannot be avoided (i.e. risk assessment)
- Combat risks at source (i.e. at the design and planning stages)
- Adapt work to the individual including the way the workplace is designed, the type of equipment provided and the choice of production methods
- Give priority to collective measures of protection (e.g. roof edge protection rather than individual safety harnesses)
- Take account of technological innovation (such as composite roofing panels that provide a safe working platform as well)
- Make sure that employees are properly instructed

Line managers should be supported by a health and safety department along with in-company safety advisors. Safety advisors are not responsible for health and safety on site but serve in a supportive and consultative role as illustrated in Figure 12.2. Small companies may engage external health and safety expertise to help them and many large contractors have external advisors who advise the main board of directors especially regarding their legal liabilities.

12.6 The safety of site operations

Under the HSWA, contractors (as employers) have general duties to their employees, and to others not in their employment, as regards their health and safety. Section 2(1) of the Act requires employers to *ensure the health and safety at work of all his employees* and Section 2(2) extends this to:

(a) the provision and maintenance of safe plant and **safe systems of work**
(b) arrangements for ensuring safe means of handling, use, storage and transport of articles and substances
(c) the provision of information, instruction, training and supervision
(d) provision of a **safe place of work** and provision and maintenance of **safe access and egress** to that place of work
(e) provision and maintenance of a **safe working environment** and **adequate welfare facilities**

HSWA section 3(1) requires employers and the self-employed to *conduct their undertaking in such a way that persons not in their employment are not exposed to risks to their health or safety.* This imposes a duty of care on contractors, for instance, for their subcontractors and for passers-by and visitors to the site, etc.

Safe place of work

The idea of a safe place of work is derived from the law and, in particular, section 2(2)(d) of HSWA referred to above and regulation 26 of CDM 2007 which requires such places of work to:

- be safe and have properly maintained access and egress
- be made and kept safe and without risks to the health of workers
- have sufficient working space taking into account any work equipment present

Regulation 2 of CDM 2007 states that a place of work is *any place which is used by any person at work for the purposes of construction work.* However, a truly safe place of work is very difficult if not impossible to achieve, especially in construction. Even on the best run projects, there will be hazards to deal with and therefore safe systems of work will have to be developed to deal with them.

Safe systems of work

A safe system of work is simply a defined method for doing a job in a safe way and both Hughes and Ferrett (2005) and St. John Holt (2001) explain that it is *a formal procedure that results from a systematic examination of a task in order to identify all the hazards and assess the risks.* Where risks remain, a safe system of work will be required.

Wallace (2000) suggests that a safe system of work is made up of five key components:

- **Personnel** – they must be competent to do the job, have suitable training and be properly supervised
- **Supervision** – the direct supervisor (ganger or foreman) must have a knowledge and understanding of the work to be done, must make sure that the defined safe system of work is adhered to, must be capable of motivating the personnel doing the job, and must have the ability to take corrective action when necessary
- **Working environment** – at the outset, this is defined by the safe place of work standard but, as the work operations unfold, this may well change. Hazards such as noise, vibration, physical strain and moving plant and machinery must be monitored and managed during the work activity. Further safe systems of work may need to be developed as the job progresses
- **Procedures** – should be well designed and clearly define how the task shall be done correctly, safely and efficiently
- **Permits** – may be necessary to prevent unauthorised personnel entering a particular place of work and may be part of a safe system of work where hot working is to be carried out or where entry into a confined space such as a manhole, sewer, or tunnel is required

Any changes to a safe system of work must be:

- Carefully considered with risks correctly assessed
- Written down and any changes to the method statement carefully noted in writing
- Properly communicated to the workforce via a task talk or similar discussion or, if necessary, through additional training
- Monitored to make sure that they are working correctly

Some safe systems of work can be verbal, perhaps where the risks are low and the work is of short duration, so long as the hazards and control measures are explained to the operatives, preferably by their immediate supervisor or manager. St. John Holt (2001) suggests that operatives must not be allowed to devise their own method of working as this is not a safe system of work and, more usually, safe systems of work are documented in the form of method statements.

Method statements

Method statements which define a safe system of work are often referred to as **safety method statements** and they are often confused with method statements used in construction planning. Best practice suggests that they are one and the same thing because safety is both an integral aspect of the planning of work operations and a legal requirement with respect to the provision of a safe place of work.

Method statements may be written down in a tabular or prose format but in both cases they should include consideration of:

- Time needed to do the job
- How the job will be done
- The plant, machinery and temporary works required (e.g. earthwork support, scaffolding, formwork)
- The order or sequence of work
- Means of access and egress to and from the work place
- Lifting equipment and the means of getting materials in place
- Means of preventing falls of people and objects
- Means of preventing unauthorised access to the place of work (e.g. children at night)
- Interface with other trades working in the same place

A sample safety method statement, using a pro forma style, is illustrated in Figure 12.3. Note that the construction method or work sequence is included together with appropriate risk control measures.

12.7 Risk assessment

The process of risk assessment is a legal requirement:

- Implied by the HSWA – section 2
- Specifically required by the Management of Health and Safety at Work Regulations 1999 – regulation 3

The basic idea is to look at planned activities to see what hazards (if any) are present and what risk of harm they might pose to workers and others affected by the work activity. The intention is to eliminate risk, where possible, and to control any residual risk by effective management.

Losses/accidents are caused by exposure to risk. This may be the result of unidentified hazards or residual risks that have been left unmanaged. The Health and Safety Executive (HSE) suggests that risk assessment should not be overcomplicated and that there are five basic steps to follow HSE (2006). We have suggested additional measures at step 3:

(1) Identify the hazards
(2) Decide who might be harmed and how
(3) Evaluate the risks and decide on precautions
    (a) Determine the control measures required
    (b) Evaluate remaining risks
    (c) Make contingency plans for the residual risks
(4) Record your findings and implement them
(5) Review your assessment and update if necessary

SAFETY METHOD STATEMENT

Contract	Contract No.	Prepared By	Date	Checked By	Date
Canal renovation	C2314	GHM	26 May	APH	27 May

Operation	Plant
Construct sheet piled cofferdam	22 RB crane BSP 900 pile hammer Komatsu 380 escavator Two 30T wagons 150mm diesel pump

Work sequence	Supervision and monitoring
Construct hardstanding for piling rig in lock mouth using imported quarry waste Erect guide frame and install Frodingham 3N piles Pump out water between hardstanding and piles Fill behind piles for access	Site engineer Piling foreman Operation to be monitored daily by the site agent Banksman to work with mobile plant Daily check on crane equipment, load indicators and operation

	Controls
Remove piling frame Construct top bracing with steel wallings and struts Remove hardstanding and install bottom bracing Erect secure ladder access to bottom of cofferdam	Authorised personnel area only When working over water to remove piling frame, life jackets and safety harness must be worn Area to be fenced off at night with chestnut paled fence Warning notices to be displayed either side of cofferdam 'Danger deep excavation' and 'Danger deep water'

Emergency procedures	First Aid	PPE schedules
Send for site first aider and/or call emergency services where necessary Rescuers must not put themselves in danger Follow first aid drill if appropriate Do not remove evidence Notify site agent	Dinghy to be moored adjacent to cofferdam Two Lifebuoy in wooden locker First aid box in site office 2-way radios	Safety harnesses Hard hats Gloves Welding goggles High visibility vests Life jackets Ear defenders

Figure 12.3

There is no standard way of doing risk assessments but, in firms of over five employees, risk assessments must be recorded, and they must also be kept under review, because risks may change according to changes in circumstances. The best idea is to keep risk assessments simple and understandable so that they are easy to communicate. This sort of approach is illustrated in Figure 12.4.

CONTRACTOR'S RISK ASSESSMENT

Contract		Contract No	Prepared by	Date of assessment
Canal renovation		C2314	GHM	23 May

Operation	Potential hazards	Risk assessment			Control measures
		H	**M**	**L**	
Construct sheet piled cofferdam	Trapping and crushing by moving plant		M		Warning device/banksman
	Work at height	H			Guard rails along lock
	Falls of materials	H			Leave piles proud for edge protection
	Confined space working			L	
	Working over water	H			Safety harness and life jacket to be worn

Programmed for	Training/certification	Action	PPE Required (specify)

16 June	All operatives to be CSCS certified
	
	Plant operators to be CTA certified
		Method statement ☑
		Work permit ☑
		Assessment :
		  COSHH ☐
		  Noise ☑
			Safety harnesses
			Hard hats
			Gloves
			Welding goggles
			HV vests
			Life jackets
			Ear defenders

Figure 12.4

12.8 Hazard and risk

The main emphasis in UK health and safety legislation is the control of risk and hazard identification is an *essential first step in risk control* (HSE 1997). Hazards give rise to risks and *if there are no hazards, there are no risks* (HSE 2000). However, construction work is, by definition, hazardous and so the best that can be done is to identify the hazards and either eliminate them or control the ensuing risks.

A **hazard** is *something with the potential to cause harm* (HSE 2000) and might be the result of a work activity, process, plant or machine or substance, etc., such as:

- working at height
- working from a ladder
- deep excavations
- moving plant or vehicles
- mixing mortar or plaster
- machining wood
- asbestos
- solvents or toxic substances

A **risk** is *the likelihood of potential harm from [a] hazard being realised* (HSE 2000). The extent or level of the risk is determined by:

- the likelihood of harm occurring
- the potential severity of the harm and
- the number of people who might be exposed to the hazard in question.

Falls from height is often identified as a hazard, but this is not correct:

- **Working at height** is *something with the potential to cause harm* and is therefore the hazard
- When someone is working at height, there is a ***chance*** of falling with the ***possible consequence*** of a serious injury and this person is therefore ***at risk***
- The extent of the risk will depend on the likelihood or chance of falling and the severity of the fall
- For example, falling 20 m to the ground is likely to lead to a fatality or, at best, a very severe injury and therefore the risk must be high unless measures are put in place to prevent a fall.

There are two principal categories of hazards – generic and specific – and these are illustrated in Figure 12.5 which gives some practical examples. You can always spot a hazard if you ask the question ***'does it have the potential to cause harm?'*** If the answer is 'yes' then it is a hazard.

Falling is just one of the risks faced by operatives when working at height and it is frequently the case that there are other hazards to consider in any given situation. For instance, **chasing out brickwork** for pipes or wiring is a hazardous activity because it creates dust, noise, vibration and flying debris, etc., depending on how the work is done. Hazards can be designed out (e.g. hiding pipes or wires behind dry lining) but using full body protection and respiratory

HAZARDS	
Generic	**Specific**
DEFINITION – Hazards that are present on virtually every site	DEFINITION – Hazards that are peculiar to a particular site

EXAMPLES
 	
• Dust and fumes
• Noise
• Vibration
• Harmful substances or materials
• Electricity
• Fire
• **Work at height** ───────┐
• Heavy weights
• Moving plant and vehicles
	EXAMPLES
	
	• Nearby school, pedestrian crossing or bus stop
	• Busy road junction
	• Railway line nearby
	• Presence of water (e.g. river or canal)
	• Overhead high voltage cables
	• Underground services (e.g. gas main, electricity cables)
	• Underground railway tunnel
	• Existing buildings to be demolished
	   o Asbestos
	   o Unstable/insecure structural elements
	   o Sharp objects such as waste hypodermic needles and syringes
	   o Unprotected lift shafts or pits
	   o Petrochemicals, solvents, acids
	• Deep excavations
	• Contaminated ground
	   o Chemicals, solvents, toxic metals
	   o Toxic/flammable gases
	   o Bacteria/viruses from clinical waste

Work at height examples	
Roofwork • Fragile surfaces (e.g. rooflights, roof sheeting) • Weather (high wind, rain, hot sun) • Sharp-edged materials (e.g. cladding panels)	
Steelwork • Heavy weights (trapping/crushing risks) • Large or awkward components • Hot work	
Brickwork and pointing • Bricks and tools that could fall • Cement-based materials and chemicals • Abrasive wheel-type hand tools (for raking out joints)	
High-pressure jet cleaning of buildings • Falling materials • High pressure liquids • High velocity waterborne debris	

HAZARDS IN CONSTRUCTION

Figure 12.5

apparatus are risk reduction measures because the hazard (i.e. chasing out) is still there. For further examples refer to Figure 12.5.

Legally speaking, it is not necessary to identify every possible hazard or assess every risk because UK health and safety legislation is largely founded on the principle of **so far as is reasonably practicable** or similar words to that effect. This means (with some exceptions) that, so long as everything reasonably possible has been done to seek out hazards and avoid risks to those who might be exposed, this is enough to satisfy the law (the leading case of *Edwards* v. *National Coal Board* [1949] refers).

Joyce (2007) reports that this principle has been established as 'not unlawful' in the European Court of Justice and therefore *measures to reduce risks do not have to be grossly disproportionate.* It is a matter of balancing the effort and cost involved. Consequently, risks are reduced to the level of ALARP (as low as reasonably practicable) although the possibility of low likelihood but high-severity risks must be guarded against as must the possibility of overlooking hazards that might give rise to unmanaged risks.

Common hazards in construction

Finding or 'spotting' hazards is not an easy or obvious process and requires considerable knowledge and experience. Useful guidance may be found in:

- Legislation (e.g. COSHH, Confined Spaces, CDM 2007, Part 4)
- Approved Codes of Practice (ACoP)
- HSE guidance
- British and International standards
- Product information and data sheets
- Industry guidance
- Research and CIRIA publications
- Company accident/ill health/near miss data

Another useful source of information is the HSE annual statistics of work-related injuries and ill health in construction available at the HSE website. Once you are aware of the common causes of accidents and ill health in construction, you can look for the hazards that might lead to these consequences:

- Falling from a height
- Being struck by moving object
- Being struck by moving vehicle
- Suffering a slip, trip or fall on the same level
- Being injured while handling, lifting or carrying

Persons at risk

Part of the risk assessment process is to consider the persons at risk from the various hazards identified. This includes:

- passers-by
- directly employed workers and site staff
- subcontractors
- official visitors (e.g. architect, engineer, factory inspector)
- other visitors (e.g. invited guests, students, etc.)
- unauthorised visitors (e.g. children, people 'up to no good!!', etc.)

It can easily be forgotten that construction sites pose hazards for members of the public as well as workers and typically around five or six members of the public are killed each year as a result of construction work – many more are injured. Passers-by can be killed or injured by falling materials and equipment, overturning plant or scaffold collapse, by contact with site vehicles and plant or even by being forced into the road by poorly planned diversions. Children are also at risk because construction sites provide tempting playgrounds.

Risk evaluation

There are two main types of risk assessment:

- Quantitative risk assessment
- Qualitative risk assessment

Quantitative evaluation *produces an objective probability estimate* based on known facts and statistical data (St. John Holt 2001), whereas **qualitative** risk assessments are *based upon personal judgement backed by generalised data on risk* (Hughes and Ferrett 2005).

Most risk assessments in practice are qualitative and normally they are perfectly adequate and simple to do. Many are also generic – i.e. standard risk assessments for commonplace hazards where the risks are well known and the control measures the same each time (e.g. manual handling, COSHH assessments, etc.). This can make sense provided the assessments reflect specific circumstances, but generic assessments should never be used for site specific or high-risk activities.

Qualitative risk assessments can give the impression that they are quantitative because numbers are sometimes used in the assessment. This methodology is still qualitative because the numbers are based on personal judgement and not on statistical data. The HSE suggests a simple 3 × 3 matrix approach as illustrated in Figure 12.6, whereas the other extreme offered by St. John Holt (2001) suggests that a qualitative risk rating can be given by taking consequence × (number of persons exposed + probability of harm).

Control measures

The ACoP to the Management of Health and Safety at Work Regulations 1999 says that the risk assessment should include:

(a) a record of the **preventative and protective** measures in place to control the risks and
(b) what further action, if any, needs to be taken to reduce risk sufficiently.

Preventative and protective measures are often referred to as **control measures**, the purpose of which is to reduce the risks to ALARP. This is not always achievable, however, and therefore further measures may need to be put in place to control any residual risks. For instance, a control measure might be to provide operatives working at height with a safety harness for a particular task. However, there is still the risk that operatives will not use the harnesses correctly or that they will not clip on to a safe anchorage. This residual risk could be reduced

3x3 matrix

Likelihood		Severity	
High	certain/near certain	Major	death or major injury
Medium	often occurs	Serious	causing short term disability
Low	seldom occurs	Slight	other injuries/illness

The resulting matrix is as follows where risk is given by severity of harm x likelihood of occurrence:

		SEVERITY		
		Low	Medium	High
LIKELIHOOD	Low	L	L	M
	Medium	L	M	H
	High	M	H	H

Risk = Likelihood x Severity

Some practitioners like to put numbers to the risk evaluation where:	Example
• Low = 1 • Medium = 2 • High = 3	If the likelihood is 2 and the potential severity is 3 the resultant risk is 6 out of a maximum possible of 9. This is at the lower boundary of the high risk category. The numbers do not really mean anything – they just indicate a level of risk – and it is sometimes better to think high, medium or low risk.

Many other hazard rating systems have been developed including the 6x6 matrix below:

6x6 matrix

Likelihood		Severity	
1	Remote	1	Minor injury
2	Unlikely	2	Illness
3	Possible	3	Accident
4	Likely	4	Reportable injury*
5	Probable	5	Major injury*
6	Highly probable	6	Fatality*

*RIDDOR category

QUALITATIVE RISK ASSESSMENT

Figure 12.6

to an acceptable level by providing a foreman to supervise the task who will ensure that the safe system of work operates properly.

Control measures are precautions to protect those at risk from work activities and are equally relevant to the site set-up as the day-to-day construction operations taking place. Effective site safety management is a matter of good planning and this includes making sure that the site itself is well organised, well run and kept tidy and that unauthorised access is prevented.

Figure 12.7 illustrates some issues that the site manager may consider when planning the site set-up together with examples of control measures for a variety of work activities.

12.9 Planning the work

Construction sites present many hazards and even the best-managed sites cannot be considered 'safe'. When planning for safety, it is tempting to list all the hazards you can think of, deal with them and 'tick the box' – job done! – but there are better ways to make sure that risk is properly managed and nothing important is overlooked.

One method is to use a **task-based approach** such as the **job safety analysis** suggested by St. John Holt (2001):

(1)   select the job/task
(2)   break it down into logical steps
(3)   identify the hazards in each step
(4)   devise ways of eliminating the hazard or reducing the risk from it
(5)   record the analysis
(6)   review and update as necessary

This is a logical approach which requires careful consideration of what is involved in the job and a good understanding of construction methods is needed. Figure 12.8 illustrates the principles with an example of the demolition of an old terraced stand at a major football ground.

One of the disadvantages with this method is that it is possible to miss hazards that may be present at the interface with other tasks. For instance, running concurrently with the demolition activity may be some drainage work to divert an existing sewer away from the new stand.

As an alternative, a **hazard-based approach** may be taken. This involves looking at the project and deciding whether specific hazards will arise and, if so, how and where. Examples of such hazards include work at height, manual handling, workplace equipment such as power tools, saw benches and welding equipment (see Figure 12.9). This method involves knowing what the hazards are and recognising them in different circumstances. It's not as easy as it sounds!

An advantage of this approach is that the ensuing risk assessment will be linked to hazards which can be easily recognised by most people from ACoP, published guidance and standard texts, and therefore the process becomes familiar to those involved. This is likely to ensure a consistent approach to particular hazards across all sites within a contracting organisation. A disadvantage is that individual hazards may appear repeatedly making the analysis repetitive and overlong and, of course, it is perhaps easier to overlook hazards than the task-based approach.

Some contractors like to include health and safety considerations in their short-term contract planning. For instance, in a 6- or 8-week detailed programme of work activities, health and safety actions may be included on the bar chart, or health and safety specific plant and equipment may be identified on the programme. The Engineering and Construction Contract specifically requires the contractor to show provisions for health and safety requirements on

SITE LAYOUT PLANNING	
Site perimeter • Plywood hoarding, proprietary interlocking metal panels or mesh fencing • Covered walkways • Protective fans and debris netting • Pedestrian diversions • Clear directional and warning signs	
Access • Pedestrian security gates • Vehicular barriers/gates • Security office • Measures to protect children o Immobilise plant, prevent access to heights, cover or barrier open excavations and drainage systems, store heavy materials safely	
Site accommodation and welfare facilities • Office and rest rooms • Sanitary and washing facilities • Drinking water • Drying facilities • Fire fighting and first-aid equipment • Statutory notices • Materials storage	**Traffic control** • Separation of pedestrians and vehicles • Signed traffic routes • Loading/unloading area • Car parking area • Wheel washers **Overhead power lines** • Disconnection/diversion • Goalpost barriers • Height restrictions
Waste disposal • Skips Refuse chutes or hoists • Hazardous/special waste containers	**Lighting** • Site lighting • Access lighting • Individual task lighting

WORK ACTIVITIES	

Excavations	• Collapse
	  o Battered sides to excavations
	  o Trench boxes, drag boxes, manhole boxes
	  o Hydraulic waling frames
	  o Shoring and sheet piling
	• Excavation guard rails/barriers
	• Plant stop blocks
	  o Use of detectors and 'dial before you dig'

Work at height	• Access
	  o Full scaffold
	  o Mobile scaffold towers
	  o Scaffold stairways
	  o Use of permanent stairways
	  o Mobile elevating work platforms (cherry pickers, scissor lifts, etc.)
	  o Passenger and materials hoists
	  o Mast climbers
	• Fall prevention
	  o Leading edge protection
	  o Harnesses with 'distance' restraints
	  o Purlin trolleys
	• Fall arrest
	  o Scaffolded and boarded 'crash deck'
	  o Harnesses and running lines
	  o Safety nets
	  o Air bags

Manual handling	• Mechanical lifters, e.g. A-frames, scissor lifters • Hoists • Elevators, conveyors • Rough terrain fork lift trucks

CONTROL MEASURES

Figure 12.7

Demolition of terracing at a major football ground

Task	Task breakdown	Hazards	Risk control measures
Demolish stand	1. Form access	a. Moving traffic	Traffic management plan with pedestrian segregation
		b. Moving plant	Banksman with machine
	2. Break up terracing	a. Moving plant	Banksman with machine
		b. Dust	Hose down regularly with water
		c. Noise	Muffled hydraulic breakers
	3. Remove safety barriers	a. Post-tensioned cables	Relieve stresses before removing
	4. Soft strip roof	a. Work at height	Safety harnesses attached to running line
		b. Wind	Check wind speed with anenometer: no work above Beaufort Scale 5
		c. Falls of materials	Exclusion zone beneath work area
	5. Dismantle steelwork	a. Work at height	Use man baskets and cherry pickers
	6. Remove earth banking	a. Moving plant	Banksman with machine
		b. Dust	Hose down regularly with water
		c. Contaminated ground	Laboratory analysis + PPE for all operatives
		d. Moving vehicles	Traffic management plan
	7. Demolish rear wall	a. Unstable wall	Progressive demolition by hand from MEWP* using temporary shoring system
		b. Dust	Hose down regularly with water

*Mobile elevating work platform (e.g. cherry picker, genie boom or scissors lift)

JOB SAFETY ANALYSIS

Figure 12.8

the accepted programme (see core Clause 31.2), and an example of how this might be done is given in Figure 12.10.

12.10 The Construction (Design and Management) Regulations (CDM) 1994

The CDM regulations were introduced in 1994 as a consequence of the Temporary or Mobile Construction Sites Directive of the European Union (92/57/EEC). The regulations were accompanied by the ACoP and were later revised in 2000 mainly to reflect changes due to case law concerning the duties of designers. A revised ACoP was published in 2004.

WORK ACTIVITY	HAZARDS
Demolitions	• Work at height • Falling debris • Premature collapse • Asbestos and other hazardous substances • Noise from heavy plant • Vibration from hand-held breakers, etc.
Bulk excavations	• Moving plant, e.g. excavators, dumpers, wagons • Contaminated ground • Underground services • Overhead power lines

Groundworks	• Collapse of excavations
	• Falling materials
	    o Into trenches
	    o From work above (e.g. roofing)
	• Excavation plant (striking/crushing/trapping, etc.)
	• Confined spaces
	• Waterborne diseases (e.g. Weil's disease)
	• Wet concrete (dermatitis)
	• Dust/debris from disc cutters

Steelwork	• Work at height • Temporary instability • Cutting and welding equipment (burns, sparks, toxic fumes) • Manual handling (trapping, cuts) • Weather (overexposure)
Brickwork & blockwork	• Work at height • Fragile materials (roof sheets, roof lights) • Power tools for cutting and nailing • Manual handling (e.g. heavy blocks) • Dust/debris from disc cutters • Cement dust/mortar/plasticisers (dermatitis)
Woodwork	• Power tools and woodworking machinery • Electricity • Sharp hand tools • Wood dusts • Noise and vibration • Glues, paints, solvents, chemicals
Cladding and roofing	• Work at height • Manual handling (heavy rolls of roofing felt, heavy/large sheet materials) • Sharp edged/heavy materials (cuts/abrasions) • Wind and weather

GENERIC CONSTRUCTION HAZARDS

Figure 12.9

Although revolutionary at the time, CDM 1994 gradually became the subject of increasingly heavy criticism including:

• Unnecessary bureaucracy and paperwork
• Difficult to enforce especially as regards designers' duties

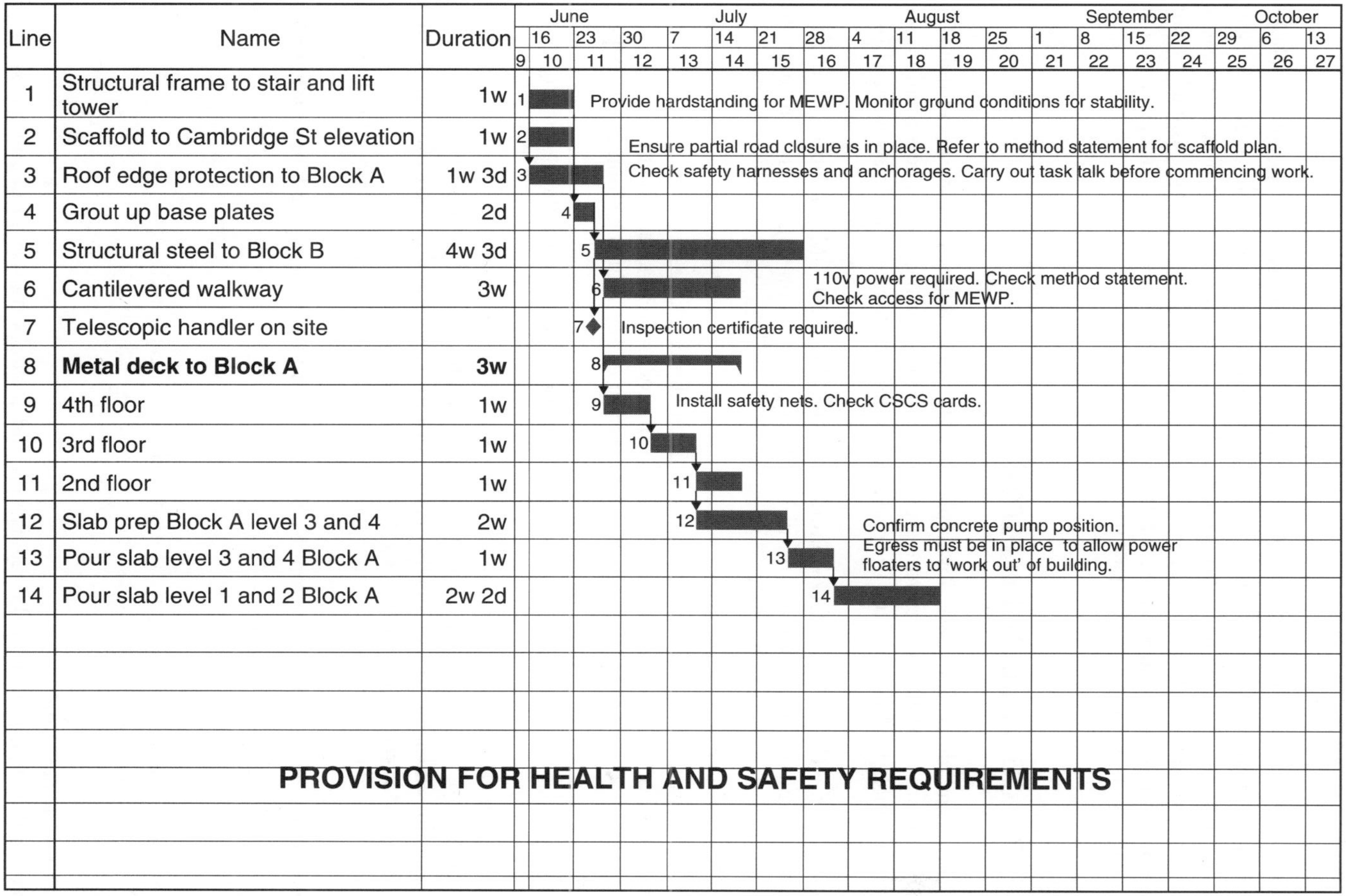

Line	Name	Duration
1	Structural frame to stair and lift tower	1w
2	Scaffold to Cambridge St elevation	1w
3	Roof edge protection to Block A	1w 3d
4	Grout up base plates	2d
5	Structural steel to Block B	4w 3d
6	Cantilevered walkway	3w
7	Telescopic handler on site	
8	**Metal deck to Block A**	**3w**
9	4th floor	1w
10	3rd floor	1w
11	2nd floor	1w
12	Slab prep Block A level 3 and 4	2w
13	Pour slab level 3 and 4 Block A	1w
14	Pour slab level 1 and 2 Block A	2w 2d

Timeline columns: June 9, 16, 23, 30; July 7, 14, 21, 28; August 4, 11, 18, 25; September 1, 8, 15, 22, 29; October 6, 13, 27

Bar annotations:

1. Provide hardstanding for MEWP. Monitor ground conditions for stability.
2. Ensure partial road closure is in place. Refer to method statement for scaffold plan.
3. Check safety harnesses and anchorages. Carry out task talk before commencing work.
5. 110v power required. Check method statement.
6. Check access for MEWP.
7. Inspection certificate required.
9. Install safety nets. Check CSCS cards.
12. Confirm concrete pump position.
13. Egress must be in place to allow power floaters to 'work out' of building.

PROVISION FOR HEALTH AND SAFETY REQUIREMENTS

Planned by Asta Powerproject

Figure 12.10

- Ineffective planning supervisors
- Clients allowed to delegate their statutory duties to an agent

However, criticisms of CDM 1994, whilst to a certain extent justified, cannot entirely be blamed on the regulations themselves and much of the excess paperwork blamed on CDM was not in fact demanded by the regulations but was largely the product of an over enthusiastic industry. Admittedly, designers' duties were very poorly drafted (vague at best would be a fair description), and planning supervisors, whilst having onerous duties, had no real power under the regulations. These were serious defects.

12.11 The CDM Regulations 2007

The CDM Regulations 2007 have now replaced CDM 1994 following widespread industry consultation, and CDM 2007 is therefore, arguably, what the industry wants. The regulations are certainly significantly different from the previous version and there is also a new ACoP (L144, 2007).

The aim of the Health and Safety Commission with the new CDM regulations is to drive the further reduction of accidents and ill health in construction with a set of regulations that:

- Are clearer with more specific expectations of duty holders
- Are flexible enough to accommodate a variety of procurement arrangements
- Emphasise the need to plan and manage work rather than producing mountains of paperwork
- Emphasise better communication and coordination through duty holders working in integrated teams
- Are simpler as regards the assessment of duty holders' competence
- The planning supervisor role has been replaced with the CDM coordinator
- The health and safety plan (which covered both the design and construction stages of a project) has been replaced with the construction phase plan
- Both clients and designers have more extensive duties to supply information about the site and the design (pre-construction information)
- The provisions of the Construction (Health, Safety and Welfare) Regulations 1996 are now included within CDM (except the bit on 'falls' which is now covered by the Work at Height Regulations 2005).

The emphasis is now very much on the management of health and safety.

CDM 2007 is in five main parts and there are also five schedules:

- Part 1 – Introduction
  Defines the applicability of the regulations and includes important definitions such as 'client', 'construction work', 'construction phase plan', 'contractor', 'designer', 'principal

contractor' and numerous others which are required in order to properly interpret the detailed regulations

- Part 2 – General management duties
  Sets out the duties that apply to all construction work, irrespective of size, including requirements as to competence, cooperation, coordination, the duties of clients, designers and contractors and the general principles of prevention, etc.
- Part 3 – Additional duties
  Sets out the duties that apply where the project is notifiable to HSE as defined in regulation 2(3), as well as clients' duties, additional duties of designers, contractors and principal contractors, CDM coordinators duties and requirements of the construction phase plan and the health and safety file
- Part 4 – Health and safety on construction sites
  This part contains detailed regulations concerning safe places of work and site security, etc., together with requirements for excavations and inspections, traffic routes and vehicular movements on site, fire detection and fire fighting, etc.
- Part 5 – General
  This part deals with rights to take legal action in the civil courts where there has been a breach of the regulations and also transitional provisions and revocations and amendments of existing legislation
- Schedules:
  (1) Contains details to be notified to the HSE on the F10 notification
  (2) Specifies legal requirements for the provision of welfare facilities on site
  (3) Lists the details to be recorded for inspections of excavations, cofferdams, etc.
  (4) Lists details of regulations revoked as a consequence of CDM 2007
  (5) Lists amendments to existing legislation as a consequence of CDM 2007

Application

The regulations apply to all construction work, which is defined in regulation 2(1), irrespective of the size of project. 'Domestic clients' have no duties under CDM but designers and contractors acting for them do. Projects have to be notified to HSE (on an F10 form) if they are planned to last longer than 30 days or exceed 500 person days of construction work. If this is the case, additional duties apply.

Duty holders

The client

Clients have to make sure that the arrangements for carrying out the project will be carried out without risk to the health or safety of anyone and that the welfare arrangements for the project are suitable.

For notifiable projects, the client must appoint a CDM coordinator and a principal contractor, and he/she must make sure that they are competent and have allocated adequate resources to the project including sufficient time.

The client has a further duty to give all relevant information to any designers and contractors he may appoint. This is called the 'pre-construction information' and corresponds to information in the client's possession and information that he could reasonably obtain.

The CDM coordinator

A CDM coordinator must be appointed for any notifiable project and must advise and assist the client with respect to his/her duties under the regulations. He/she must also make sure that appropriate arrangements are in place for planning and coordinating health and safety measures during both the design and construction stages of a project.

The CDM coordinator does not prepare the construction phase health and safety plan (that job belongs to the principal contractor) but must liaise with the principal contractor with regard to pre-construction information requirements.

The CDM coordinator has onerous duties (but still no real powers) including:

- collecting pre-construction information
- presenting the pre-construction information in a convenient form
- ensuring that designers comply with their duties
- making sure that designers and the principal contractor cooperate with regard to design issues during the construction phase
- prepare the health and safety file in conjunction with the principal contractor

The designer

Designers' duties under CDM apply irrespective of the size of the project or whether or not the project is notifiable. This illustrates the central role that designers play in reducing risks both during construction and when the building or structure is in use.

Consequently, designers must:

- avoid foreseeable risks to the health and safety of any person
- eliminate hazards that give rise to risks
- reduce risks from remaining hazards

They must also make sure that the client is aware of his own duties before starting work on the design and share information with the client and other designers and contractors. For notifiable projects, designers must not go beyond the initial stages of design unless a CDM coordinator has been appointed and he/she must provide relevant information about his design to the CDM coordinator.

The principal contractor

The principal contractor role is a statutory appointment for notifiable projects only. It is usually the main contractor who undertakes the role, but it can be another contractor or even the client.

The principal contractor must plan, manage and monitor the construction phase so that there is no risk to health or safety. This duty is subject to the test of 'reasonable practicability'.

Preparation of the construction phase (health and safety) plan is also the principal contractor's responsibility and this must be developed in sufficient detail before work is started.

The client must make sure that the construction phase does not start unless the construction phase plan complies with regulations 23(1)(a) and 23(2) and also that the principal contractor's welfare arrangements comply with regulation 22(1)(c).

Other duties of the principal contractor include:

- liaising with the CDM coordinator
- preventing unauthorised access to the site
- providing information to contractors and workers
- consulting with others
- drawing up site rules
- making sure that workers are provided with a site induction

Other contractors

Under CDM 2007, contractors (usually 'subcontractors' in common parlance) are required to cooperate with the principal contractor and comply with directions given by him. They are also required to provide information relating to the management and prevention of health and safety risks relating to their own work.

Contractors must also comply with the construction phase plan and provide RIDDOR (Reporting of Injuries, Diseases and Dangerous Occurrences Regulations 1995) information to the principal contractor as appropriate.

Principles of CDM

The 4 Cs

In common with HSG65, the guiding principles of CDM 2007 are the 4 Cs, i.e.:

- Cooperation   The duty to cooperate applies to everyone involved in the project from the client to site workers
- Coordination   All duty holders must coordinate their activities with each other to ensure the health and safety of all those affected by the construction work
- Communication   Whilst not an express duty, communication is vital to the operation of CDM and includes:
  - Pre-construction information
  - The construction phase plan
  - The health and safety file
  - Information and training
  - Consultation and directions
  - Display notices
- Competence   No duty holder may be appointed unless they are competent and all workers must either be competent or supervised by a competent person

Prevention

In addition, all duty holders under CDM must apply the principles of prevention during both design and construction. These are the principles laid down in the Management of Health and Safety at Work Regulations 1999 and included in the ACoP to CDM 2007, Appendix 7 (see also Section 12.5).

Welfare facilities

The standard of welfare facilities provided on site are given central importance in CDM 2007 and are clearly regarded as being vital for the health and well-being of people working on site. Contractors must price the required standards of sanitary, washing, drying, changing and resting facilities into their tenders and make sure that allowances have been made for the provision of drinking water and first aid.

In addition:

• The client must be satisfied that welfare arrangements comply with the regulations and that they are maintained throughout the project
• Contractors must ensure that minimum statutory standards of welfare are provided (CDM Schedule 2)
• For notifiable projects, the principal contractor must make sure that Schedule 2 is complied with

Pre-construction information

As far as the contractor is concerned, his ability to plan and manage health and safety on site is very much dependent on the quality of information provided to him. This is recognised in CDM 2007 which requires the client to provide information to both designers and contractors to help them to manage risk. Designers are required to add to this information by highlighting significant hazards implicit in their design so that contractors and others may be alerted to risks that they may not have been otherwise aware of. Pre-construction information should refer to:

• The site and any associated hazards or risk issues
• Existing services and hazardous substances
• Existing information such as surveys or a health and safety file
• Significant risks which may not be obvious to contractors and others

The information supplied should be provided early on in the procurement or tendering process and must be project specific. It may include notes on drawings or suggested construction sequences where this would be helpful to the contractor.

Pre-construction information is required to be presented to contractors and others in a convenient form by the CDM coordinator. This may be done in practice by means of a pre-construction information pack which should cover topics contained in the CDM 2007 ACoP, Appendix 2. Figure 12.11 summarises matters that should be included in the pack.

PRE-CONSTRUCTION INFORMATION

TOPIC	EXAMPLES
1. Description of project	• Project description • Programme and key dates • Details of client, designers, CDM coordinator, etc.
2. Client's considerations and management requirements	• Health and safety goals • Communications and liaison • Site planning and security, e.g. hoardings, permits, emergency procedures
3. Environmental restrictions and existing on-site risks	• Safety hazards, e.g. existing buildings or services, hazardous materials, unstable structures • Health hazards, e.g. asbestos, contaminated land
4. Significant design and construction hazards	• Design assumptions and erection sequences • Design coordination during construction • Design risks
5. The health and safety file	• Description and format • Requirements regarding content

CONSTRUCTION PHASE PLAN

Description of project	• Description of work and site details, programme and key dates • Contact details of duty holders
Management of the work	• Organisation structure and responsibilities • Arrangements for communication and consultation o Meetings, site induction, training • Arrangements for dealing with information flow, design changes • Selection and control of subcontractors • Production and approval of risk assessments and method statements • Reporting and investigation of accidents and near misses • Site rules • Fire and emergency procedures
Arrangements for controlling significant site risks	• Delivery, storage and removal of materials and waste • Plant management • Services • Falls, lifting operations, excavations, traffic management, etc. • Health risks including asbestos, hazardous substances, noise and vibration
The health and safety file	• Format and layout • Data collection • Storage

Figure 12.11

Construction phase plan

Since they first came into force in 1995, the CDM regulations have distinguished between 'the principal contractor' and other 'contractors'. The purpose of this distinction is to elevate one of the contractors on site – usually the main contractor – to the statutory role of principal contractor with overall responsibility for the management of health and safety on the project.

As part of this responsibility, regulation 23 of CDM 2007 requires the principal contractor to prepare a construction phase plan which is defined in regulation 2(1) as *a document recording the health and safety arrangements, site rules and any special measures for construction work.*

This plan must be prepared before the start of the construction phase of a project and it is the client's duty under regulation 16 to make sure that this happens. The purpose of the construction phase plan is to ensure that the construction phase is planned, managed and monitored effectively so that the work is carried out, so far as is reasonably practicable, without risk to health or safety. The plan is intended to be developed over the course of the project.

Regulation 23(2) stipulates that the plan must:

- Identify the risks arising from the construction work
- Include site-specific risks
- Include suitable and sufficient measures to address the risks
- Include site rules

Appendix 3 of the CDM 2007 ACoP sets out what should be contained in the construction phase plan and this is summarised in Figure 12.11.

Site rules

Site rules are part of the principal contractor's safety management system and form the basis for the organisation, planning and control of site activities. They lay down standards for safe working practices, fire and emergency procedures and behaviour on site. They also include issues such as restricted areas, permits-to-work and hot work.

Under CDM 2007, regulation 22(1)(d), it is the duty of the principal contractor to draw up a set of site rules which should be in writing and readily available to anyone working on the site.

Health and safety file

Regulation 20(2)(e) of CDM 2007 requires the CDM coordinator to prepare a health and safety file for projects that are notifiable. This is intended to contain information that might be useful to anyone who carries out construction work on the finished structure in the future or is involved with the cleaning, maintenance, alteration or demolition of the structure. The idea is to alert others to risks that only people involved in the original project know about. Residual hazards, design principles, special construction methods, location of services, the presence of hazardous materials, etc., should be included in the health and safety file.

The client's duties in relation to the health and safety file are stipulated in regulation 17. The CDM coordinator must agree on the structure and format of the file with the client and liaise with the principal contractor regarding the contents of the file as stipulated in regulation 20(1)(c)(i). The health and safety file is a legally required document that follows the building or structure for which it was intended, just like the deeds to a house.

12.12 Health and safety training

Training in the construction industry is the responsibility of a sector skills council – Construction Skills – which is a partnership between CITB-Construction Skills, the Construction Industry Council (CIC) and Construction Skills-Northern Ireland. The scheme is employer-led and operates under a levy and grant system. The objective of Construction Skills is to tackle the skills and productivity needs of the industry.

A large number of training courses are available to the industry, for both operatives and management, aimed at improving competence and addressing the huge skills shortage in construction. They include NVQs, card schemes and experienced worker assessments. Different card schemes exist for demolition workers, scaffolders and plant operators, etc.

Card schemes include:

- CSCS – Construction Skills Certification Scheme which includes a health and safety competence test for over 200 trades, general operatives, supervisors and managers
- CTA – Certificate of Training Achievement for plant operators now replaced by CPCS (Construction Plant Competence Scheme) for plant operators based on a test of professional competence and health and safety awareness

Training may take place on or off site. Site-based training includes site induction, toolbox talks and task talks.

Site induction

In common with any employee starting a new job, people arriving at a construction site for the first time should undergo a site induction. This applies equally to workers, site management, client representatives, subcontractors and visitors.

On notifiable projects (lasting more than 30 days), it is the principal contractor's statutory duty to ensure that every construction worker on site is provided with a suitable site induction under regulation 22(2) of CDM 2007. This is in addition to any information and training required by statute.

The objective of the site induction is to make sure that the person concerned is fully aware of:

- The company health and safety policy, organisation and arrangements
- The site safety management system and who to contact
- Site rules

- Personal responsibilities
- Accident and unsafe work practices reporting procedures
- Fire and emergency procedures
- Welfare facilities
- Site hazards
- Risk assessments and safe systems of work

Induction training will need to be repeated on a regular basis because there is considerable 'churn' of personnel on construction sites.

Toolbox talks

Toolbox talks are part of the contractor's safety management system and take place in the site cabin, usually between the supervisor or foreman and the operatives.

A toolbox talk affords an opportunity to raise awareness and promote a two-way discussion about day-to-day health and safety matters such as:

- General site health and safety
- Traffic management
- Roofwork and ladders
- Manual handling
- Noise and dust
- Abrasive wheels
- Cartridge-operated tools
- Buried services
- Asbestos

Task talks

Task talks are similar in principle to toolbox talks, but they deal with specific site operations rather than general matters.

Typical task talk topics would include proposed construction methods for:

- Deep excavations
- The use and handling of drag-boxes for deep drainage
- Installation of sheet-piled cofferdams
- Hoisting and fixing external precast concrete cladding panels
- Demolition work by hand
- Fixing roof sheeting from a leading edge

The task talk gives the operatives who will be doing the work the chance to listen to the contractor's method statement and make suggestions for improvements on the basis of their own previous experience.

12.13 Measuring performance

Monitoring

Health and safety performance on site should be regularly monitored in accordance with guidance in HSG65. This can either be done before things go wrong or after:

- Active (or pro-active) monitoring involves regular inspections and checks to make sure that the safety management system is working
- Reactive monitoring looks at what has gone wrong, why and how to prevent a recurrence of the event.

It is common practice to measure accidents by calculating the incidence rate or frequency rate either on a particular site or in the company as a whole. This can be compared with industry standards or 'best-in-class' for different sized contractors:

$$\text{Incidence rate} = \frac{\text{Total number of accidents}}{\text{Average number of persons employed}} \times 1000$$

$$\text{Frequency rate} = \frac{\text{Total number of accidents}}{\text{Total number of hours worked}} \times 100\,000$$

It is usual to have a rate for:

- Fatalities and major injuries
- Over 3-day injuries
- Reportable injuries

'X-days without an accident' boards are frequently seen in practice. This, however, is a measure of failure, and it is much better to measure the success of the organisation in making sure that its safety management systems are working properly before accidents happen. Active monitoring includes making direct observations of site conditions and peoples' behaviour (or unsafe acts), talking to people to elicit their views and attitudes and looking at documentation, reports and records.

Audit

Audit and review is part of the health and safety 'control loop' aimed at seeing how well the health and safety management system is working and where the defects are. HSG65 (HSE 1997) gives guidance on the subject.

Individual site managers and senior management in the company need to know how sites are performing with regard to management arrangements, risk control systems and workplace precautions. The audit process collects information about what is happening on the site and makes judgements about the adequacy and performance of the safety management system.

Audits are often carried out by managers or directors from other parts of the company or other regions or areas where the company works. The idea is to audit without any preconceived notions or prejudices.

Some companies have long lists of 'compliance' issues and give the site a score out of 100, whilst other companies merely investigate whether, say, 15/20 key aspects of the management system are working or not and score the site 'compliant' or 'non-compliant'. As part of the audit process, some contractors conduct a 'stop-shift' audit by stopping individual workers and asking them questions about health and safety issues.

12.14 Enforcement of legislation

For the vast majority of construction work, the HSE factory inspectorate is the enforcement authority and it is the factory inspector who the construction site manager will meet should his/her site be subject to an inspection.

The powers of an inspector include:

- Rights of entry at reasonable times, etc., without appointments
- Right to investigate, examine
- Right to dismantle equipment, take substances/equipment
- Right to see documents, take copies
- Right to assistance (from colleagues or police)
- Right to ask questions under caution
- Right to seize articles/substances in cases of imminent danger

If there is a serious accident or incident on site, this will normally be investigated by the HSE who will determine the cause of the accident, what needs to be done to prevent a recurrence, what lessons can be learnt and whether a prosecution is relevant. The prosecution may be brought by the HSE or the Crown Prosecution Service (CPS). Where there is a fatality, it is the police who will investigate, in conjunction with the HSE who will provide technical support. Any ensuing prosecution will be made by the CPS.

Death or injury is not a prerequisite for a prosecution and, where there is an incident that could have resulted in serious harm, but did not, a prosecution may follow. The collapse of tunnels under construction at Heathrow Airport in 1994 is a case in point where multi-million pound fines were handed out by the courts but no one was even scratched!

Enforcement action

The factory inspector has the right to issue legal notices which are written documents requiring a person to do or stop doing something which might be dangerous or in contravention of the law. Ignoring notices is a criminal offence.

Improvement notice

- this says what is wrong and why
- it says how to put it right
- it gives a set time to take action

Prohibition notice

This prohibits the carrying out of unsafe practices or the use of unsafe equipment with immediate effect. A prohibition notice may refer to a particular site operation, for example excavation work or scaffolding or it might require the immediate suspension of all work on site.

Warnings

Factory inspectors may also give written or verbal warnings or advice so that potentially unsafe practices may be averted without the need for a formal notice. They may also explain regulations and available guidance, etc., in confidence. This is quite important because formal notices frequently lead to prosecution and any responsible contractor will wish to avoid this at all costs.

Prosecution

Both employers and employees face the possibility of prosecution and this could result in a court appearance in either:

- The Magistrates' Court, where there is the possibility of a maximum fine of £5000 if found guilty of a summary offence
- The Crown Court, where the fines are unlimited and there is the possibility of a custodial sentence for certain offences (e.g. manslaughter, breach of asbestos licensing legislation or obstruction of an inspector in the performance of his duties)

Corporate Manslaughter and Corporate Homicide Act 2007

Following countless corporate manslaughter cases that have failed in the courts, including the Herald of Free Enterprise, Clapham Junction and Piper Alpha, this long-awaited legislation means that companies cannot escape their liabilities for management failures that lead to fatal accidents.

The Act introduces a criminal offence of corporate manslaughter in England, Wales and Northern Ireland (homicide in Scotland) where an organisation, through its senior managers, conducts its activities in such a way as to lead to a person's death. Senior managers who play a significant role in the making of decisions about how the whole company's activities are to be managed or organised will be subject to the existing health and safety legislation and to the common law offence of gross negligence manslaughter.

At the time of writing, sentencing guidelines for corporate manslaughter are yet to be decided, but fines of up to 10% of turnover could be imposed (e.g. with a company turnover of £200 million, the fine for a single fatality would be $\leq$ £20 million).

Naming and shaming

HSE prosecutions database

Health and safety offences are a 'black mark' against the otherwise good name of a contractor, not least because they remain on the contractor's record for 5 years, a bit like penalty points on your driving licence. However, the advent of the Internet has enabled the HSE to make this information public in the form of a register of convictions, which can be found 'on the web' at the official website of the HSE.

All prosecutions which have resulted in successful convictions are included except those cases which are either subject to appeal or applicable in Northern Ireland or the Republic of Ireland. The database is often referred to as the 'name and shame' list.

The HSE database can be searched either geographically or by industry and there is a special section for construction. The data can be searched either by individual cases or by the particular breach or breaches of the law and can be sorted in A–Z or Z–A order.

HSE enforcement notices database

The HSE also has a public register of notices, similar to that for prosecutions, which makes publicly available on the internet the details of improvement notices and prohibition notices issued. This information is kept 'on the record' for 5 years.

The register can be searched geographically or by industry and particular types of notice can be filtered if required. The notice reference number is hyperlinked which then leads to the details of the particular notice concerned, for example the type of notice, when served and upon whom and the specific breach of the law involved.

12.15 Accidents and incidents

The reason why it is so important to discover the hazards present in any project is so that the risks arising from the various hazards can be properly managed. If hazards are not dealt with, they may manifest themselves in such a way as to cause injury, ill health or damage to persons and property or they may result in a 'close shave' incident of some sort.

Reporting of Injuries, Diseases and Dangerous Occurrences Regulations 1995

Reporting accidents and ill health at work is a legal requirement under the RIDDOR. The legal duty is on employers, self-employed people and people who control premises and employers must report any work-related deaths, injuries, cases of disease, or near misses involving their employees irrespective of where they are working. This must be done by the quickest means possible either by calling the HSE Incident Contact Centre (ICC) or, alternatively, by submitting form F2508 either online, by email or by post.

The following must be reported:

- deaths and major injuries
- over 3-day injuries
- injuries to members of the public or people not at work who have to be taken to hospital
- certain work-related diseases
- dangerous occurrences – where something happens that might have resulted in an injury (i.e. 'near miss')

Reportable major injuries include fractures (but not fingers, thumbs or toes), amputation, dislocation of the shoulder, hip, knee or spine, loss of sight, etc., and must be reported without delay. Over 3-day injuries are those which are not 'major' but result in an employee or self-employed person being absent from work, or unable to work normally, for more than 3 consecutive days. They must be reported within 10 days of the incident occurring.

Detailed records of any reportable injury, disease or dangerous occurrence must be kept including when and how the incident was reported, the date, time and place of the event and details of those involved and a brief description of what happened. Records can be kept in a file, on computer or in the accident book.

Accident book

Not all accidents are reportable but every employer is required to keep an accident book (where there are more than ten people employed) under the Social Security (Claims and Payments) Regulations 1979. Details of every accident must be recorded, including how it happened.

The cost of accidents

Research indicates that the uninsured costs associated with accidents represent a factor of 11 times those costs which are covered by insurance. This is illustrated in Figure 12.12 which shows that the overall cost of an accident to the contractor where the injured party is awarded £10 000 damages in the civil courts. Bearing in mind that the insurers will pay out less than £10 000 (i.e. there will be an excess or uninsured loss of say £500), this is a loss to the contractor of 2.2% of turnover. The calculation does not take into account any fines that the contractor may have to pay for breach of health and safety law (which cannot be insured against) or the court costs (which can).

With typical contract margins of between 3 and 5% of turnover, half of which is normally regarded as risk, this simple example illustrates just how costly an accident might be to a contractor particularly as £10 000 is not a large insured loss by modern standards.

12.16 Human factors

Health and safety cannot be effectively managed without appreciating the impact of human behaviour and other related factors in the workplace (HSE 1999). However, whilst behavioural

Accident	Any unplanned event that results in injury or ill health of people or damage or loss to property, plant, materials or the environment or a loss of a business opportunity (HSE)	• fall from roof while working at height • contact with live power line • crushed by overturning vehicle or crane • trip or fall on uneven surface • nail through foot from waste timber
Disease	A work-related medical condition arising out of contact with harmful substances which may be reportable under RIDDOR 1995*	• dermatitis from repeated contact with wet cement • asbestosis or mesothelioma from asbestos fibres
Dangerous occurrence	Any 'near-miss' incident as defined under RIDDOR which could have led to serious injury or loss of life and must be reported to the enforcing authorities	• collapse of scaffolding • overturning of crane • unintended collapse of a building or falsework • fire from 'hot work' such as welding
Near miss	Any other incident that could have resulted in an accident but did not - non-reportable under RIDDOR but should be notified to employer or principal contractor for further investigation	• tower crane lifting a panel of wall formwork swings, narrowly missing overhead power cable • wet concrete spills from crane skip when filling column shutters at edge of 5-storey building – concrete delivery truck driver and some passers-by splattered with concrete but no one hurt • bricklayer pointing house chimney inadvertently drops trowel which bounces off scaffold and falls onto pavement just missing a passer-by

*RIDDOR 1995: Reporting of Injuries Diseases and Dangerous **Occurrences** Regulations 1995

Insured Costs	**Uninsured Costs**
• Employer's liability • Public liability • Fire/physical damage • Legal costs • Business interruption • Product liability	• Fines and legal costs • Sick pay • Medical costs • Investigation time • Loss of key staff • Loss of morale • Delay/disruption • Client dissatisfaction • Loss of reputation • Loss of future business

Example

A contract value of £5 million with an insured loss of £10 000 due to an accident and an excess of £500. The financial cost of the accident to the contractor is therefore:

$$\frac{\text{Insurance excess of £500} + \text{Uninsured losses of £10 000} \times \text{a factor of 11} \times 100}{\text{Contract value of £ 5 000 000}}$$

$$= 2.2\%$$

ACCIDENTS

Figure 12.12

scientists and psychologists help us to understand the human factors operating within complex organisations, it must be acknowledged that there are no perfect explanations as to why people behave the way they do in the work situation.

It is too simplistic, however, to accept 'the commonly held belief that incidents and accidents are the result of a "human error" by a worker in the "front line"... which is beyond the control

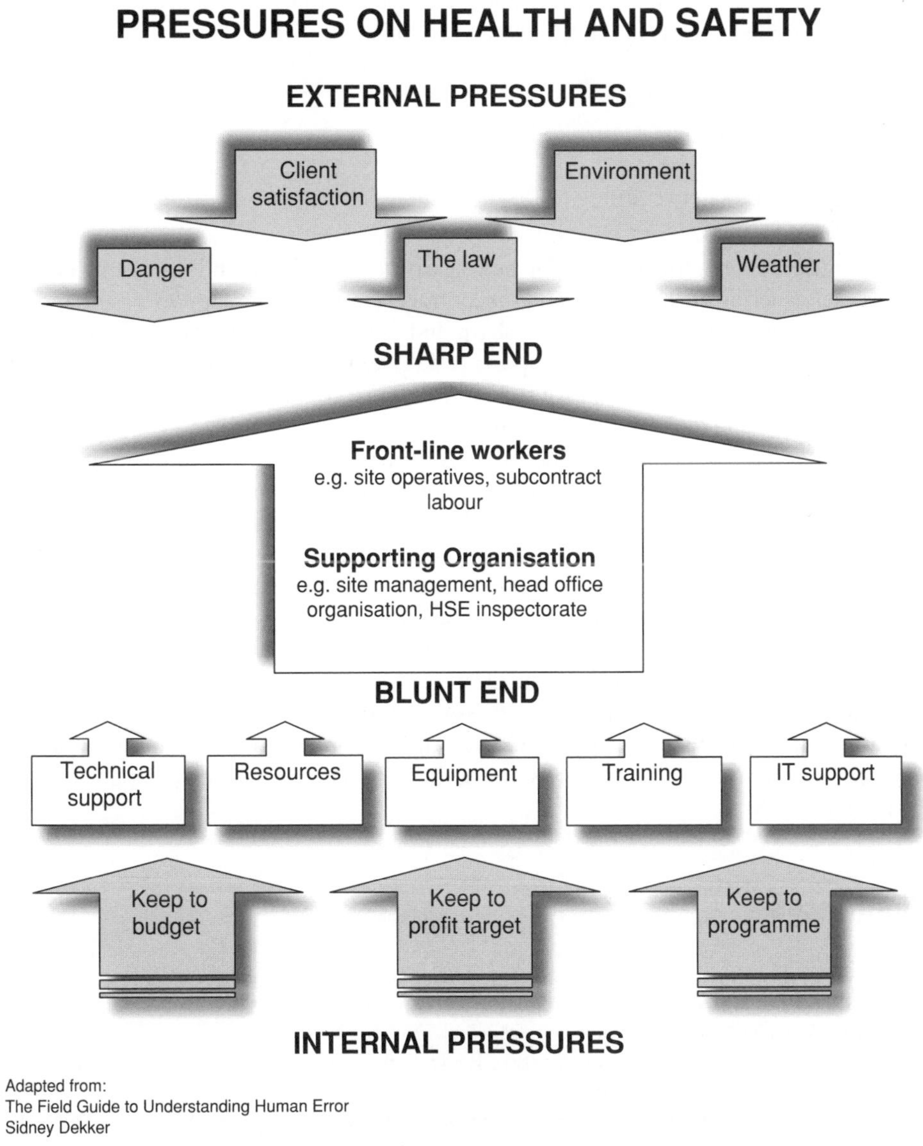

PRESSURES ON HEALTH AND SAFETY

EXTERNAL PRESSURES

Client satisfaction

Environment

Danger

The law

Weather

SHARP END

Front-line workers
e.g. site operatives, subcontract labour

Supporting Organisation
e.g. site management, head office organisation, HSE inspectorate

BLUNT END

Technical support

Resources

Equipment

Training

IT support

Keep to budget

Keep to profit target

Keep to programme

INTERNAL PRESSURES

Adapted from:
The Field Guide to Understanding Human Error
Sidney Dekker

Figure 12.13

of managers' (HSE 1999). What is more believable is that 'a human error problem is actually an organisational problem' (Dekker 2006).

Accidents happen to people at work due to their involvement in the work activity and the HSE (1999) estimates that up to 80% of accidents are, at least, partly attributable to the actions or inactions of people. This is not surprising considering the extent to which people 'at the workface' are exposed to danger. Human failures in health and safety are called 'errors' or 'violations' where 'errors' are unintended actions and 'violations' refer to deliberate actions, inactions or deviations from rules. However, blaming the individual for accidents ignores the less obvious causes of accidents which may lie deeper within the supporting organisation. These are identified in HSG48 (HSE 1999) and include:

- poor work planning
- lack of safety systems
- poor communications
- poor health and safety culture

Dekker (2006) views the organisation as a 'sharp end' and 'blunt end' consisting of 'front-line' workers, such as site operatives and subcontract labour, and the supporting site and head office organisation which underpins the 'sharp-end' activities. This view is depicted in Figure 12.13. In this model, the blunt end supports and drives the sharp-end activities and it is the blunt end that provides the necessary technical support, resources and training, etc., which enables the sharp end to deliver the end product.

However, despite the support that the internal organisation offers, the blunt end also imposes pressures on the sharp-end activities such as keeping on budget and programme and making sure that the desired profit target is achieved. When these 'internal pressures' are added to the 'external pressures' imposed by legislation, client expectations and the dangers presented by the working environment, etc., it is little wonder that the behaviour of those engaged in physical production sometimes manifests itself as 'errors'.

HSE (1999) explains that 'human factors refer to environmental, organisational and job factors, and human and individual characteristics, which influence behaviour at work in a way which can affect health and safety'. It also suggests that human factors can be viewed simply by considering three issues: the job, the individual and the organisation. Attitudes, risk perception, culture and leadership are some of the factors that impact on people's health and safety-related behaviour.

References

Dekker (2006) *The Field Guide to Understanding Human Error*. Ashgate Publishing.

HSE (1997) *Successful Health and Safety Management*, HSG65. HSE Books.

HSE (1999) *Reducing Error and Influencing Behaviour*. HSE Books.

HSE (2000) *Management of Health and Safety at Work, Approved Code of Practice*. HSE Books.

HSE (2006) *Five Steps to Risk Assessment*, INDG163 (revised). HSE Books.

HSE (2007) *Managing Health and Safety in Construction, CDM 2007, Approved Code of Practice*. HSE Books.

Hughes, P. & Ferrett, E. (2005) *Introduction to Health and Safety at Work*. Elsevier.

Joyce, R. (2007) *CDM Regulations 2007 Explained.* Thomas Telford.
St. John Holt, A. (2001) *Principles of Construction Safety.* Blackwell Science.
Wallace, I.G. (2000) *Developing Effective Safety Systems.* Institution of Chemical Engineers.

Web references

http://www.hse.gov.uk/riddor/
http://www.hse.gov.uk/enforcement/prosecutions database/
http://www.hse.gov.uk/statistics/industry/construction.htm

13 Planning the project

13.1 Introduction

The tendering period is a busy time for the contractor and getting the price finalised and submitted on time is a high-pressure process. Once the tender has been delivered, there is a short period of anticlimax for the estimator in particular and the process starts again with the next enquiry.

The client team, on the other hand, is very much focused on the project in hand, and they will be anxious to compare the tenders received with the budget and cost plan. If the tenders are too high, changes will have to be made to the design or specification in order to make savings. When the preferred tender has been checked for mistakes or qualifications, the successful tenderer will be notified and contract documents prepared.

The time between contract award and taking possession of the site is frequently very short and clients often apply pressure to start 'yesterday'. Clients' advisors and contractors alike find this pressure hard to resist and often all concerned bend over backwards to get things moving.

This chapter concerns the contractor's processes in planning for the construction stage and in making sure that the contract proceeds in a timely and efficient fashion.

13.2 Planning procedures within a large organisation

The reasons for undertaking planning have been outlined in Chapter 8. During a major construction project, the contractor will need to implement appropriate procedures in order to keep the master programme under constant review. Figure 13.1 shows the relationships between the master programme, stage programmes and the 4–6 weekly and 1–2 weekly short-term programmes at site level.

The organisation structure of the planning department within a large contracting firm is shown in Figure 13.2. This shows the relationship between the pre-tender planning and the contract planning functions. The contract planning staff may be site-based on the larger projects and on smaller projects may be serviced by planning staff from an adjacent major site.

The contractor's planning procedures for the construction phase are discussed under the following headings:

- Pre-contract planning
- Contract planning

RELATIONSHIP BETWEEN PLANNING UNDERTAKEN DURING THE CONSTRUCTION STAGE

Personnel involved:

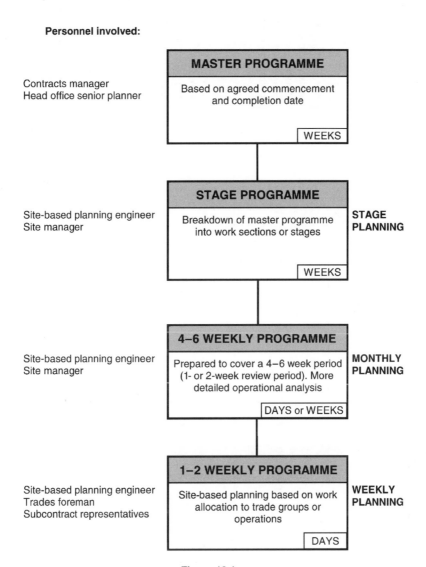

Contracts manager
Head office senior planner

MASTER PROGRAMME

Based on agreed commencement
and completion date

WEEKS

Site-based planning engineer
Site manager

STAGE PROGRAMME

Breakdown of master programme
into work sections or stages

WEEKS

**STAGE
PLANNING**

Site-based planning engineer
Site manager

4–6 WEEKLY PROGRAMME

Prepared to cover a 4–6 week period
(1- or 2-week review period). More
detailed operational analysis

DAYS or WEEKS

**MONTHLY
PLANNING**

Site-based planning engineer
Trades foreman
Subcontract representatives

1–2 WEEKLY PROGRAMME

Site-based planning based on work
allocation to trade groups or
operations

DAYS

**WEEKLY
PLANNING**

Figure 13.1

13.3 Pre-contract planning

On award of the contract, the pre-contract planning process commences. The contractor may have up to 6 weeks in order to plan the commencement of works on site, or he may simply have 6 days. The commencement date will normally have to be negotiated with the client's

ORGANISATION OF THE PLANNING FUNCTION WITHIN A LARGE CONTRACTING ORGANISATION

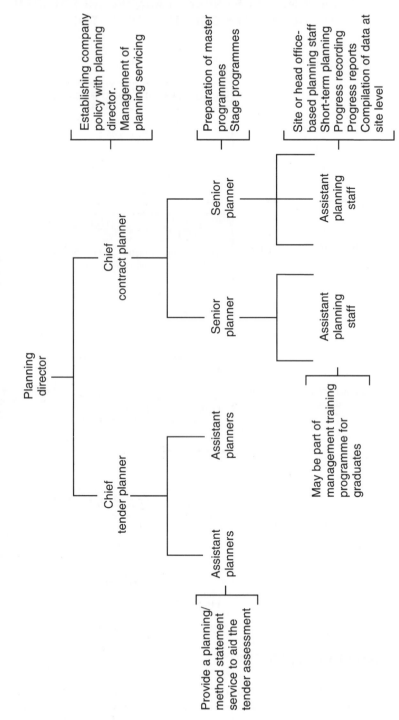

Figure 13.2

representative or may be specified in the tender documents. On a major contract of, say, £3 000 000 in value, the following stages will be involved in the pre-contract planning:

- The pre-contract meeting and arrangements for commencing work
- Placing orders for subcontractors and suppliers
- Site layout planning
- Construction method statement
- Master programme preparation
- Preparation of requirement schedules
- Preparation of contract budgets
- Preparation and approval of the construction health and safety plan

It must be pointed out that no two companies undertake exactly the same procedures at the pre-contract stage. Procedures depend on the policy of the company and the need for establishing standard routines which may be followed on every new contract. The procedures outlined below could be considered normal in a larger company or in a more organised medium-sized organisation. Within the larger organisation, a more formal approach will be taken to the arrangements for commencing work.

13.4 Pre-contract meetings

Prior to the commencement of the project, a series of pre-contract meetings will be held. For instance, a pre-contract meeting will be called by the client's representative in order to coordinate the start of works on site. It is important to establish lines of communication between the client's team and the contractor that are clear and transparent so as to avoid confusion and disagreement. Apart from obvious contact details, the contractor needs to know, for instance, who has authority under the contract to issue instructions, to certify payment and to agree on the value of variations.

The pre-start meeting is set up to organise a multiplicity of matters including:

- Organisational and contractual details
- Dates of progress and valuation meetings
- Safety management arrangements
- The commencement of work on site

The contractor will also hold his own internal pre-start meetings in order to hand over the tender documentation and make arrangements for commencing work on site. It is also an opportunity to delegate responsibility to those entrusted with the duty of organising and managing the project.

The pre-contract arrangements outlined here relate to those undertaken by the contractor on a traditional contract and will require consideration of the following:

- The pre-contract planning meeting
- The registration of drawings and distribution of information

- Preparation of the construction-phase health and safety plan (in compliance with the CDM regulations)
- Arrangements for commencing work

The contractor's pre-contract meeting

A meeting will be held to announce the award of the contract and allocate responsibility to the construction team responsible for undertaking the work. All documentation produced at the tender stage will now become available to the construction section. The following personnel will normally attend the meeting which may be chaired by the chief estimator or the contracts manager.

Chief estimator

The chief estimator will probably act as meeting chairman in the first instance as he is fully aware of all decisions taken at the tender stage. He will be responsible for handing over all the estimating data to the contracts section including:

- Estimate summary and analysis
- Build-up of the all-in rates and net bill rates
- Summary of subcontractors' and suppliers' quotations
- Pre-tender method statements
- Preliminaries build-up
- Pre-tender safety assessment
- Tender adjudication report

Contracts manager

The contracts manager will be responsible for organising the commencement of work. At this stage, it is not always possible for the construction manager to be appointed or available as he may be tied up on another contract. On the larger multi-million pound projects, it would prove advantageous, however, to have him allocated to the project team at the pre-contract stage.

The contracts manager is therefore responsible for all pre-contract activities within the medium- and smaller-sized organisation. He would be responsible for finalising the operational method statement and assisting with preparing the master programme.

Since 1995, it has been a statutory requirement to prepare a health and safety plan for the construction phase of most projects and the client must be satisfied that this plan complies with the Construction (Design and Management) Regulations prior to allowing the commencement of work on site. This is called the 'construction-phase plan' under CDM 2007 (HSE 2007).

The contracts manager's responsibilities will also include making arrangements for the transfer of key staff to the project or the recruitment of additional staff to manage the project. Arrangements will have to be made for the delivery of site accommodation and the mobilisation of plant and equipment for the initial site operations.

Company buyer

The buyer, or quantity surveyor in the medium-sized organisation, is responsible for placing orders for subcontractors and suppliers on the basis of the information received at the tender stage. It may be the policy of the company to ask subcontractors and suppliers to requote for the work now that the contract has been secured. Smith (1995) makes an interesting reference to the practice of 'bid peddling' once the contractor has been awarded a contract. This practice is often frowned upon by the subcontractor fraternity and rightly so.

In many of the medium-sized construction firms, the contract quantity surveyor under-takes the responsibility for placing subcontract orders and preparing contracts. This practice enables the surveyor to become familiar with the project from the outset. The placing of sub-contract orders will, however, continue throughout the project period. The scheduling of key subcontract dates will need tying in with the master programme and relating to the project requirement schedules.

Chief quantity surveyor

The chief quantity surveyor or managing surveyor is responsible for allocating surveying personnel to the project and arranging for the checking and signing of the contract. The surveyor will also be responsible for preparing the contract cumulative value forecast on the basis of an analysis of the contract bills and master programme (see Chapter 14).

Office manager

It will be necessary to establish communication channels for the distribution of project infor-mation as it is received from the architect, and this will involve establishing procedures for the circulation of drawings and correspondence. This is normally the job of the office man-ager, who will also be responsible for allocating office staff to deal with wages and material invoicing.

He will also be responsible for:

- sending a copy of the F10 notice of appointment of the principal contractor to the site manager so that it can be displayed on site for workers to read,
- issuing any statutory building notices to the local authority and
- making applications for hoarding licences, footpath crossings and the provision of tempo-rary service connections.

The necessary insurance requirements for contract specific cover will also need organising.

The client's pre-contract meeting

The client's pre-contract start-up meeting is usually chaired by the architect or engineer, and the design team, client's representative and the contractor will be present in order to establish initial contact between the parties. This enables channels of communication to be set up for the issue and distribution of project information.

At the pre-start meeting, the contractor may be asked to present his outline programme. This may indicate the requirements for key subcontract dates in order that realistic information requirements may be assessed. Outstanding matters in relation to the contract commencement dates may be discussed, together with arrangements for the signing of the contract if this has not been done.

A typical agenda for such a meeting might contain some or all of the following items:

(1) Introductions
(2) Apologies for absence
(3) Employer's organisation and delegated powers
(4) Contractor's organisation
(5) Tax exemption matters (New CIS)
(6) Insurances
(7) Notices
(8) Commencement date
(9) Safety management arrangements
(10) Programme and method
(11) Site boundaries and access
(12) Setting out arrangements
(13) Working hours
(14) Contractor's tip and cleaning roads
(15) Communications and correspondence
(16) Progress meetings
(17) Valuations and payment
(18) Emergency procedures
(19) Any other business
(20) Date of next meeting

13.5 Placing orders for subcontractors and suppliers

As indicated above, it will be necessary to consider company policy with regard to procedures for placing orders for subcontractors and suppliers. The practice of offering work on the basis of a Dutch auction should not be encouraged. Chapters 6 and 18 deal with subcontractor selection and control criteria.

Within larger organisations, there will be standard procedures to award subcontracts. This may be on the basis of 'in-house' subcontract conditions or using standard forms such as the JCT 05 Standard Building Sub-Contract. Some main contractors engage subcontractors on letters of intent, with a lump sum price or schedule of rates, presumably to avoid the provisions of the Construction Act.

Within small- and medium-sized organisations, subcontracts may simply be awarded on the basis of a letter of appointment with no formal written contract entered into. Alternatively, some contractors have their own bespoke conditions which subcontractors should read very carefully.

13.6 Site layout planning

Site layout planning is an essential part of pre-contract planning. The contractor is often required to submit his proposals for approval by the client's representative prior to commencing work on the project. It is important to consider the allocation of preliminaries facilities allowed in the original estimate as this establishes the basis of the preliminaries budget.

Overspending on the preliminaries is a common problem on contracts and one must learn to work within the monies allocated at the tender stage – no matter how inadequate.

Where a keen price has been put in for the work at the tender adjudication stage, there will no doubt have been some reductions in the site overhead allowances.

Site layout plan checklist

Location of offices and site accommodation

- On open sites, the site accommodation should be located close to the entrance in order that vehicles and personnel may be readily observed entering the site. Under the CDM regulations, the principal contractor is responsible for preventing unauthorised access to the site. Site notices instructing personnel entering the site to report to the office should be displayed. Site compound areas should be adequately stoned up or surfaced and consideration should be given to the site parking of vehicles for staff and operatives. The site office area may be fenced and provided with secure gates.
- The space available on site may be at a premium and on restricted sites, consideration may be given to stacking site accommodation on top of each other or locating the offices on a gantry over the pavement. Alternatively, the contractor may be forced to rent some space adjacent to the site. The contractor may also think about locating his offices in the building being refurbished, if this ties in with the sequence of work or moving the accommodation into completed parts of the building as it is being constructed. On restricted sites, problems also relate to the storage of materials and the location of plant and equipment.
- The location of offices and accommodation will also be influenced by the location of access roads and site services. Careful attention is required with regard to site security and arrangements may need to be made to have the site well illuminated at night. Powerful security lighting may be used for this purpose in order to deter theft and break-ins. On a site in a rough district of Manchester, this was recommended by the police, but during the first week of the project the security lighting was stolen! Consideration will have to be given to the siting and location of the site signboards and any sample panel areas (e.g. facing brickwork).

Location of site services

- Unless the contract period is short, the site will need to be serviced with permanent water, power and sewerage facilities and existing services will need to be located. Water and drainage connections will need to be made and the site will have to be hooked up to electricity and telephone services. Services crossing the site may have to be diverted.

- The statutory authorities will need to be contacted in order to establish existing locations and entry points for new services. The contractor will be required to establish the proposed location of temporary standpipes for site mortar mixing or facilities for the washing down of site vehicles. On open, greenfield sites, permanent services may have to be laid early in the project to provide temporary service connections.
- Power supplies to major items of plant such as tower cranes will need consideration, together with the provision of electrical service connections for site power tools and site temporary lighting.

Temporary roads, hardstandings and access

- Reference will again have to be made to monies allocated in the contract preliminaries for these items. Temporary hardstandings may be required for the location of cranes and piling rigs or for the unloading of materials such as ready-mixed concrete.
- It may be necessary to provide a temporary access road around part of the building for mobile lifting or access equipment during the cladding stage of the works.
- It may prove opportune to lay hardcore to car park areas early in the contract in order to use them as temporary hardstandings for the works.
- Foundations for tower crane bases or tracks will need preparing early in the contract prior to the erection of the cranes.

Location of plant and equipment

- The contractor will have to consider the storage and security of plant during evenings and weekends. A locked compound may be provided. The location and working radius of cranes will need to be marked on the site layout plan in order to establish the best location for delivery points.
- Consideration will have to be given to the rights of adjoining owners with regards to the swinging of tower crane jibs and booms over land adjacent to the site. The contractor has no rights to enter the air space of an adjacent site without permission from the owner, and may be sued in trespass if he does so. Likewise the contractor has a duty of care to persons using footpaths, streets and highways adjacent to the works. The main contractor is responsible for the deeds of his subcontractors when working in the vicinity of an adjacent highway. The contractor may be sued for negligence in the case of an accident involving the public.
- Consideration will also have to be given to site space requirements around the building for the provision of scaffolding.

Material storage areas on site

- One of the major considerations is the safe and secure storage of materials and components, in order to reduce waste and ensure safe working procedures.
- Materials are an expensive commodity and care must be taken in their handling, storage, protection and placing in position. Considerations for material storage (lay down) areas should be shown on the site layout plan.

- Bricks and blocks should be stored on a firm clean surface or on pallets suitable for handling with a fork lift truck. Blocks should not be stacked too high. Bricks should be stored in locations which do not become waterlogged and where the brick stockpiles will not become contaminated by mud.
- Structural timber should be ordered in cut or stock lengths to avoid cutting waste and may be banded for ease of handling. Timber must be stored clear of the ground and protected with polythene or tarpaulin sheets to avoid changes in moisture content.
- Finishing timbers such as architraves, skirtings and door linings should be ordered in door sets. Skirtings and timber mouldings must be stored internally on horizontal racking. On housing projects and the like, garages should be constructed early in the project where possible and used for storage purposes.
- Roof trusses may be stored on purpose-built, timber or steel storage racks, which support the trusses clear of the ground. It is preferable to deliver the trusses to site in phased deliveries so that they may be lifted directly into position.
- Scaffolding and props may be stored on horizontal racking with the fittings stored in drums or bins. These are expensive components to lose, as the contractor will be charged for losses by the hire firm.
- Drainage goods are fragile and easily damaged if mishandled. Drainage fittings may be delivered to site in a crate. Pipes may be banded and stored on timber separators. Materials may be stored adjacent to the works in a small compound.

13.7 The master programme

The contract master programme is an important management tool. It is an essential requirement in the coordination and control of the many integrated tasks to be undertaken during the project and is also used by the client's contract administrator to monitor the contractor's progress. On a major project, many different programmes are required to cover each stage of the work in order to ensure the smooth flow of information during the project. This includes consideration of:

- A contract master programme covering the major phases or sections of work and clearly indicating the planned sequence of construction
- A programme indicating the key dates for the release of design team information in order to meet the requirements of the contract master programme
- A programme to coordinate the requirements of subcontractors, material supplies and the resources of the main contractor
- Separate detailed programmes relating to the various project phases or stages highlighting the links between each work stage

The master programme forms the basis of the contractor's budgetary control and financial forecasting procedures and aids the client in assessing his cash-funding requirements at the monthly payment stages. It is also important in relation to the contract.

Despite its importance, the master programme is not usually part of the contract documents. Under the JCT 05 Standard Building Contract, for instance, these normally consist of:

- the articles of agreement
- the contract drawings
- the bills of quantities (or the specification in the without quantities version)

Including the master programme or a method statement in the contract documents is fraught with problems. This was discovered in the case of *Yorkshire Water Authority* v. *Sir Alfred McAlpine & Son (Northern) Ltd.* (1985) in which the contractor was relieved of his obligations to carry out the works because they were physically impossible. This is an entirely different situation, however, to a contractual obligation to provide copies of the master programme.

For instance, under JCT 05 Standard Building Contract Clause 2.9.1.2, 'the contractor shall, without charge, provide the architect/contract administrator with two copies of the master programme for the execution of the works'. The contractor is also obliged, 'within 14 days of any decision by the architect/contract administrator in relation to an extension of time', to provide two copies of any amended programme. However, this clause applies only where the contractor has actually prepared a programme, which, unlike ICE Clause 14, he is not obliged to do.

If the master programme was included as a contract document, it would somewhat impair its flexibility and usefulness as a management tool. It would involve both parties having to strictly adhere to the programme and the contractor would be obliged to start and finish each operation by the programmed dates, and in the programmed sequence, or risk being in breach of contract.

Consideration should also be given as to whether or not the master programme constitutes effective notice of the contractor's information requirements to the architect. In the case of *London Borough of Merton* v. *Stanley Hugh Leach* (1985), for instance, the judge held that the programme, if it gave sufficient detail, would constitute effective notice of the contractor's requirement for further information. It follows that a post-contract programme is a unilateral declaration of notice by the contractor of when he requires information, provided that the notice is not unreasonably premature.

Notwithstanding contractual obligations, it is good practice for the contractor and design team to agree amicably on a realistic contract programme prior to the commencement of the project. Copies of the master programme should be circulated to the architect for courtesy approval and also to the major subcontractors to indicate their approximate commencement dates and periods on site. This is particularly important when nominated subcontractors are involved and will avoid recourse to legal action which can only lead to strained relationships, which everybody can do without.

The master programme may be presented in bar chart, network or precedence format depending on the programming techniques adopted by the contractor. Contractors tend to use the programming technique which best suits their mode of operations and which works for them.

The programme should show the contractual possession and completion dates for the project, together with the main work activities on site. The master programme indicates the sequence of operations, and relationships may be shown between related operations by introducing links in the form of a linked bar chart.

The programme is used to record progress weekly and monthly throughout the contract period in order to achieve the planned completion date.

Figure 13.3 illustrates an extract from a master programme showing three stages of work on a steel-framed building project:

- Site establishment = 7 weeks
- Substructure = 9 weeks
- Superstructure = 14 weeks

A single bar line distinguishes the overall duration of each stage. The programme also shows the subtasks within each main or summary activity. Using project management software, the subtasks can be 'rolled-up' to produce a simpler overview of the master programme. Finish-to-start and start-to-start links have been shown to link the various stages of work.

13.8 The target programme

During the pre-contract period, it is common practice for contractors to develop a target programme as well as a master programme. The target programme effectively compresses the master programme into a shorter time period, thereby saving on time-related costs and releasing resources sooner for the next contract. For instance, the master programme may show completion in 22 weeks but the target programme will indicate completion in 17 weeks, thereby saving 5 weeks (see Figure 13.4).

In order to achieve the target programme, it must be realistic and not just hopeful. It therefore needs to be based on the informed opinion of the construction team, and especially the contracts manager or site agent, that the project can realistically be completed faster than the time shown on the master programme.

The implications are far reaching in terms of the contractor's own supervision and resources, the procurement and management of subcontractors and the ability or willingness of the design team to provide the necessary drawings and other information in time.

Following the judgment in the case of *Glenlion Construction* v. *The Guinness Trust* (1987), a programme showing a shorter completion period than that required under the contract, even though it may be annotated with information deadlines, will not be successful in putting pressure on the architect or engineer to produce drawings, etc., to a tight timescale.

Any claim by the contractor based on the failure of the contract administrator to provide information, and thereby causing delay, will not be valid unless the contract is delayed beyond the *contractual* completion date. The architect or engineer is obliged only to furnish drawings and instructions 'within a reasonable time of the conclusion of the contract', and Keating (Furst and Ramsey 2000) takes the view that the employer is not under an implied obligation 'to enable the contractor to complete by the earlier date'. Therefore, 'provided that the contractor can still complete within the contract period, he cannot recover prolongation expenses' and 'the employer is under no obligation to pay compensation if the contractor is unable to achieve an accelerated programme'.

Target programmes place added pressure on the site management team and subcontractors alike and a careful balance needs to be struck between the demands of faster working and the health, safety and quality standards expected.

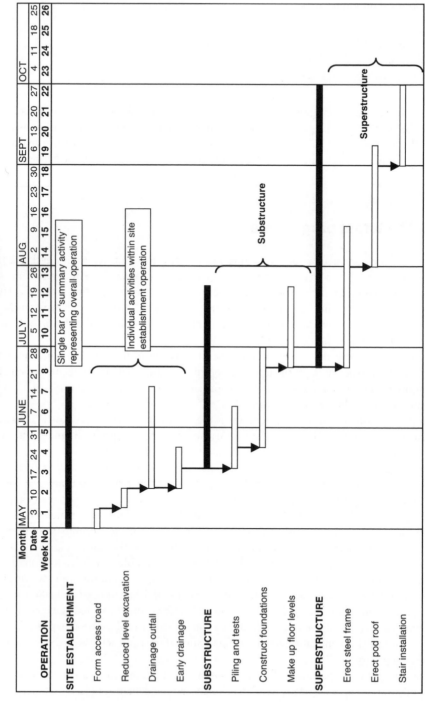

Figure 13.3

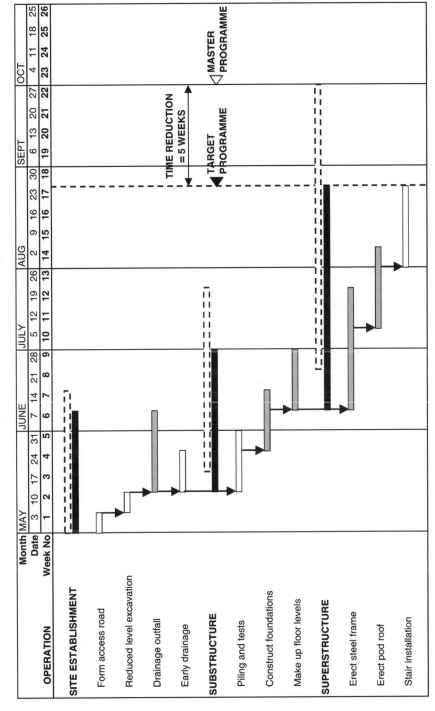

Figure 13.4

13.9 Subcontractor programmes

The master or target programme frequently shows subcontract packages as a single bar line simply because it is impossible to show every site activity in detail – the programme would be too complex and unwieldy. This principle is demonstrated in Figure 13.4 which shows, for example, 'piling and tests' as a single bar on a target programme with a duration of 3 weeks. However, each subcontract package is effectively a subproject with its own list of activities and events and the contractor needs to exercise control over them.

Contractors, therefore, commonly prepare subcontractor programmes for this purpose and an example is given in Figure 13.5. In this example, the piling and tests package is shown in detail with several activities and events happening within the overall duration of 3 weeks indicated on the target programme.

13.10 Procurement programmes

Figure 13.6 illustrates a procurement programme where the bar lines represent the procurement periods for each major component supplier or subcontractor. Each bar line has been flagged with three milestone symbols relating respectively to information requirements, placing the order and the commencement of the operation on site.

The following examples illustrate how the procurement programme may be developed into specific programmes for the procurement of the 'steelwork' and 'finishes' packages.

Steelwork

Figure 13.7 shows a detailed bar chart display indicating the procurement requirements for the steelwork activity using a simple linked bar chart and early warning symbols. In practice, the steelwork activity would be shown as a single bar on the main procurement bar chart. However, by using a project management software package, the detailed programme may be accessed simply by clicking on the bar line or, on some packages, by clicking on an icon to expand the activity and show the detail.

Finishes

Figure 13.8 shows a procurement bar chart for the finishes operation depicting lead-in times incorporated within the bar lines. Lead-in times are for the benefit of both the contractor and those providing the information. This might be the project architect or other designer or the design and build designer engaged by the contractor.

Each of the above examples clearly indicates 'must have' or 'must do' information reminders. It is no use producing key requirement dates if nobody keeps to them, but for this to happen dates must be realistic and not simply established to strengthen contractual claims. Management action must be taken when key dates indicated by the milestone symbols are not met or if the data is not available on the specified date. Delays resulting from the late release of

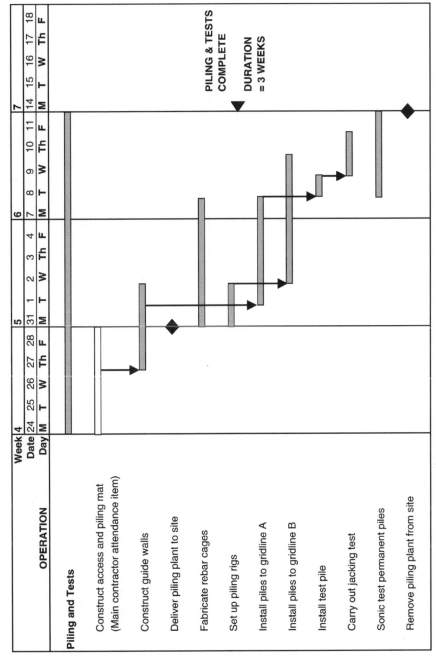

SUBCONTRACTOR PROGRAMME – PILING AND TESTS

Figure 13.5

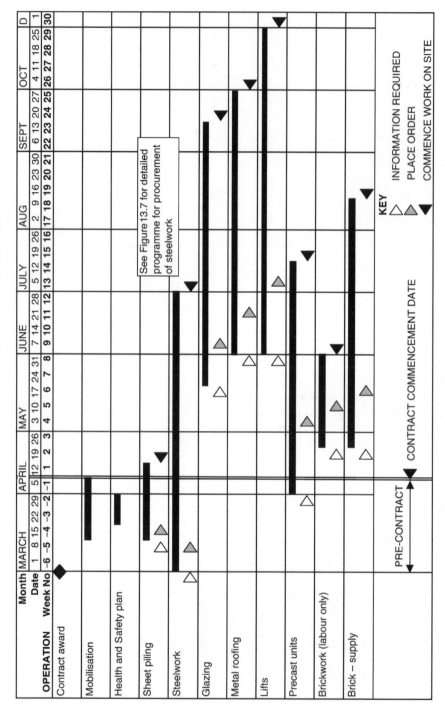

Figure 13.6

PROCUREMENT PROGRAMME – STEELWORK

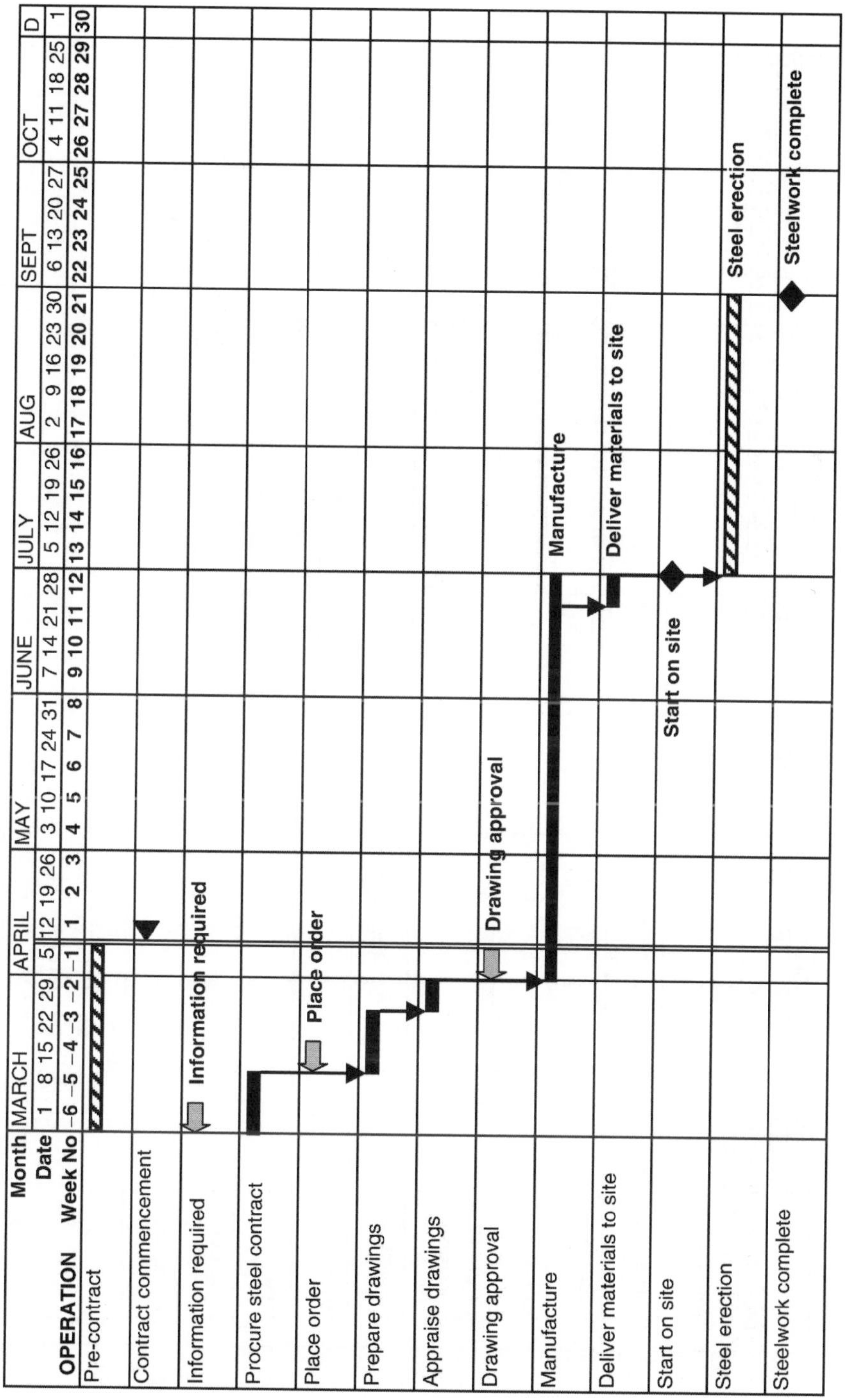

Figure 13.7

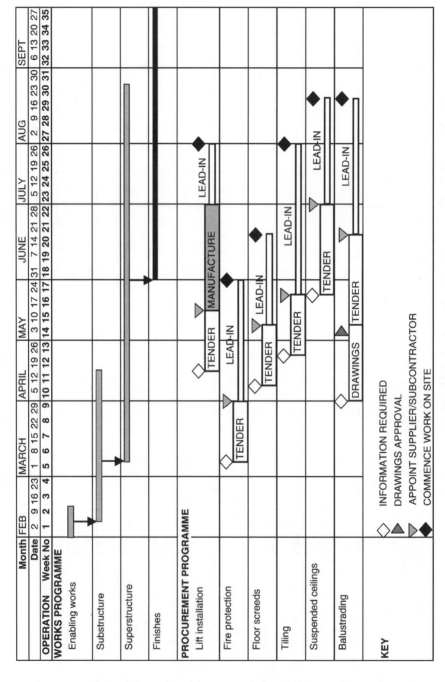

Figure 13.8

information from whatever source must be confirmed in writing at the appropriate time in order to have contractual effect.

13.11 Requirement schedules

A long-standing bone of contention between contractors and contract administrators is the thorny issue of the provision of information and this has been the cause of many claims and disputes. The problem is especially prominent with conventional contracts where the design is incomplete at the tender stage.

The purpose of requirement schedules is to aid the smooth running of the contract by providing a programme for the preparation and release of information by the architect to the contractor throughout the project.

The importance of this is emphasised in the JCT 05 Standard and Intermediate Building Contracts which make specific provision for an information release schedule. In JCT 05 Standard Building Contract, the articles of agreement provide for an option (Recital 5) which binds the employer (client) to provide such a schedule saying what information the architect will release and when. This is also referred to in Clause 2.11. Where the contract does not provide for a schedule; however, the onus is on the contractor to give notice to the architect sufficiently in advance of the date when the information is required (Clause 2.12.3). Furthermore, Clause 2.12.2 also allows the architect to use his own judgement as to when information will be released, but no contractor worth his salt would be happy with this arrangement.

In any event, there is an important linkage between information release and contractual claims whether or not there is an information release schedule. Under the JCT 05 Standard Building Contract, for instance, failure of the architect to provide timely information may be a default, and therefore a relevant event under Clause 2.29.6, thereby justifying adjustment of the completion date. Similarly, such default may give rise to a claim (under Clause 4.24.5) for loss and expense.

Figure 13.9 illustrates an information requirements schedule requesting information in respect of work packages. The responsibility for providing the information on behalf of the client's design team is indicated on the schedule.

However, it is pointless asking for information relating to painting schedules at week 1 on a 70-week project and expecting to formulate a claim at week 20 simply because the information has not been received.

Many contractors may consider that 'the claims start here' as far as requirement schedules are concerned, but this should not be the reason behind requesting information requirements at the pre-contract stage of the project. Surveyors, however, remain mesmerised by information requirement schedules!

Certain schedules relate to the contractor's internal requirements, while others concern key dates for the release of information from the client or client's representative. The following examples illustrate the types of requirement schedules in use.

Contractor's internal schedules

- Key materials schedule
- Plant schedule
- Domestic subcontractor requirements

INFORMATION REQUIREMENTS SCHEDULE

DETAILS REQUIRED	Contract Newcastle							Contract No			
	Design team								Workpackage requisition		
Work element/package	Arch	Str. eng	Serv. eng	Date required	Lead-in period	Start on site			Requisition required by	Last date for order	Start on site
Demolitions	✓	✓		15 Dec	3 weeks	19 Jan					19 Jan
Foundations and lift pit	✓	✓	✓	5 Jan	4 weeks	2 Feb					2 Feb
Precast units	✓	✓	✓								
Raised access floor	✓		✓								
Suspended ceilings	✓		✓								

Figure 13.9

Client-based information requirements

- Subcontract schedule (nominated subcontractors and suppliers)
- Drawings and information release schedule
- Request for information sheets

Key materials schedule

A materials schedule should be prepared in conjunction with the contractor's planning or buying section showing the key materials scheduled to meet key operational programme dates. The schedule should show ordering information referenced to the bills of quantities, delivery required by dates and supplier contact details. The material requirements are assessed from the contract drawings, with the quantities in the bills being used as a check.

Phased deliveries may be arranged for materials such as bricks and precast floor units. On sites with restricted space, or inner city refurbishment projects, just-in-time deliveries to suit the contract programme may be preferred.

It is important that the material schedule contains full information with regard to what is expected of the supplier. Additionally, site management personnel must be encouraged to use and update the schedules as it is important that material suppliers are kept informed of the progress position with regard to any amendments to planned delivery dates.

Plant schedule

The plant schedule indicates key dates for the major items of plant required on the contract. The schedule may be presented in tabular format or as a bar chart programme. Alternatively, plant requirements for each construction activity may be indicated on the master programme and most linked bar chart software packages allow notes to be added for such purposes. The plant schedule enables the plant department to plan its resources between the various projects operating at any one time and order hired-in plant in good time (e.g. large mobiles or tower cranes are not available 'off-the-shelf'). It also enables the right balance to be struck between using contractor-owned plant and hired-in plant from specialist firms.

A careful check must be kept on plant expenditure compared to the allowances for plant within the contract preliminaries. Overspending on plant is common at site level mainly because site managers can always find an excuse to keep plant on-hire for a bit longer than intended.

Subcontract schedule

The coordination of subcontractors and suppliers is an essential part of project control. The schedule illustrated in Figure 13.9 enables the site manager to overview key contract dates with respect to progress, to chase up the architect for information outstanding and to make sure that subcontractors are 'on message' as regards their start date. Details are also shown on the schedule relating to the subcontract order, contact address and notification dates. Reference should be made to the subcontractor's programme prepared at the contract stage. The late release of key information relative to subcontractors and suppliers may lead to an extension of time claim when the delay is caused by the architect.

Drawings and information requirement schedule

It is essential to monitor and record the receipt of contract drawings and other information issued by the architect. It is the contractor's responsibility to give adequate notice of information requirements such as:

- Setting out dimensions and measurements to site boundaries
- Reinforcement details for pile caps, foundation beams, etc.
- Details of ground floor services and pockets or fixing bolts
- Fixing details for cladding panels
- Door, window and ironmongery schedules
- Colour schedules for internal decoration
- Service layout details

When the architect fails to release information on time, the contractor may be entitled to an extension of time and loss and expense. The schedule shown in Figure 13.9 allows the contractor to record the date of the information request and compare this with the information release date.

Key information requirements may also be highlighted in the form of milestone events on the master programme. Alternatively, the early warning system developed by a major contractor in the mid-1960s is still a valid method of highlighting project requirements today. This is discussed in Chapter 16.

Request for information sheets

Figure 13.10 shows a written request for information of a more specific and detailed nature, i.e. drainage layout details relating to holes, pockets and ducts prior to pouring a ground floor slab area. This format allows the information requested to be matched with the information received on the same request form.

13.12 Contract planning

Once the contractor has commenced work on site, planning is required at regular intervals to determine when and how particular site activities are to be carried out. This type of planning is called 'contract planning' and is carried out monthly or weekly (short-term planning).

Contract planning involves monitoring the master programme and updating it 'as built', reporting progress to management and making sure that the health and safety plan and all safety method statements for specific activities are up to date. Planning at this stage of the project is carried out at a fairly detailed level in order to flesh out the master programme and provide a basis for the detailed day-to-day arrangement of work on site. This helps to prevent

REQUEST FOR INFORMATION SHEET

Contract	Wigan	Request date	15 Jan
Contract No	84/1	Prepared by	AB

Issued by	Issued to	Distribution
C. Wooton (Site Agent)	Architect – DBS Associates	Quantity surveyor Contracts manager File

INFORMATION REQUIRED	INFORMATION SUPPLIED
Drainage layout for ground floor slab to include: 1. Gully positions 2. Main service ducts in laboratory unit 3. Connections to toilets 4. Floor duct for waste water from machine hall	
CONTRACTOR'S COMMENTS	

Information required by	6 Feb
Date of response	
Further action date	

Signed	Signed
Date	Date

Figure 13.10

inefficient working or duplication of effort on site and enables the contractor to make the best use of his resources.

During the contract stage, the site manager or planner may break the master programme down into 'bite-size' pieces in the form of a **stage programme**. A further 'drilling down' into the programme may be undertaken in the form of a **short-term programme** and **sequence studies** may be carried out to compare the cost and time implications of alternative construction methods for important or cost-sensitive programme activities. An important means of communicating the contractor's detailed plans to the workforce is via **toolbox talks** and **task talks** which also give operatives the opportunity to put over their views about proposed methods of working and safety issues (see Section 12.12).

Stage programme

The contractor's master programme is frequently developed as the project progresses in order to show a finer level of detail. Figure 13.11 shows a stage programme for part of the substructure activity shown on the master programme above.

The activity comprises three sub-activities: piling and tests, construct foundations and make up floor levels. The construct foundations activity is shown in expanded form with logic arrows indicating dependency. Some activities (e.g. fix rebar) are continuous and others (e.g. blind bases) are intermittent.

Where alternative construction methods are being considered, the contractor might ask his quantity surveyor to prepare a cost comparison of the alternatives so that cost can be compared to the time allowed on the programme. There may be time benefits from choosing the more expensive alternative, thereby saving on preliminaries and possible liquidated and ascertained damages.

Short-term programme

When the contract stage is reached, the level of planning becomes more detailed and many contractors prepare short-term programmes on a monthly or weekly basis. For example, a programme covering, say, 2 weeks' work might be produced, and at the end of the first week of this programme, another 2-week programme would be prepared for the following fortnight reflecting progress, problems and any changes. On refurbishment-type projects of a relatively short duration, planning schedules may be prepared on a daily basis for each trade gang employed on the project.

Short-term programmes afford the contractor a much better means of controlling day-to-day operations on site and act as a useful method of communication between the site manager and the foreman, gangers or work package contractors. They can also be used in toolbox talks in order to discuss the health and safety implications of the programme with the workforce.

The principles of short-term planning based on a 2-week review period are illustrated in Figure 13.12 which shows the construct foundations activity from the stage programme. The programme for weeks 5 and 6 is prepared at the start of the project and, at the end of each weekly period, the short-term programme for the next 2-weekly period is prepared (weeks 6 and 7).

The short-term programme may be prepared by the site-based planning engineer or by the construction manager. Coordination meetings are usually held on a Friday afternoon with trades foremen and subcontractors' representatives and the work for the forthcoming 2-weekly period can be reviewed and the programmes prepared.

Responsibility for short-term planning may also be delegated to a site engineer when he is in charge of external works operations or specific concrete pours. This may be incorporated in the management-training programme developed by the company.

Objectives of short-term planning

- To assist the coordination of operations in the short term, especially when considering the continuity of work for trade gangs and subcontractors

STAGE PROGRAMME – FOUNDATIONS

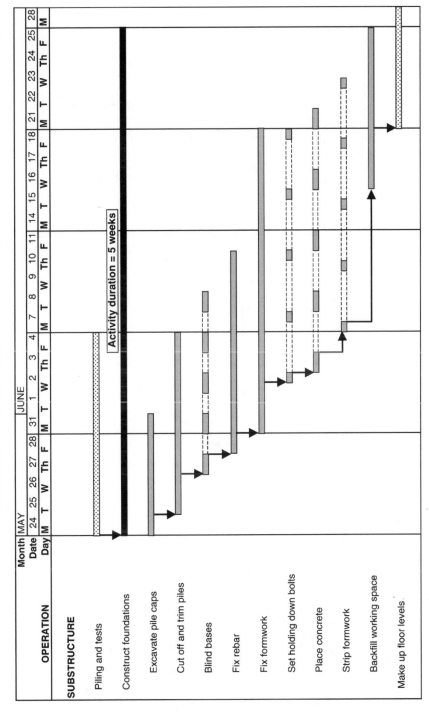

Figure 13.11

SHORT-TERM PROGRAMME

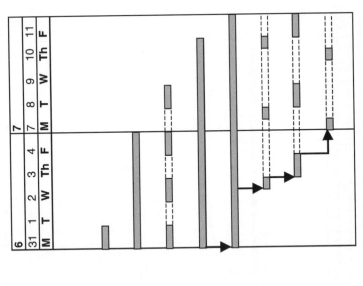

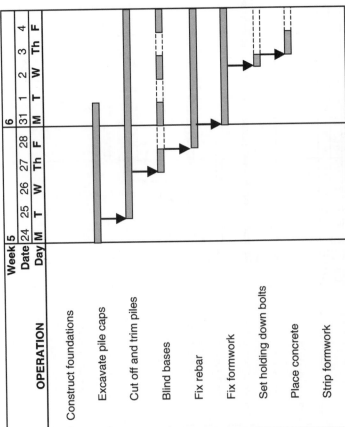

Figure 13.12

- To keep the master contract programme under constant review in the short term
- To highlight information requirements in the short term in order to meet planned completion dates for each stage of the work
- To assess key material requirements in the short term
- To keep senior site management informed of the progress position in the short term

References

Furst, S. & Ramsey, V. (2000) *Keating on Building Contracts*, 7th edn. Sweet and Maxwell.
HSE (2007) *Managing Health and Safety in Construction, Approved Code of Practice*. HSE Books.
Smith, A.J. (1995) *Estimating, Tendering and Bidding for Construction*. Macmillan.

14 Planning cash flow

14.1 Introduction

The construction industry has a bad reputation for its high levels of corporate and individual insolvencies, especially among contractors. Surprisingly, many of the companies and sole traders that get into financial difficulties are generating satisfactory profits but fail because they simply run out of cash. An insolvent company is unable to pay its debts as and when they fall due, and if cash is not available at the right time, the company ceases trading.

Cash flow may be defined as 'the movement of money in and out of the firm'. Consequently, payments made for contract work represent negative cash flow (money out) for the client and positive cash flow (money in) for the contractor.

In the case of *Gilbert Ash (Northern) Ltd* v. *Modern Engineering (Bristol) Ltd.* (1973), Lord Denning, Master of the Rolls, famously said that 'there must be cash flow in the building trade – it is the very lifeblood of the enterprise'. Construction firms get their 'cash' from monies received on contracts or, for house builders from sales revenue, and some firms enjoy both sources of income stream.

However, reliance on 'money coming in' is not enough to run a business and remain solvent. To do this, 'working capital' is required which helps to pay the bills before normal revenues are received and also helps the business to survive when the cash dries up for some reason (e.g. bad debts on a contract, insolvency of a debtor). This working capital is normally obtained through a bank overdraft, bank loans or from funds invested in the company by its shareholders.

Cash flow is a problem not only for contractors but also for developers and construction client organisations. Clients can also run into financial difficulties and therefore contractors need to be as sure as possible that they will be paid in full and on time by their clients if they are not to find themselves with 'cash flow problems'.

14.2 Client's cash flow

Developers and other clients to the industry provide the capital investment required for construction projects to go ahead. This money may be either borrowed from banks, provided by shareholders' investments or generated from profits, or a combination of all three.

The developer or client has a different view of cash flow from the contractor because their cash position is always negative until sales income or revenue from the completed building is forthcoming. An example of a client's cash flow is given in Table 14.1.

Table 14.1 Client's cash flow.

Money in	Money out
Housing development	Land purchase
Deposits	Interest on borrowings
Sales completions	Planning and legal fees
Rental income	Professional fees
Production revenues	Infrastructure costs (e.g. roads and sewers)
Sales of completed buildings (e.g.	Site remediation (e.g. removal of contamination)
speculative offices and factories)	Building costs (monthly or stage payments)

The cash flow position of two types of client is shown in Figure 14.1, which shows the cash requirements of a developer client building three houses and a client building a speculative office block project.

14.3 Contractor's cash flow

The contractor's cash flow position is somewhat different from that of the client. The contractor relies on interim or stage payments from the client to provide money in and this helps to pay for the money out payments for wages, materials, subcontractors, etc. However, because the contractor has to wait for perhaps 2 or 3 months for his money to come in, reliance also has to be placed on the credit provided by suppliers and subcontractors in order to reduce the negative cash flow effect of contract payments. An example of a contractor's cash flow is given in Table 14.2.

14.4 Cash flow forecasting

As part of the financial planning for the business, a prudent contractor will prepare a **cash flow forecast**. This will be used for negotiating banking facilities, such as an overdraft, for anticipating cash shortages, to aid the financial control of contracts and to avoid 'overtrading' (taking on too much work with too little working capital).

Cash flow forecasting is not easy because the contractor is never sure exactly how much money will be received from his portfolio of contracts. The forecast needs to be constantly updated within the contractor's reporting system because the timing and management of cash flows between receiving payment and paying creditors can mean the difference between insolvency and survival.

Cash flows for the company are usually prepared on a contract-by-contract basis and accumulated for the company as a whole so as to give a complete picture of what is happening. By doing the calculations, a company can predict the minimum and maximum cash required over the trading year so as to arrange a comfortable working capital facility.

Figure 14.2 is a simple illustration of how such a cash flow forecast may be constructed for a small contractor with an annual turnover of around £3 million. The forecast and graph are prepared using a popular spreadsheet programme. The net cash flow for each month of trading

CLIENT CASH FLOW

Small housing development project

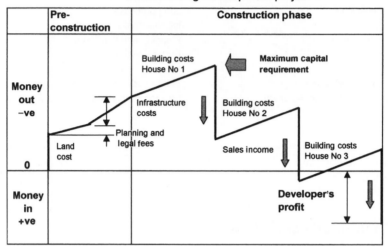

4-month office block development

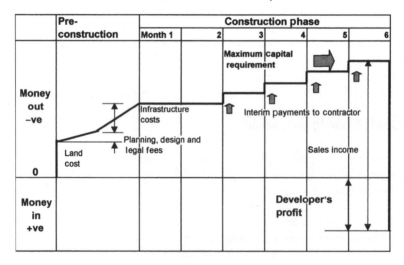

Figure 14.1

is given simply by deducting the cash out from the cash in. Negative figures illustrate potential problems, which need to be investigated. The cumulative cash flow is derived by adding the current month's figure to the previous month.

The graph beneath the spreadsheet illustrates the 'roller coaster' of cash flow forecasting. This is not necessarily a problem but if, in this case, the contractor's overdraft facility was £50 000, the directors would be getting a little 'twitchy'!

Table 14.2 Contractor's cash flow.

Money in	Money out
Monthly payments on contracts	Head office running costs
Final account payments	Staff salaries
Retentions released on practical completion	Company cars and expenses
Retentions released on issue of the final certificate	Payments to suppliers
Returns on investments (e.g. land and property)	Payments for plant hire
Medium–long-term bank borrowings	Contract payments to subcontractors
Shareholders' funds invested in the business	Wages and labour-only payments

14.5 Credit terms in construction

Construction is largely a credit-based industry where customers expect a period of deferment before having to pay for goods or services received. Traditionally, the construction client expects the contractor to carry out a month's work which is then valued and certified and the certificate is later honoured by cheque. The average period of credit under a JCT 05 contract is illustrated in Figure 14.3, but this period can be much longer under other forms of contract. Contractors enjoy even better credit facilities from their subcontractors and suppliers.

The industry culture of late payment of subcontractors and suppliers was a central theme in the Latham Interim Report 'Trust and Money'. Suppliers are commonly kept waiting for 60 or 90 days and may be forced to suspend supplies or even withdraw credit facilities from persistent offenders. Recovery action through the courts may well ensue. Some subcontractors never see their retention monies despite years of waiting.

Typical credit terms in the industry are given in Table 14.3.

The construction industry operates a system of retentions where a percentage of the payments made for work in progress is held back by the client from the contractor (or by the contractor from subcontractors). Typically, this is 3% or 5% but can be 10% or more where non-standard forms of contract are used.

The money held back is effectively capital lock-up as far as the contractor (or subcontractor) is concerned, and this has a cost effect in terms of interest payments on working capital requirements. This adds to the cost of building and is counterproductive from a client value point of view.

The practice of retentions has been widely criticised over the years, including by Banwell (1964) and Latham (1994), but latest feedback seems to be that retention bonds have not gained significant popularity (RICS 2006).

14.6 Forecasting contract value

The first step in preparing a cash flow forecast is to determine what the contractor's income will be. This is a matter of assessing the likely income from the various projects that the contractor has. The total income of the company will be the accumulated income from each of the contracts. A bit of 'crystal ball gazing' is required here as some of these projects may not

6-MONTH CASH FLOW FORECAST

	JAN	FEB	MAR	APRIL	MAY	JUNE
CASH IN						
CONTRACT A	100 000	100 000	50 000			
CONTRACT B	50 000	50 000	60 000	100 000	120 000	100 000
CONTRACT C			50 000	100 000	150 000	150 000
CONTRACT D				50 000	100 000	200 000
TOTAL CASH IN	150 000	150 000	160 000	250 000	370 000	450 000
CASH OUT						
SALARIES	20 000	20 000	20 000	20 000	20 000	20 000
RENT	1000			1000		
TELEPHONE		500			500	
MATERIALS	40 000	60 000	90 000	150 000	200 000	200 000
SUBCONTRACTORS	50 000	80 000	100 000	100 000	150 000	200 000
TOTAL CASH OUT	111 000	160 500	210 000	271 000	370 500	420 000
NET CASH FLOW	39 000	−10 500	−50 000	−21 000	−500	30 000
CUM CASH FLOW	39 000	28 500	−21 500	−42 500	−43 000	−13 000

CASH FLOW FORECAST

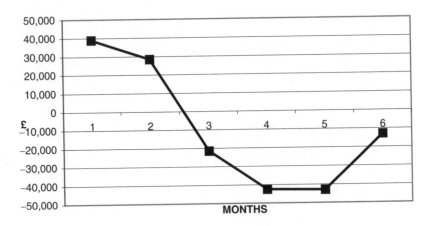

Figure 14.2

have started and some will be finished but awaiting the outcome of the final account which may be in dispute. The income from other ongoing projects will be derived from the contractor's internal reporting system.

When assessing the financial requirements of a project, or a number of projects to be undertaken over an annual period, the contractor must ask the following questions:

CLIENT CREDIT TERMS (JCT05)

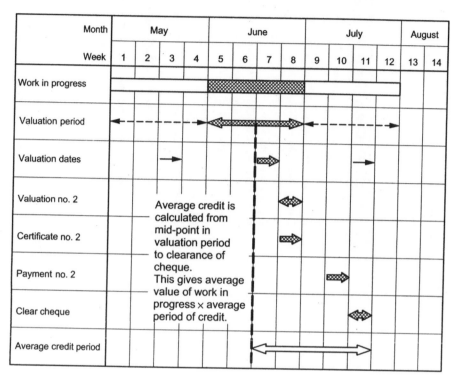

Figure 14.3

- What is the basis of the forecast?
- What is the basic information required?
- How does timing affect the monetary requirements?
- How accurate does the forecast need to be?
- How does the contractor monitor the cash requirements during the project?
- Will the budget information help to convince lenders that the cash funding requirements have been realistically thought through? A slick presentation at this stage may help, but this is doubtful

Various techniques are available to help but their usefulness depends on what stage has been reached and the information available. An overview of these methods of forecasting funding requirements will now be outlined.

S curves

Various types of S curves are available on which to base a cumulative value forecast. The S curve is best applied to projects where the provisional sums are evenly spread though the contract,

Table 14.3 Typical credit terms.

Contract	Payment terms	Overdue payments
JCT 05 Standard Building Contract	Monthly, 14 days from architect's certificate	Simple interest on overdue payments = Bank of England base rate + 5%
JCT 05 Design and Build Contract	14 days from contractor's application	Simple interest on overdue payments = Bank of England base rate + 5%
JCT 05 Major Project Contract	14 days after issue of payment advice	Simple interest on overdue payments = Bank of England base rate + 5%
ICE7 Conditions	28 days from contractor's statement	Compound interest at 2% over bank base lending rate
New Engineering Contract	3 weeks from project manager's assessment date	Compound interest at the rate stated in the contract data (minimum 2%)
JCT Standard Building Sub-Contract	21 days after the appropriate interim certificate for the main contract is payable	Simple interest on overdue payments = Bank of England base rate + 5%
Civil Engineering Contractors Association Form (Blue Form)	38 days after the main contractor's application to the employer	Interest at the same rate specified in the main contract
Non-standard forms	Varies enormously but between 50 and 70 days from invoice is not unusual	Statutory rate applying at the time of court judgement
Typical plant hire contract	15 or 30 days from invoice. Invoice is usually submitted when plant is off-hired	Statutory rate applying at the time of court judgement
Typical suppliers' terms (e.g. builders merchants)	30 days from end of month of delivery of materials	Statutory rate applying at the time of court judgement

as the technique does not allow for high-expenditure provisional items early in the project period. The basis for this method of forecasting is the presumption (established by research) that the cumulative expenditure on any construction project normally approximates to an S-shaped curve. Consequently, if we take each monthly payment on a project and add it to the total of previous payments, a cumulative S curve will be described.

The quarter–third approximation is a geometric curve which has been found to give a reasonable assessment of value, while other approximations based on the ogee curve give a similar range of cumulative values. The $\frac{1}{4}$:$\frac{1}{3}$ rule gives a good approximation of this curve. The basis of the rule is that:

- $\frac{1}{4}$ of total expenditure will be made during the first $\frac{1}{3}$ of the project timescale
- a further $\frac{1}{2}$ of total expenditure will be made during the middle $\frac{1}{3}$ of the project
- the final $\frac{1}{4}$ of total expenditure will be made during the final $\frac{1}{3}$ of the project timescale

It works by plotting a graph and Figure 14.4 shows the basic principles of the quarter–third approximation. Figure 14.5 gives a cumulative value forecast using this method for a contract lasting 6 months with a value of £160 000.

Cumulative percentage value

This method involves preparing a forecast of the cumulative percentage value per month on the basis of data analysed from similar types of project. The purpose is to produce a reasonable S-curve approximation at the feasibility or tender stage of the project. This method is illustrated in Figure 14.6, which shows a tabular display of the cumulative percentage value forecasts for projects lasting up to 12 months.

The table in Figure 14.7 shows data abstracted from Figure 14.6 for a contract with a value of £160 000 and lasting 6 months. The appropriate contract duration is chosen along the horizontal scale and the estimated percentage of the total contract value for that month of the project is read off vertically.

This approximation gives a slightly lower cumulative value forecast than the quarter–third method, as shown by the comparison of values and curve profiles in Figure 14.7. The tabular display in Figure 14.6 is ideal for displaying as a spreadsheet, as this allows the analysis to be speedily undertaken and displayed in line graph form.

The idea of percentage value analysis relative to time may be developed by collecting empirical data for different types of projects – schools, housing, factories, office developments, etc. In this way a series of different S-curve profiles may be developed for forecasting the cumulative project values at feasibility and tender stages.

Using the bar chart programme

This programme may be used to express the rate of value accumulation during the project. This may be achieved by analysing the contract estimate into an operational format in order to allocate the monies to a linear timescale on the bar chart.

S-CURVE APPROXIMATION
1/4–1/3 RULE

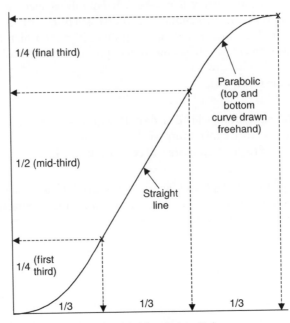

PRINCIPLES

1/4 value expended in first 1/3 of project period

1/2 value expended in middle 1/3

1/4 value expended in final 1/3 of project period

Figure 14.4

    As the project develops and further information becomes available, a more realistic cumulative value forecast may be prepared from the pre-tender programme or master programme. This enables the value forecast to be directly linked to the sequence of construction operations. The cumulative value forecast is therefore better prepared at this stage of the project as it provides a tool for controlling the project. At monthly intervals, the actual cumulative and forecast values may be matched as part of the company's monthly cost-reporting procedures. The cumulative value forecast may also be used for assessing the client's and contractor's cash funding requirements. It is often a requirement of the tender submission that a cumulative value forecast is provided at the tender stage of the project in order to assist the client's assessment of his funding requirements.

S-CURVE PRESENTATION

Project value £160 000

Contract period 6 months

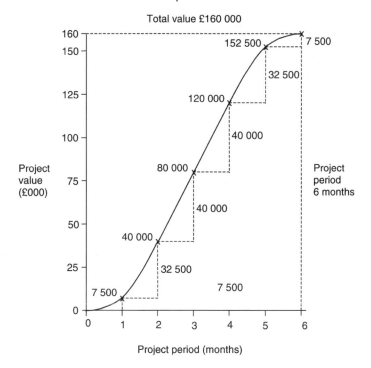

Total value £160 000

VALUE FORECAST (Cumulative)			
Month	Cumulative value	Month	Cumulative value
1	7 500	4	120 000
2	40 000	5	152 500
3	80 000	6	160 000

Figure 14.5

Figure 14.8 illustrates a bar chart for a factory development project, showing the weekly and cumulative forecast values at each of the planned valuation dates. A cumulative value–time forecast based on the programme is shown in Figure 14.9. If the relationship between value and time is represented by a nearly straight line, as in this case, then so be it. The cumulative value forecast normally follows the basic S curve, except where high values are expended on such items as erecting steelwork early in the project period.

PROJECT PERIOD (months)

Cumulative % value complete forecast

MONTH	1	2	3	4	5	6	7	8	9	10	11	12
0	0	0	0	0	0	0	0	0	0	0	0	0
1	100	45	24	18	12	9	7	6	5	5	4	3
2		100	76	45	30	24	18	14	12	11	9	7
3			100	79	60	45	35	27	23	20	17	13
4				100	82	67	55	45	37	31	27	22
5					100	85	72	63	53	45	38	33
6						100	88	77	67	59	51	45
7							100	90	80	72	64	56
8								100	92	83	75	66
9									100	92	85	76
10										100	93	85
11											100	93
12												100

HOW TO USE

Read off cumulative % value relative to contract duration, i.e. 6 month contract

CUMULATIVE VALUE

Month	Cumulative %
1	9
2	24
3	45
4	67
5	85
6	100

Figure 14.6

CUMULATIVE VALUE FORECAST

Contract value £160 000
Contract period 6 months

Cumulative value forecast (from % graph)

Month	Cumulative %	Cumulative value	Compared with 1/4–1/3
1	9	14 400	7 500
2	24	38 400	40 000
3	45	72 000	80 000
4	67	107 200	120 000
5	85	136 000	152 500
6	100	160 000	160 000

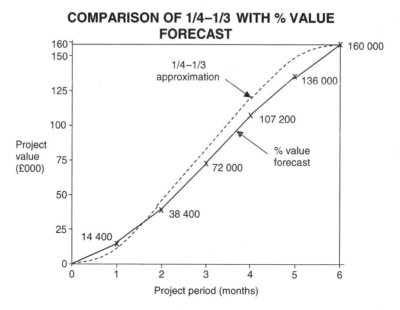

COMPARISON OF 1/4–1/3 WITH % VALUE FORECAST

Figure 14.7

It must be pointed out at this stage that the cumulative value forecast is only as good as the accuracy of the planning forecast. If the planning is too ambitious, then so will be the resulting value forecast. It is the responsibility of management to overview the budget forecasts and ensure that they are attainable. All levels of management thrive on achieving success and meeting targets. The achievement of budget forecasts during a project helps establish a team approach to ensuring a successful project.

FACTORY PROJECT – BAR CHART PROGRAMME – CUMULATIVE VALUE FORECAST

OPERATION	BUDGET	1	2	3	4	5	6	7	8	9	10	11	12	13	14	15	16	17	18	19	20
Establish	5 000	5																			
Piling	10 000		10																		
Caps & g.beams	21 000			7	7	7															
Drainage	6 000		2	2	2																
Erect frame	20 000						10	10													
Roof cladding	14 000								7	7											
Ext. brickwork	30 000									6	6	6	6	6							
Floor slab	12 000													4	4	4					
Internal services	20 000															5	5	5	5		
External works	16 000																	4	4	4	4
Preliminaries	10 000	0.5	0.5	0.5	0.5	0.5	0.5	0.5	0.5	0.5	0.5	0.5	0.5	0.5	0.5	0.5	0.5	0.5	0.5	0.5	0.5
Weekly value (£000)		5.5	12.5	9.5	9.5	7.5	10.5	10.5	7.5	13.5	6.5	6.5	6.5	10.5	4.5	9.5	5.5	9.5	9.5	4.5	4.5
Cumulative weekly value		5.5	18.0	27.5	37.0	44.5	55.0	65.5	73.0	86.5	93.0	99.5	106.0	116.5	121.0	130.5	136.0	145.5	155.0	159.5	164.0
Monthly value					37.0				73.0				106.0				136.0				164.0
Valuation periods					1				2				3				4				5

Time in weeks

20-week project period

Project value £164 000

Figure 14.8

VALUE FORECAST FROM PROGRAMME

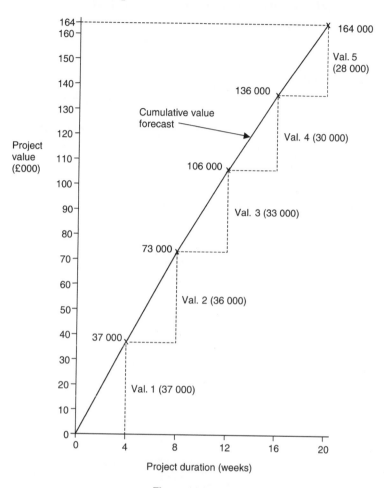

Figure 14.9

The good-guess method

This method is not to be scoffed at. An experienced forecaster will have accumulated a great deal of 'know-how' over the years and will probably use a formal method simply by intuition. With practice and experience, forecasts based on intuition can produce acceptable results, but in any case, all the calculations and formulae in the world cannot militate against late payments and bad debts.

14.7 Forecasting the contractor's income

Figure 14.10 shows the principles of assessing the payments to be made to the contractor at the end of each payment period. The final payment is released to the contractor at the end of

PRINCIPLES OF FORECASTING MONTHLY PAYMENTS DUE TO CONTRACTOR

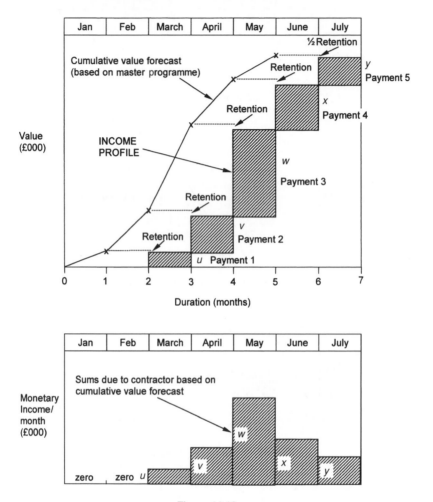

Figure 14.10

the defects period. The histogram, shown below the line graph, allows the monthly payment sums to be more clearly presented.

The forecast of income from the contract valuations and the release of the interim payment certificate to the contractor depend on the payment terms contained in the various forms of contract.

JCT 05 Standard Building Contract

Under JCT 05 conditions, the payment terms are stated as 14 days from the date of issue of the interim certificate. The overall period, however, between the date of the interim valuation and the payment of the money into the contractor's bank account, may cover some 28 days (see Figure 14.3). This time delay is made up of:

- 7 days between the valuation date and the issue of the interim certificate
- 14 days' payment period
- 7 days to receive and clear the payment

The interim payment is subject to retention, as stated in the contract particulars, which is 3% unless otherwise stated. The retention is reduced by one half on issue of the practical completion certificate and the balance of the retention fund is released on issue of the certificate of making good defects at the end of the defects rectification period. The defects rectification period is 6 months unless otherwise stated but this is likely to be 12 months where there are significant mechanical engineering installations.

JCT 05 Design and Build Contract

Under the JCT 05 design and build conditions, the payment terms may be either stage payments (Alternative A) or periodic payments (Alternative B). Periodic payments are normally monthly, whereas stage payments are based on agreed cumulative values linked to stages or milestones as stated in the contract particulars. The payment period from the payment request date is specified in the contract as 14 days. Similar terms to JCT 05 Standard Building Contract apply to the issues of retention and defects liability.

ICE Conditions of Contract (7th Edition)

For civil engineering contracts where the 7th edition is used, valuations are agreed between the engineer and the contractor and payment becomes due within 28 days of the issue of the certificate.

Worked example

The cumulative value forecast shown in Table 14.4 has been obtained from a master programme following the principles explained in Section 14.6 above. The contract period is 6 months and retention is 3%.

Figure 14.11 shows the cumulative value forecast presented graphically, together with the forecast cumulative and monthly income profile.

14.8 Movement of money

In the first 2 months of a typical project, the contractor will be paying out money for wages, materials, subcontractors, etc., and this represents negative cash for the contractor. Some of this money will be recovered when the client pays the first interim certificate at the end of

Table 14.4 Cumulative value forecast.

Month	Cumulative value forecast (£)	
1	45 000	
2	124 000	
3	198 000	Payments = monthly
4	265 000	Retention = 3%
5	320 000	
6	380 000	

month 2. This is unlikely to be sufficient to put the contractor in the black, however, because by this time another month's work will have been done and more costs will have been incurred.

The pattern of money in and money out on a contract may be illustrated as a saw tooth diagram as shown in the top part of Figure 14.12. This shows that the contractor is in a negative cash position until month 6, which means that he will have to finance the project at his own expense. His finance will normally be borrowed money, perhaps from loan or overdraft facilities, and it will have a cost in terms of interest payments. Therefore, even though cash is coming in from the contract, more money is going out due to the delay in contract payments.

A further negative cash flow consideration is that the client will keep some retention money back from the contractor to cover for possible defects in the work. It can therefore be seen that the contractor is in a negative cash position for a considerable period of the contract, until such time as the contract payments begin to outweigh the monies being spent.

Figure 14.12 illustrates the contractor's cash position assuming that there is no delay in paying for the wages, materials and services required for a contract. Realistically, however, the contractor will use the credit facilities offered by his suppliers and subcontractors to offset the negative cash flow effect of having to wait for the client's payments. The payment delay situation is shown in the bottom part of Figure 14.12. Here, the contractor makes no payments out until money is due in from the contract. Consequently, the contract will be 'in the black' more quickly and the contractor will have to find less working capital for the job. This means lower interest payments and more profit for the contractor.

Harris and McCaffer (2006) suggest that a weighted average payment delay may be calculated to demonstrate the effect of using credit facilities to improve cash flow. An example of this is given in Table 14.5.

Because most construction work is carried out on credit, a delicate balance exists in the supply chain between income and expenditure. For instance, under the standard forms of contract (e.g. JCT 05 and ICE7), the contractor works for a month but the client is not required to pay for work in progress until a valuation has been carried out and a certificate issued by the architect or engineer. This means that the contractor will have to wait for a month or more before getting paid and he will have to find the money to pay wages and salaries in the meantime.

On the other hand, the contractor does not pay straightaway for subcontract services, plant hire or materials supplied until a 'credit' period has expired. Subcontractors will be paid according to the terms of their contract (2–4 weeks) and suppliers perhaps 30 days from the end of the month when the materials were delivered. Plant hire companies usually invoice

FORECAST OF CONTRACTOR'S MONTHLY PAYMENTS

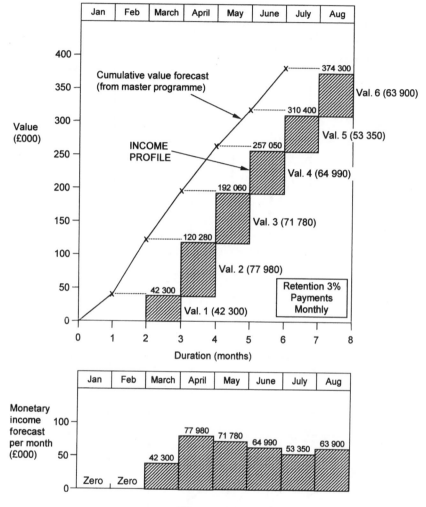

Figure 14.11

when the plant has been off-hired from site, and payment is normally due a month later (see Section 14.5).

The credit system works well unless the chain is broken and one party or the other fails to pay on time. When this happens, cash flow problems can arise for contractors or subcontractors with inadequate working capital. The longer the contractor can defer paying his creditors, the better it will be for his cash position, but care needs to be exercised to avoid withdrawal of credit facilities or possibly court action for recovery of monies outstanding.

Creditors cannot be paid out of profits if the cash is not available and so delaying payment to creditors, while awaiting payment on contracts, has become an art form in construction.

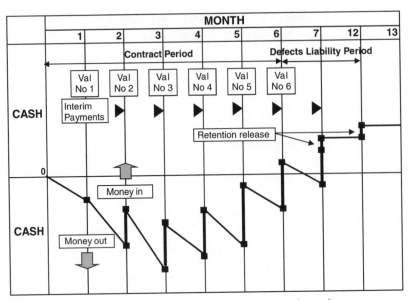

1 – No payment delay situation

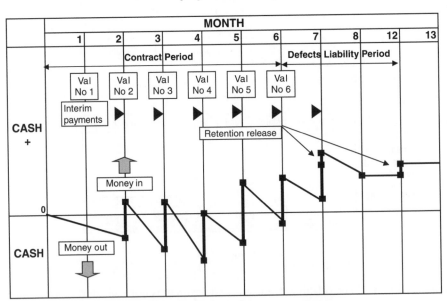

2 – Payment delay situation

CASH FLOW PRINCIPLES - 1

Figure 14.12

Table 14.5 Weighted average payment delay.

	Percentage of contract value	Payment delay or credit period (weeks)	Weighted average payment delay (weeks)
Labour	10	1	0.1
Materials	20	6	1.2
Plant	20	6	1.2
Subcontractors	50	3	1.5
Total	**100**		**4.0**

However, when small firms are squeezed between a large debtor (e.g. a main contractor) and a creditor (e.g. a supplier wanting payment), with one controlling the money and the other controlling credit facilities for materials, serious problems can occur. The Housing Grants, Construction and Regeneration Act 1996 was enacted to address these problems but in practice payment problems are still common in the industry.

14.9 Working capital

Working capital is the money a firm needs to fund its day-to-day operations in order to pay bills on time. In order to fund the negative cash situation, the contractor needs access to working capital in the form of shareholders' funds, bank borrowings or overdraft facilities. Contractors who rely solely on overdrafts rather than long-term funding, however, are vulnerable to insolvency because overdrafts can be withdrawn without notice. This usually happens just when the contractor has reached his overdraft limit!

Figure 14.13 illustrates the contractor's cash position and working capital requirements. In the top part of Figure 14.13, the area between the zero cash line and the minimum cash line indicates the least amount of working capital needed by the contractor assuming that contract payments are received on time. If this is not the case, the maximum cash line would apply and the area between the zero cash position and maximum cash line would represent the working capital needed to fund the project.

The areas between the zero cash line and the minimum and maximum cash lines represent the contractor's minimum and maximum capital lock-up on the contract. It is this area which needs to be funded out of working capital.

In the bottom part of Figure 14.13, a payment delay of 1 month (4 weeks) is indicated which means that the contractor is in the black by month 2 rather than month 6. As a rule of thumb, a 1-month payment delay improves cash flow by approximately 50% with a consequent reduction in the contractor's working capital requirements.

14.10 Improving cash flow

Despite relatively low interest rates at the time of writing, the cost of borrowing is still a matter of concern for contractors because profit margins are so low and banks lend at a premium over the base rate. Consider the situation for a contractor with an annual turnover of £20 million,

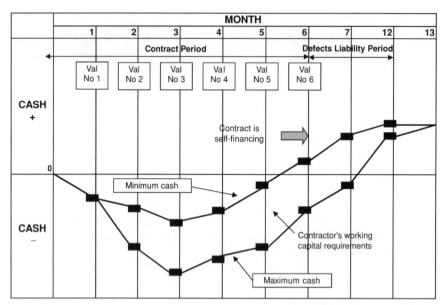

1 – No payment delay situation

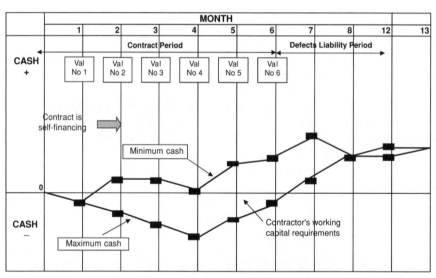

2 – Payment delay situation

CASH FLOW PRINCIPLES - 2

Figure 14.13

annual profits of £600 000 and bank borrowings of £3 million with a lending rate of 2% over base:

Actual cost of money = base rate (say) 5% + 2% = 7%

Cost of borrowings = 7% × £3 000 000 = £210 000

Cost of borrowing expressed as a percentage of turnover = 1%

With profits at only 3% of turnover, the contractor is vulnerable should interest rates rise or profits fall. Profit in contracting is always at risk because of the uncertain nature of the work and the likelihood of mistakes in tenders and inaccurate forecasts on contracts. For these reasons, and because contractors invariably make more money by keeping other people's money for as long as possible, new ways of reducing negative cash flows are always attractive. Some common methods are listed below.

At tender stage

These methods will bring in early money but must be done before submitting the priced bills:

- Load money into undermeasured items
- Load money into early items such as excavation and substructures
- Load money into mobilisation items in the preliminaries

During the contract

These methods will reduce working capital requirements:

- Submit interim application on time
- Overmeasure the work in progress
- Overvalue materials on site
- Agree on the value of variations as soon as possible
- Keep good records and submit claims early
- Deal with defective work quickly to avoid delayed payment
- Make maximum use of trade credit facilities

Post-contract

These methods will increase profit levels:

- Submit all documentation as soon as possible
- Ensure timely release of retentions by submitting health and safety file information on time
- Agree on final account as soon as possible
- Collect outstanding retentions on time

Table 14.6 Tender analysis.

Phase		Item	Cost (£)	Total (£)
Phase 1	Two-storey extension Value = £150 0000 Duration = 10 months	Operations • Substructure • Superstructure • Finishings and services	150 000 600 000 750 000	1 500 000
Phase 2	Refurbished canteen Value = £750 000 Duration = 8 months	Operations • Demolitions • Superstructure • Finishings and services	37 000 225 000 488 000	750 000
Phase 3	Office refurbishment Value = £1 500 000 Duration = 10 months	Operations • Demolitions • Superstructure • Finishings and services	75 000 450 000 975 000	1 500 000
Contract preliminaries				250 000
Total project value				4 000 000

14.11 Forecast value and cash flow example

Project brief

A contractor has obtained a contract for a project to be undertaken in three phases. Details of the phases with approximate durations and monetary values are shown in Table 14.6. The overall project duration is 24 months.

Project task

Produce a cumulative value forecast for the complete project and an assessment of the contractor's working capital requirements for the first 6 months of the project period. This is to be based on the following:

- The above values include a 5% contribution to profit and overheads
- 5% retention is to be applied to the payments
- Costs are to be paid at the end of the month in which they are incurred (i.e. no delay in meeting the cost situation)
- Interim payments are to be made monthly, payable 1 month after the valuation date (JCT 05 contract)

Worked solution

The approach to the assessment of the cumulative value forecast and the working capital requirements involves the following stages:

CASH FUNDING FOR THREE-PHASE PROJECT

Time in months

Phase I must be complete before Phase II and Phase III start

PHASE I · PHASE II · PHASE III

OPERATION		VALUE (£'000)	1	2	3	4	5	6	7	8	9	10	11	12	13	14	15	16	17	18	19	20	21	22	23	24
Subst.	PH.I	150	75	75																						
Superst.		600		100	100	100	100	100	100																	
Finishings		750						150	150	150	150	150														
Demolition	PH.II	37											37													
Superst.		225												45	45	45	45	45								
Finishings		488															122	122	122	122						
Demolition	PH.III	78															78									
Superst.		450																90	90	90	90	90				
Finishings		972																			162	162	162	162	162	162
Prelims	ALL	250	20	10	10	10	10	10	10	10	10	10	10	10	10	10	10	10	10	10	10	10	10	10	10	10
Monthly value		95	95	185	110	110	110	260	260	160	160	160	47	55	55	55	255	267	222	222	262	262	172	172	172	172
Cumulative monthly value		95	95	280	390	500	610	870	1130	1290	1450	1610	1657	1712	1767	1822	2077	2344	2566	2788	3050	3312	3484	3656	3828	4000

Figure 14.14

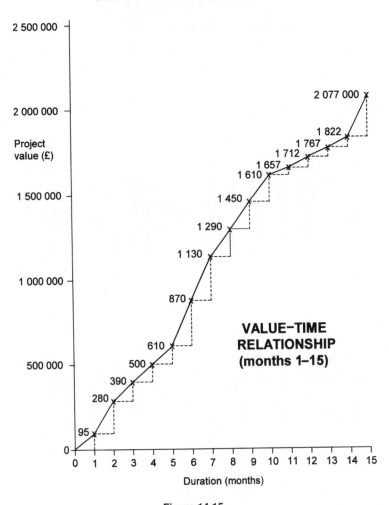

Figure 14.15

- *Step 1*

  Assess the cumulative value forecast for the three phases of the project by allocating project values to a bar chart programme.

  Figure 14.14 shows a bar chart display for the three phases of the project with monies allocated to project operations. The value for contract preliminaries has been allocated separately throughout the contract period with an extra sum of £10 000 for establishing the site at week 1 of the project. The cumulative value forecast is presented along the bottom of the bar chart and displayed as a value–time graph in Figure 14.15.

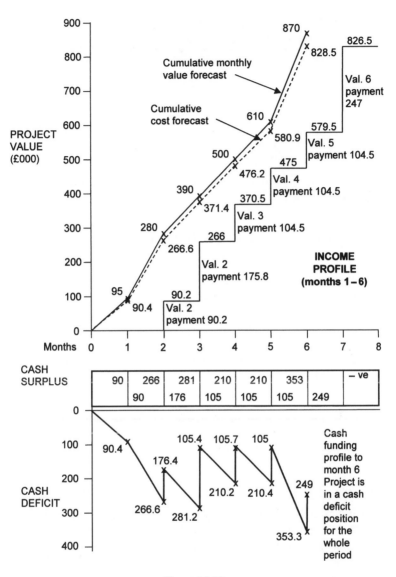

Figure 14.16

- *Step 2*

  Establish the cumulative cost forecast remembering that value is cost plus margin.

  $$\text{Cost} = \text{value} \times \frac{100}{(100 + \text{margin})}$$

  Therefore, where cumulative value is £390 000 and margin = 5%

  $$\text{Cost} = £390\,000 \times \frac{100}{(100 + 5)} = £371\,428$$

Table 14.7 Working capital requirements.

Month	Maximun cash requirement (£)
1	90 400
2	226 600
3	281 200
4	210 200
5	210 400
6	353 300

Calculate remaining costs and plot the cumulative cost–time relationship on the graphical display (see Figure 14.16).

- *Step 3*
  Calculate the contractor's actual income allowing for a 1-month payment delay and 5% retention.

Interim payment no. 1
Forecast value	£95 000
Less 5% retention	£4 750
	£90 250

Calculate remaining payments and plot them on the display making sure to allow a 1-month delay (see Figure 14.16).

- *Step 4*
  Plot a saw tooth diagram for the first 6 months of the project using the cumulative cost and payment figures calculated above. Money out (cost) is plotted downwards at an angle and money in (payment) is plotted upwards (see bottom of Figure 14.16).

- *Step 5*
  Display the maximum and minimum working capital requirements in the form of a table.

Figure 14.16 indicates the relationship between value–time, cost–time and income–time for the first 6 months of the project period. The cash funding profile is shown in the form of a saw tooth diagram under the graphical display. This is based on the no delay situation in meeting the cost as stipulated in the project brief.

The cash funding profile indicates that during the first 6 months of the project period, cash requirements peak at the end of months 3 and 6 respectively. The maximum working capital requirements at the end of months 1–6 are shown in Table 14.7.

References

Banwell, H., Sir (1964) Report of the Committee (Chairman Sir Harold Banwell). *The Placing and Management of Contracts for Building and Civil Engineering Work.* HMSO.

Harris, F. & McCaffer, R. (2006) *Modern Construction Management*, 6th edn. Blackwell Publishing.

Latham, M., Sir (1994) *Constructing the Team.* HMSO.

RICS (2006) *Contracts in Use.* Royal Institution of Chartered Surveyors.

Part C Control

15 Project control procedures

15.1 Introduction

Like any other business, builders, contractors and developers have to plan and organise their day-to-day activities in order to manage effectively. Unfortunately, it is a fact of life that the best-laid plans often go wrong, and when they do, they often go wrong in the worst possible way. This is called Murphy's law.

Managers who can anticipate a problem before it becomes a crisis have more chance of making a success of their business than someone who takes the Mr Macawber approach of simply looking at excess income over expenditure as a measure of success.

Effective management requires control, but all contractors have different ideas on the degree of control necessary for the projects they undertake. Many factors need to be considered, including the size and organisation of the firm and the scale and complexity of the projects in hand. However, in order to monitor performance, information needs to be collected within a structured reporting system, however simple, so that appropriate action can be taken if and when things start going wrong.

There are many aspects to the control of a business, but in terms of construction projects, three essential areas stand out for special consideration. These are the control of:

- Time
- Money
- Resources

15.2 Control procedures within organisations

As companies change size, they need to review control procedures. Denyer (1972), in his glossary of management terms, defines 'control' as the setting up of standards, making regular comparisons of actual events with those standards and taking corrective action. Figure 15.1 illustrates these principles diagrammatically in the form of a control cycle. This illustrates the need for communication channels to be established between all levels of management. Denyer (1972) outlines various methods of control which may be applied to a construction situation, as follows.

CONTROL APPLIED TO CHECKING DEVIATION FROM PROGRAMME

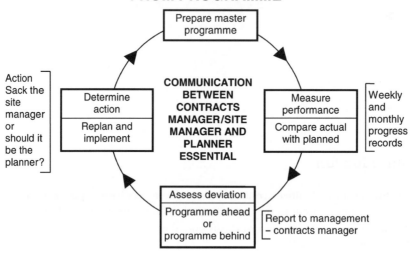

Figure 15.1

Visual and personal

Checks applied by a supervisor or site manager. The regular visits to a project by the principal of a small organisation are a means of applying control to a project. The principal gets the feel that all is right or it quickly becomes apparent that all is not well – call it management intuition, if you like.

Reporting

The submission of regular reports during a project provides control information to management. This applies to weekly site reports on progress and monthly cost–value reports on the profitability position. Monthly reports on progress and cash funding also aid management decision-making. As companies change size, the value of internal reporting procedures needs to be considered.

Control by exception

This is a well-established management principle. It enables management to consider reporting only on serious deviations, otherwise it is assumed that performance is up to standard. Management by exception saves management time in analysing irrelevant reports and unnecessary data.

Control policy and the establishment of the necessary information in order to control the business must be established by senior management. The management must lay down policy statements in order to establish standards for controlling finance, time and quality. Control systems may be established for reporting on project performance and profitability. These will require strict monitoring and enforcement in order to succeed.

15.3 Definitions

Financial plans are called budgets and the process of making, monitoring and adjusting them is called **budgetary control**. Hall (1974) identifies budgetary control as an important management technique used for the purpose of controlling income and expenditure. Control is achieved by preparing budgets relating to the various activities of the business and these provide a basis for comparison with actual performance.

Alternatively, Harper (1976) defines a budget as a monetary cost plan relating to a period of time, while Pilcher (1992) indicates that a budget is a financial plan for the contract as a whole and is a financial version of the programme.

The budget may be expressed in tabular or graphical format such as S curves, line graphs or histograms, although in practice managers tend to favour graphs as a way of expressing data.

15.4 Types of budgets

Harris and McCaffer (2006) identify the various types of budgets applicable to contracting organisations. These include:

- **Operating budget** – staffing, overheads, labour, materials, subcontractors
- **Annual sales budget** – anticipated turnover including existing contracts
- **Capital expenditure budget** – spending on plant and equipment, development land, etc.
- **Cash flow budget** – a forecast of the movement of monies in and out of the business in order to determine company borrowings (working capital requirements)
- **Master budget** – a forecast balance sheet and profit and loss account

The budgets produced at head office level will be based on budgets relating to specific contracts or developments, built up to show the picture for the company overall. Developers or contractor/developers need to anticipate future borrowings very carefully as the working capital requirements for schemes relying on sales of completed units are particularly heavy.

A housing developer will prepare an annual sales budget in order to monitor actual sales against a predetermined plan, but a general building contractor may be more concerned about forecasting his cash flow. In this case, cash flow budgets would directly help with cash control and also identify working capital requirements. Budgets may also relate to turnover forecasts or the value of cumulative tender enquiries received. Capital budgets lay down the planned requirements for the long-term survival of the business.

At project level, contractors need to forecast the amount of work they expect to carry out each month (the value) and what the expenditure on wages, materials, plant, etc., is likely to be (the cost). For forecasting purposes, the difference between value and cost is margin (overheads and profit).

This process facilitates control over payments made by the client under the contract and enables the contractor to assess how the contract is performing financially. These budgets will relate to the contractor's tender using the priced bills of quantities and taking account of any tender adjudication adjustments.

15.5 Preparing budgets

The procedures involved in the preparation of a budget are outlined by Harper (1976) who suggests five stages of development:

(1) Prepare the forecast
(2) Consider the company policy
(3) Compute the budget by expressing resources, quantities or values in monetary terms to form the initial budget
(4) Review the forecast, policies and the initial budget until an acceptable budget emerges
(5) Accept the budget at all levels of management. Ensure that it has the company's backing from the top down

The budget is now established.

Figure 15.2 indicates the stages of preparing a budget in diagrammatic form and illustrates the principles of the budgetary control cycle as applied to a sales, cumulative value forecast or cash flow budget.

15.6 Contract budgets

It is an essential function of management to prepare forecasts in order to establish a plan for the future of the business. Without a plan against which to monitor performance, management has no control and no business can be run successfully without keeping close control over the day-to-day finances. This involves providing the information necessary for keeping managers informed so that decisions can be made about how the business should react to current circumstances.

Businesses are living entities (going concerns) and therefore money is coming in and going out all the time. There are wages and suppliers' and subcontractors' bills to pay and monies are being received for contract work. This activity cannot be allowed to happen willy-nilly because the business will soon be in a mess. Contractors, therefore, have to prepare forecasts and these are required for many aspects of their business as well as the contract work undertaken. For instance, the company has to plan for the next year's trading in order to make sure that sufficient work is obtained to enable the company to meet its commitments and keep going. This is very much tied in to the estimating, planning and quantity surveying functions within a construction company.

Budgets may also be developed for the labour, plant and preliminaries expenditure on contracts. These will be based on an analysis of the contractor's net estimate, in other words excluding profit and overheads. The estimator's allowances in the priced bills of quantities will be extracted into a total for each budget category.

BUDGETARY CONTROL PRINCIPLES

STAGES IN PREPARING A BUDGET

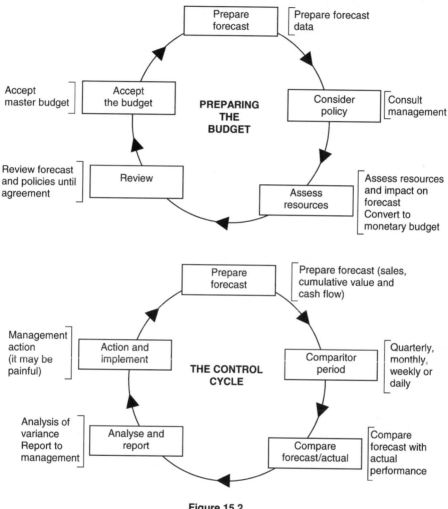

Figure 15.2

15.7 Labour, plant and preliminaries budgets

Budgets can be presented in a useful and interesting way by plotting money or resource allocations against time on a bar chart and this can be extended into a graphical display in the form of a value–time curve forecast, for instance. During the project, the contractor may then plot actual expenditure or resource consumption and use the analysis to monitor variances as part of his monthly cost control and reporting procedures. The subject of cost reporting at monthly intervals forms part of the discussion in Chapter 17.

Labour budget

Figure 15.3 indicates a bar chart display for labour demand expressed in man-hours/man-weeks. The graphical display in Figure 15.4 has been presented as a cumulative forecast of man-weeks plotted against time. This allows the actual man-weeks used on the project to be matched with the forecast and any variance monitored as the project proceeds. The bar chart and cumulative graph could equally well be expressed in money.

This simple approach is highly effective as an aid to resource and cost control, but needs to be reconciled with the contract progress position in order to be an effective management tool. The reason for the variance of two man-weeks could be due to the contract being behind programme. This could have resulted from insufficient labour being available to the contractor or due to unexpected delays on site. Management should investigate such variances.

Plant budget

Figure 15.5 shows a plant budget for a project in bar chart form and Figure 15.6 is the graphical display of the forecast cumulative plant expenditure. Plant cost allocations from the estimator's priced bills of quantities are shown on the bar chart display and plotted on a weekly cumulative graph for comparison with actual expenditure.

Plant costs on site frequently exceed expenditure allowances in the tender and site managers are notorious for keeping plant on site for too long just 'for convenience'. Whether plant is hired or owned by the contractor, strict control over budget variances needs to be exercised and plant and equipment must be off-hired as soon as it is not required.

Preliminaries budget

Similar principles may be applied to the preparation of a contract preliminaries budget by analysing the preliminaries allowances in the estimate, preparing a bar chart, and allocating money to time.

Figure 15.7 indicates a tabular display for a preliminaries budget. This has been prepared in bar chart format with the preliminaries expenditure forecast presented to a timescale. Fixed costs relating to the establishment of the site and its consequent dismantling are also indicated. Figure 15.8 shows the cumulative expenditure forecast presented as a line graph.

Contract preliminaries is one of the most common areas of overspending on construction projects. This may be due to management failing to allow sufficient monies to cover this section of the tender or over-elaborate site organisation. It is also the area most subject to adjustment at the tender adjudication stage, either to make the tender more competitive or because the contractor decides to tender on a shorter programme period. The total of preliminaries typically represents between 5 and 15% of the contract sum, but on small contracts the percentage could be a lot higher.

Actual preliminaries costs may be matched with the forecast expenditure in tabular and graphical form as the contract proceeds, and in this way the preliminaries variance may be monitored continuously during the project. Plotting the variance as a $\pm/-$ graphical display

LABOUR EXPENDITURE BUDGET

OPERATION	Labour budget man hours	Man weeks	1	2	3	4	5	6	7	8	9	10	11	12	13	14	15	16	17
Foundations	800	20	4	4	4	4	4												
Brickwork to DPC	360	9					4	5											
Ground floor slab	200	5							5										
External brickwork	960	24							2	6	6	6	4						
Roof		S/C																	
1st fix	240	6													2	2	2		
Plaster		S/C																	
Weekly man weeks			4	4	4	4	8	5	7	6	6	6	4	0	2	2	2	0	0
Cumulative man weeks			4	8	12	16	24	29	36	42	48	54	58	58	60	62	64	64	64

Man weeks

Figure 15.3

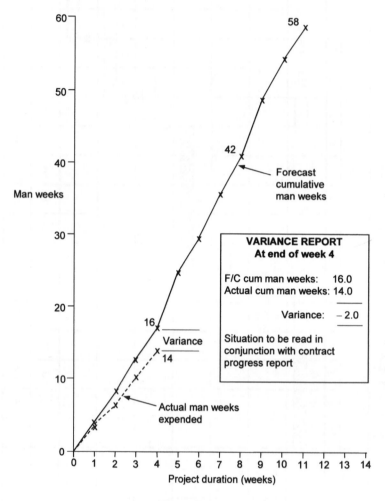

Figure 15.4

somehow focuses attention on the issues (see Figure 15.8). The preliminaries budget is also an ideal application for a spreadsheet display.

15.8 Budget control procedures

Control procedures must be established in order to monitor the budget at clearly defined time intervals. These may be daily, weekly, monthly, quarterly or annually. Management will then

PLANT EXPENDITURE BUDGET

OPERATION	Plant allocated	Budget £	1	2	3	4	5	6	7	8	9	10	11	12	13	14	15	16
Site clearance	P.Shovel 4 Lorries	2000	400	800	800													
Excavate foundations	Hyd. B/A. 2 Lorries	5000			1000	1000	1000	1000	1000									
Concrete foundations	Concrete pump	1000								1000								
Brickwork DPC	Mixer	200									100	100						
Ground floor slab	Concrete pump	1000											1000					
External walls	Scaffold mixer	4000												1000	1000	1000	1000	
1st fix	Crane	1000																1000
Weekly expenditure forecast			400	800	1800	1000	1000	1000	1000	1000	100	100	1000	1000	1000	1000	1000	1000
Cumulative expenditure forecast			400	1200	3000	4000	5000	6000	7000	8000	8100	8200	9200	10200	11200	12200	13200	14200

Figure 15.5

PLANT BUDGET

CUMULATIVE EXPENDITURE FORECAST

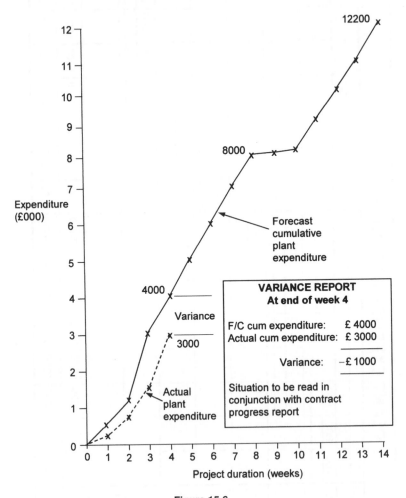

Figure 15.6

be able to take corrective action where variances from the budget occur. Management must act where variances are apparent at each review date, whether they are positive or negative. Any director can shout when it is obvious that things have gone wrong, but the astute manager is the one who analyses variance trends and applies decisive action at the appropriate time.

The application of variance analysis forms part of the budgetary control process. Examples of the analysis of contract variances are given in Chapter 17.

Variance analysis is the matching of actual performance with the forecast in order to assess the difference, whether it be positive or negative. It is the responsibility of management to highlight variances, report on the findings and implement action, the effect of which may often

CONTRACT – BOLTON FACTORY
CONTRACT VALUE – £600 000

PRELIMINARIES BUDGET

BUDGET FIGURES ARE NET PRELIMINARIES SUMS IN THE BILL

Total prelims £47 400

OPERATION	BUDGET £	1	2	3	4	5	6	7	8	9	10	11	12	13	14	15	16	17	18	19	20
														Time in weeks							
Establish site	8 000	4	4																		
Site accommodation																					
Establish	1 000		1																		
Hire	200/w			0.2	0.2	0.2	0.2	0.2	0.2	0.2	0.2	0.2	0.2	0.2	0.2	0.2	0.2	0.2	0.2	0.2	
Dismantle	1 000																				1
Site management																					
Site manager	1 000/w	1	1	1	1	1	1	1	1	1	1	1	1	1	1	1	1	1	1	1	1
General foreman	500/w								0.5	0.5	0.5	0.5	0.5	0.5	0.5	0.5					
Site engineer	500/w	0.5	0.5												0.5	0.5					
Temporary roads	2 000	0.5	0.5					0.5	0.5												
Hardstandings	2 000			1	1																
Site hoarding																					
Erect/dismantle				1	1															1	1
Weekly expenditure forecast		6	7	3.2	3.2	1.2	1.2	1.7	2.2	1.7	1.7	1.7	1.7	1.7	2.2	2.2	1.2	1.2	1.2	2.2	3
Cumulative expenditure forecast		6	13	16.2	19.4	20.6	21.8	23.5	25.7	27.4	29.1	30.8	32.5	34.2	36.4	38.6	39.8	41	42.2	44.4	47.4
Actual weekly cost		5	5	2	2	2	2	2	2	2											
Cumulative weekly cost		5	10	12	14	16	18	20	22												

Figure 15.7

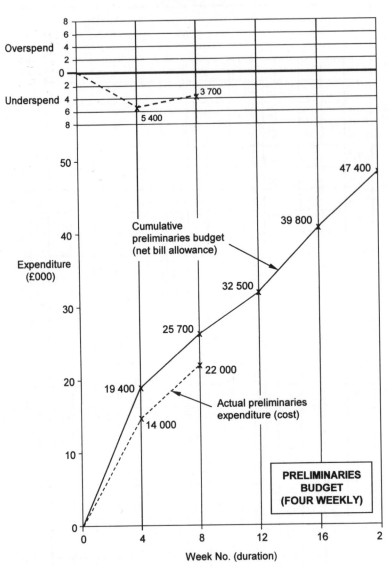

Figure 15.8

prove painful to subordinates. Failure to meet targets may have a damaging effect on morale within the organisation and questions will be asked about the initial forecasting approach and methodology. An easy way out at this stage is for management simply to admit that the forecast was wrong and to discharge the person responsible for preparing it. This is the usual escape route and gives everyone the impression that decisive action has been taken.

For budgetary control to be an effective tool in the management of the business, forecasting must be realistic and achievable.

15.9 Importance of site records

The keeping of site records during a project is imperative if a contractor is to receive his true entitlement as prescribed by the conditions of contract and also ensure that subcontractors are not successful with spurious claims for extra payment.

Records are necessary for:

* Establishing the basis of a claim for loss and expense
* Defending a counterclaim from a client
* Substantiating applications for extensions of time
* Claiming payment for extra work not included in the contract
* Reporting to progress meetings

Thomas (2001) lists 24 common types of records that contractors might keep, including site diaries, master programme updates and records of resources on site.

Records normally assume greater importance after the event than during it and contractors often fail to keep adequate records during a project unless some formal procedures are adopted and maintained. On rare occasions, it has been known for a contractor to fabricate site diaries and other records if he is in a tight corner, rather than admit that he has no records at all.

Responsibility for keeping records on site

The contractor should delegate responsibility to his site management team for keeping, circulating and maintaining records and this includes site supervisors, planning engineers and quantity surveyors.

Construction supervisory staff should be responsible for:

* Maintaining a daily site diary
* Recording time lost per day due to exceptional rain, cold or heat. The time lost per week should be included in the report to the clerk of works and minuted at monthly site meetings with the architect. This information may be used to establish a claim for an extension of time (JCT 2005, Clause 2.29.8)
* Keeping records of the contractor's own labour and that of his subcontractors. Details relating to site labour are often noted in the daily site diary. Site labour records may be used as a basis of a counterclaim against a subcontractor due to his non-performance
* Recording verbal instructions from the architect or his clerk of works. A system of CVIs (confirmation of verbal instructions) is normally implemented by the contractor with copies sent to the contract surveyor
* Records of additional excavation works, especially where underground obstructions have been encountered. The effect of delays in foundation works may lead to the issue of a variation order and hence to a possible extension of time (JCT 2005, Clause 2.29.1). Photographic

records may easily be recorded on a digital camera in order to record site conditions during foundation works. Excavation depths should be checked with the clerk of works as work proceeds and a signature obtained
- Records are to be kept in relation to the issue of drawings from the architect or the consulting engineer. Procedures should be set up in relation to the circulation of drawings to subcontractors and other specialists. Responsibility for the keeping of a drawing register should be allocated to one of the construction team
- Materials received each day or week should be recorded on goods received sheets and copies sent to head office
- Daywork records (often referred to as the 'daily liar') are the responsibility of construction management. These may be used by the contract surveyor to establish fair rates

Site planning personnel should be responsible for:

- Updating short-term programmes each week or fortnight
- Recording progress each week on all the relevant contract programmes, i.e. weekly and monthly site master programme and subcontractor programmes. Preparing weekly and monthly progress reports for submission to the contractor's project manager
- Preparing records for the contract 'as-built programme' as work on site progresses
- Recording the effect of programme delays on the final contract completion date

Site-based surveying staff are responsible for:

- Keeping records of all site measurements which relate to the final account, i.e. records of foundations and drainage remeasurement. It may prove valuable to photograph complex drainage connections for reference purposes
- Recording details and measurement of individual variations to contract. Recording any delays and disruption resulting from undertaking work in variations. This may be linked to the construction planner's records
- The pricing of site daywork sheets and obtaining the appropriate signatures
- Correlating instructions received from the architect or his representatives with the CVIs issued
- Undertaking the reconciliation of the bills of quantities items with quantities abstracted from the site working drawings, e.g. brickwork and blockwork quantities

15.10 Meetings as part of the control process

Meetings are an important means of communication throughout the various stages of any construction project and form an integral part of the control process necessary to keep a project on time and on budget and make sure that the build quality is according to specification. They have a variety of purposes, including:

- To discuss problem areas and suggest alternative solutions
- To decide on appropriate action
- To request and distribute project information

- To review the current situation and reschedule as necessary
- To consider ideas for resolving bottlenecks
- To consider the opinion of other parties
- To develop a team approach to problem-solving

It is important that meetings are short, focused and have the right people there so that the correct facts can be tabled and decisions made without constant referring back to superiors. They help to develop communication links between the parties (i.e. the client, design team, contractor and subcontractors) and to foster a team approach to decision-making and problem-solving. Examples of the various meetings which may take place during the construction process are included in Figure 15.9, which explains the purpose of the meetings and the main people involved.

Other meetings that might take place include informal meetings with the architect to discuss defects, discussions between the quantity surveyors regarding progress with the final account, weekly/daily meetings to problem solve contract emergencies and toolbox or task talks with operatives to discuss work methods and health and safety issues.

Managers should try to strike the correct balance between working and meeting as too many meetings may impede or disrupt the work pattern on a project. The value of both formal and informal meetings should also be recognised. Formal meetings such as the client's pre-contract meeting and monthly site meetings have procedures and rules, whereas informal meetings such as tender coordination meeting and weekly site meetings with the trades foreman may be of a more casual and relaxed nature.

Meetings should normally have an agenda and minutes should be recorded and circulated, but this is sometimes impractical. For instance, when the site manager and site foreman meet for 10 minutes to discuss a site problem, this would not be formalised. However, where an impromptu meeting like this leads to a change of construction method, a risk assessment should be carried out and recorded and the original method statement should be formally changed to reflect what was agreed. A further meeting to inform the operatives (i.e. a toolbox/task talk) would also be necessary.

Monthly site meetings

The meeting agenda for site meetings is the responsibility of the meeting chairman (usually the client's architect or engineer) and it should be circulated before the meeting in order that the personnel attending are fully aware of the items to be discussed. Minutes of previous meeting should also be circulated beforehand in order that they can be formally accepted as a true record.

A typical agenda for a monthly site meeting to be chaired by the project architect is indicated below:

Minute Reference[a]	Headings for the order of the meeting
1.1	Personnel attending and apologies for absence
1.2	Confirmation of minutes of previous meeting

1.3	Matters arising from minutes
2.1	Confirmation of matters raised at intermediate site visits by architect
2.2	Weather Report – record of inclement weather to date
2.3	Progress Report – normally given by the contractor or clerk of works
3.1	Drawings and information requirements
3.2	Construction queries – relating to materials or design
3.3	Design issues – anticipated future requirements
3.4	Subcontractor and specialist supplier information requirements
3.5	Liaison with statutory undertakings/utilities
4.1	Health and safety report
5.1	Variation orders – review of outstanding architect's instructions and instructions requiring confirmation
5.2	Dayworks – overview of outstanding dayworks
6.1	Financial review – quantity surveying matters for discussion
7.1	Project completion date – review of projected completion
8.1	Any other business
9.1	Date and time of next meeting

[a]to be referred to in all site minutes.

Weekly progress meetings

Weekly progress meetings are held at site management level in order to review the short-term programme from the previous week and to coordinate the activities of subcontractors and the contractor's own trade gangs. Coordination is essential for the success of the project and the weekly site meetings are used to establish the short-term or 'look-ahead' programme for the following 1- or 2-week period and to discuss interfaces between the various subcontractors and other site activities. On larger projects, the meeting may be chaired by the assistant project manager assisted by the site-based planning engineer. A meeting agenda will be prepared and minutes recorded.

Headings for the order of the meeting:

1.1	Personnel attending
1.2	Overview of progress to date
1.3	Overview of the programme requirements for the next period
1.4	Reasons for the delays – critical review
1.5	Resources to meet proposed short term programme

- Labour review
- Plant review
- Subcontractor resources
- Key material requirements

1.6	Architects instructions/information requirements
1.7	Interface between proposed operations – coordination of operations
1.8	Date of next meeting

Project stage	Type of meeting	Purpose	Who is involved
Tender	Pre-qualification	• Determine the suitability & competence of contractors • Shortlist contractors	• Client • Design team • Tendering contractors
	Pretender	• Discuss bidding strategy • Consider risk issues • Allocate responsibilities	• Director or estimating manager • Estimator • Planning engineer
	Tender coordination	• Ensure bid will be ready on time • Discuss further risk issues • Resolve problems and 'loose-ends'	• Estimating manager • Estimator • Planning engineer
	Tender adjudication	• Scrutinise the estimate • Consider risk and profit • Consider programme and construction period • Decide tender sum	• Director • Estimating manager • Estimator
Precontract	Pre-contract (internal)	• Hand over tender documents and pricing notes • Discuss materials and subcontract quotes • Discuss programme and key dates • Discuss risk issues	• Estimator • Planning engineer • Contracts director or Contracts manager • Site manager • QS • Procurement manager
	Pre-contract (external)	• Introduce participants • Hand over contract documents • Establish means of communication • Agree valuation dates	• Client • Architect • Client's QS • Contractor
	Pre-start (external)	• Finalise details, e.g. setting out, statutory approvals, existing services, etc. • Agree handover dates • Handover contractorís programme	• Client • Architect and design team • Contractor
Contract	Monthly/progress (external)	• Discuss progress • Receive site managerís report • Receive QSs report • Table problem areas • Request/receive information • Discuss variations/changes	• Architect • Site manager • Client QS/Contractor QS • Other design team members
	Monthly valuation (external)	• Agree value of work in progress • Agree value of materials on site • Discuss physical progress • Agree preliminaries • Discuss variation instructions	• Client's QS • Contractor's QS
	Health and safety	• Discuss construction phase plan • Consider accident reports • Discuss non-compliance issues • Receive information and updates • Discuss risk assessments and method statements	• Site manager • Site foreman • Health and safety adviser (head office) • Planning engineer • Subcontract manager

PROJECT MEETINGS

Figure 15.9

Subcontract coordination meetings

Most projects these days rely on subcontractors to carry out the bulk of the work. A fairly straightforward project may have 30 or more subcontractors on site and, on large jobs, there may be over 100 subcontractors and utility companies employed. The work of all these firms needs to be carefully organised and coordinated if the project is not to disintegrate into a shambles and many contractors employ a subcontract manager or coordinator to deal with these specialists.

Normally, there will be a pre-start meeting with each of the subcontractors at which the subcontract manager, quantity surveyor/procurement manager and the subcontractors' manager/director will be present. Agenda items will include:

- Subcontract details and price
- Subcontract programme and start date
- Resources required
- Common facilities to be provided by the main contractor (e.g. water and power, site accommodation and storage, waste removal, scaffolding, cranage, etc.)

As work proceeds on site, further meetings will be held with the subcontractors on a regular basis – perhaps weekly for the more complex work packages. These meetings may be held jointly or there may be separate meetings for each of the larger subcontracts. Present at the meetings will be the contractor's subcontract manager and planning engineer together with subcontractors' representatives. Items for discussion will include:

- Subcontract progress
- Problems and delays
- Resource problems
- Interfaces with other subcontractors including shared work zones, shared facilities such as scaffolding and cranage, health and safety risks and work overlaps

15.11 Key performance indicators

In order to measure overall performance on projects, key performance indicators (KPIs) may be used. They can be used to set standards at the start of a project, monitor performance as the project progresses and form a basis for improvement targets for future projects.

Through 'Rethinking Construction' and the Construction Best Practice Programme (merged into 'Constructing Excellence' in 2004 – www.constructingexcellence.org.uk) initiatives set up by the government to assist the construction industry have resulted in the development of KPIs for several aspects of project delivery including:

- Client satisfaction
- Defects
- Safety
- Time and cost predictability

- Productivity
- Profitability

By using KPIs, companies can measure project performance both internally and externally, and thereby establish benchmarks to help the organisation achieve best practice standards.

References

Denyer, J.C. (1972) *Students' Guide to Principles of Management,* 2nd edn. Zeus Press.

Hall, L. (1974) *Business Administration.* M and E Handbooks, McDonald and Evans.

Harper, W.M. (1976) *Management Accounting.* M and E Handbooks, McDonald and Evans.

Harris, F. & McCaffer, R. (2006) *Modern Construction Management,* 6th edn. Blackwell Publishing.

Pilcher, R. (1992) *Principles of Construction Management,* 3rd edn. McGraw-Hill.

Thomas, R. (2001) *Construction Contract Claims,* 2nd edn. Palgrave.

16 Controlling time

16.1 Introduction

Planning is one of the functions of management and a programme is a key tool in the management process. However, very few plans turn out as expected and this is especially the case in construction work. Unforeseen ground conditions, delays due to weather and waiting for instructions are just some of the many reasons why this may happen.

It is therefore crucial to ensure that the planning process includes an element of control so that the plan can be monitored in the light of prevailing circumstances. Additionally, activities may have to be rearranged or additional resources may be required to cope with difficulties encountered and a forecast of their likely impact may also be required. This is the role of the contract master programme which both the contractor and the client's contract administrator can use to monitor what is happening.

The contractor will normally have a contractual duty to complete the works on or before the date for completion stated in the contract and if he fails to achieve this without a valid reason he may suffer the deduction of liquidated and ascertained damages from his contract payments. This is of little benefit to the client, however, because his building will still be late and therefore the contract administrator will use the master programme to check that the contractor is progressing satisfactorily. If things are not going according to plan, the architect or engineer will point this out at the monthly meeting and ask the contractor what he is going to do about it.

In the worst-case scenario, failure to make satisfactory progress could lead to the replacement of the contractor and termination of his employment under the contract.

16.2 Time for completion

Most standard conditions of contract have similar provisions as regards time for completion and somewhere in the contract will be stated:

- Date for possession of the site
- Date(s) for possession of specific sections of the site (where appropriate)
- Date for completion of the works
- Date(s) for completion of section(s) of the works

The above information is to be found in the Contract Particulars in JCT 05, in the Appendix in ICE7 and in the Contract Data in NEC3. Time periods for deferment of possession may also be stated so that the employer may delay handing over the site to the contractor without breaching the contract. The contract will usually state the agreed rate of liquidated and ascertained damages (LAD) for the whole, or where applicable, sections of the works, so that the employer may charge the contractor in the event of late completion of the contract.

Within the main body of common conditions of contract will be found clauses enabling the employer to extend the contract completion date where the contractor has been delayed for reasons beyond his control (provided that the reasons for delay are the same as those specified in the contract). The purpose of such clauses is to preserve the employer's right to charge LADs for delayed completion where the contractor is at fault.

16.3 The contractor's programme

Under most standard forms of contract, the contractor is required to produce a programme of some sort to show how he intends to organise and schedule the project. For example:

- JCT 05 Standard Building Contract – Clause 2.9.1, master programme
  - Two copies of programme for execution of the works as soon as possible after the contract is signed
  - Two copies of programme updates following an extension of time under Clause 2.28.1
- ICE7 – Clause 14, programme
  - Within 21 days of contract award, a programme showing the order in which the contractor proposes to carry out the works
  - At the same time, a general description of the intended arrangements and methods of construction
  - Revised programmes where actual progress varies from the accepted programme
  - Further details of methods of construction, if requested by the engineer, including details of temporary works and use of plant together with related structural calculations
- NEC – Core Clause 3, accepted programme
  - Key dates including starting, possession, partial possession and completion dates
  - A method statement for each operation showing planned equipment and resources
  - The order and timing of the operations
  - Provisions for float, time-risk allowances, health and safety requirements and contractual procedures
  - Revised programmes showing actual progress for each operation and its effect on remaining work
  - The effects of compensation events (such as employer risk delays) and matters giving rise to early warnings (such as design problems or discovery of bad ground conditions) and how the contractor plans to deal with these issues

JCT, ICE and NEC conditions vary as to the detail that the contractor must include in his programme, but the expectation is nevertheless that the contractor will show sufficient detail so that the employer's contract administrator, architect or engineer, may monitor progress and make informed judgements about extensions of time and so on as the work proceeds.

Even in the NEC contract (which is much more explicit than others about the contractor's programme), the extent of detail required in the contractor's programme is very much at the contractor's discretion as is the precise style (e.g. bar chart or network) and format (paper-based programme, computer-based programme, specific software package to use, etc.).

Consequently, the contractor's master programme may be in any of the following formats:

- Gantt (bar) chart with no logic links
- Linked bar chart showing programme logic
- Precedence network
- Critical path analysis (arrow) diagram
- Line of balance
- Time-chainage diagram or
- A combination of the above (e.g. a linked bar chart for substructure and envelope in conjunction with line of balance for frame and fit out)

16.4 Milestones

Milestones are symbols shown on the contractor's programme – normally shown by using a small diamond shape (◆) – that are especially useful in programming and delay analysis:

(1) to denote key events on the contractor's programme such as:
- when the building is planned to be weathertight
- when a section of the work is due for completion
- partial handover dates when the employer hopes to take occupation of part of the building
- commissioning dates for power and other utilities
(2) to show when something actually happened such as:
- a drawing issue or revision
- a site instruction or architect's approval
- the placing of an order

Milestones have a duration of zero but can still be linked to the programme logic in the usual way (e.g. finish–start or start–start, etc.). In some software packages, such as Asta Powerproject and Microsoft Project, they are created by entering a duration of zero alongside the activity description. Project Commander uses 'drag-and-drop'.

Milestones behave in all respects just like any other activity on the programme, but they denote a point in time as opposed to an activity which has a defined start, duration and finish. In delay analysis, they can be linked to other activities in order to show the effect of a late instruction or drawing issue on the rest of the programme.

16.5 Early warning systems

The success of a project is often related to the links developed between the main suppliers, subcontractors and the contractor. Bar chart displays which represent the connection between procurement and the commencement of operations on site have now become an integral part of planning. This is in order to ensure that key dates are met with respect to design requirements, information flow and the delivery or commencement of works on site.

Early warning systems were developed in the mid 1960s by John Laing and extensively used. The system worked because of its simplicity in application and ease of monitoring; it allowed the user to develop his own symbols and letters to denote information requirements.

The NEC Engineering and Construction Contract (Clause 1 – general) highlights requirements for early warning. This places responsibility on both the contractor and project manager to give an early warning on any matter which could delay completion or impair performance of the works in use.

Figure 16.1 shows a range of early warning symbols (or milestone symbols) which may be used to denote an occurrence which affects a supply chain for a component supplier or subcontractor. The example indicates that actual data were released by the client's representative 1 week later than planned in respect of information and nomination. This resulted in the order being placed 1 week later than scheduled with a possible delay in the commencement of steel erection on site. Information requirements relative to a project may also be presented in a tabular format, which meets the requirements of the JCT contract under Clause 5.4.2. Delays in the release of information by the architect or other consultants may result in the contractor applying for an extension of time.

16.6 Progress and delay

Irrespective of the chosen format, provided that the contractor submits regular updates to the programme, and that he proceeds regularly and diligently with the works, and provided the contractor does not suffer any prevention or hindrance from the employer or contract administrator, he is free to complete the works on or before the contract completion date making necessary changes to the plan as the works proceed.

However, it is frequently the case that the contractor is delayed for one reason or another either due to:

- A contractor risk event, for example:
  - Delay by a subcontractor
  - Weather delays (other than exceptional conditions)
  - Labour or plant shortages
- An employer risk event such as:
  - Variations and design changes
  - Late possession of site
  - Exceptionally adverse weather
- Some other reason

This does not mean necessarily that completion will be late as the contractor may make up lost time either by working faster or by rearranging the work sequence more efficiently. Where, however, delayed completion is inevitable or likely, and this is due to an employer risk event, contracts commonly provide a mechanism for the contractor to be granted an extension of time. The reason for this arrangement is not out of kindness to the contractor but so as

EARLY WARNING SYSTEM

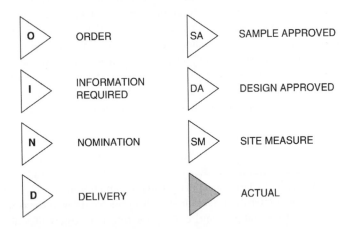

EARLY WARNING SYMBOLS FOR USE ON BAR CHARTS

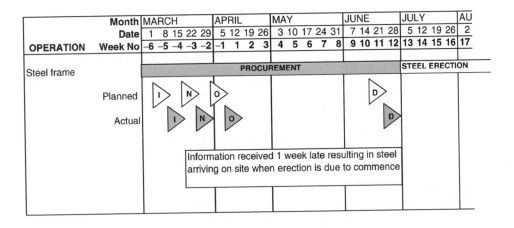

EXAMPLE BAR CHART DISPLAY

Figure 16.1

to preserve the employer's right to liquidated and ascertained damages should the contractor otherwise be late finishing the works.

The usual practice in circumstances where a project is not proceeding as originally intended is as follows:

- the contractor notifies the contract administrator that the works have been or are likely to be delayed beyond the contract completion date

- the contract administrator then has a set amount of time to consider the situation
- if the delay is the employer's fault:
  - the contract administrator will make a decision to grant the contractor an extension of time
- if the delay is the contractor's fault:
  - the contract administrator will remind the contractor of the contract completion date and his contractual obligation to proceed regularly and diligently with the works
- in either case, the contractor's master programme will be an important reference point to assist the contract administrator in coming to a decision

It is frequently the case that matters are not quite so straightforward and there will often be a number of causes of delay (delay events) – some at the employer's risk and some at the risk of the contractor. There will also be issues to untangle such as:

- whether the contractor's resources were adequate or working efficiently
- whether or not certain delays were concurrent
- whether the activities affected were on the critical path or not and
- whether or not the contractor had endeavoured to mitigate the delay (as he is obliged to do under common law)

A further problem is that most contractual obligations to provide a programme are insufficiently precise so as to produce a programme capable of being used effectively as a change control document. For instance:

- is the programme based on a form of logic which links activities together in such a way as to enable the effects of delay and/or disruption to be determined (this would normally mean making a choice between a linked bar chart and precedence diagram, as arrow diagrams are largely obsolete these days due to their inflexibility and unnecessary complexity)?
- has a method statement been provided which forms the basis of the programme?
- does the programme or method statement show details of how the contractor intends to carry out the various site operations together with the resources he intends to use?

Additional problems may include:

- the contractor's programme may be too simplistic
- the programme may contain imprecise logic or
- there may be open-ended or hanging activities in the programme, i.e. where the logic is incomplete
- the programme may not be based on a software package capable of modelling the delay/disruption scenario

16.7 Progress recording

It is of little benefit producing a bar chart programme, putting it up on the site cabin wall, and all sitting around looking at it. For the bar chart to be an effective management tool, it is necessary to record progress on it and critically analyse the operations which are ahead or behind programme. The self-progressing bar chart has not yet been invented!

The purpose of progress recording is to provide a record of weekly and monthly progress in order to assist the construction director, contracts manager and site manager and to inform the monthly site progress meeting with the architect/contract administrator.

Procedure for recording progress

As a minimum requirement, the planning engineer or site manager will monitor the progress on site on a weekly or monthly basis and produce a monthly report. The report should be completed 2 days before the site project review meeting and should consist of a visual progress report generated from the programme.

The report summary should highlight operations of concern, i.e. those which are out of sequence or behind programme. Comments should be added relating to the action to be taken in order to get the contract back on programme. Reasons for the delays may also be indicated, especially where the delays have been caused by variations to the contract, late receipt of information or delays by specific subcontractors.

The monthly progress report should be issued to the contracts director, chief planner, contracts manager, site manager, site-based quantity surveyor and the design and build coordinator where applicable.

Progress recording procedures in practice

In some medium and large contracting organisations, laptop computers are issued to senior site staff. The site manager will have access to the contract master programme and may have attended in-company training programmes in order to learn how to progress record operations on site. Thus the manager is able to progress record the master programme, operation-by-operation, and email the information to his contracts manager for a mid-Monday morning progress meeting in the head office. This is an innovative approach to getting the site manager involved in information technology procedures.

Progress recording on the bar chart

There are various ways of recording progress on the bar chart at weekly or monthly intervals. Responsibility may be delegated to the site-based planning engineer on the larger project, or the site manager on the smaller project.

As an additional safeguard, the planning department at head office may act as an independent monitor of progress, undertaking regular visits to site and preparing a separate report on the progress position. This procedure forms part of the company's overall site management reporting procedures. It also alleviates the problem of site management personnel bending the truth a little when it comes to reporting progress!

Progress recording by colour coding

It may be policy within the company to record progress by colour coding on the bar chart. For example, green could represent the actual time expended on the operation and red could be used to indicate the percentage of the operation completed.

An alternative method of recording progress is illustrated in Figure 16.2. This is based on allocating a different colour to each week of the project and shading the work undertaken in that particular week with that colour. In this way, the progress can be readily observed on the bar chart. This method may form the basis of an as-built programme.

In practice, progress is mostly expressed as a percentage, on an operation-by-operation basis. The bar chart may also show the planned and actual percentage complete at each reporting date.

Progress recording by computer

Many software packages allow the user to progress record by computer. This is usually done on the basis of the percentage completed for each operation, which involves the planner simply entering the percentages, pressing the analysis command and leaving the rest to the computer.

Most software packages display the progress indicator as a 'jagged' line as shown in Figure 16.3 which illustrates a bar chart progress analysis based on a widely used project management software package. The vertical line shows the progress review date, but if activities are ahead or behind programme, the line moves to the left or right accordingly. In this example, structural frame activities 1–4 are complete, activities 5–7 are behind and activity 8 is yet to start. Siteworks and drainage activity 1 is complete and activities 2–4 are behind programme.

Progress recording by earned value

The physical progress on a contract can also be measured according to the earned value of work carried out, using a value envelope. This is established by plotting the forecast value for the contract on the basis of both the earliest and latest times that work activities can be started. The idea is to use the float or free time in non-critical activities to create a control envelope whilst still keeping to the overall programme period as determined by the critical activities (i.e. the critical path). The boundaries of the envelope delineate the intended programme in money terms. The principles of earned value forecasting are explained in Chapter 14.

The monthly valuation is plotted on the envelope and, where the earned value is within the envelope, the contract is progressing according to the programme. Should the earned value fall outside the envelope, this signifies that the contract is either ahead or behind programme.

Figure 16.4 shows an example where the earned value is above the envelope in month 1 (generating value faster, therefore probably ahead), below the envelope in month 2 (behind) and inside the envelope in month 3 (on programme).

The monthly site progress report

It is common practice for an operational progress report to accompany the marked up bar chart. The progress report is normally written by the site planning engineer or the site manager

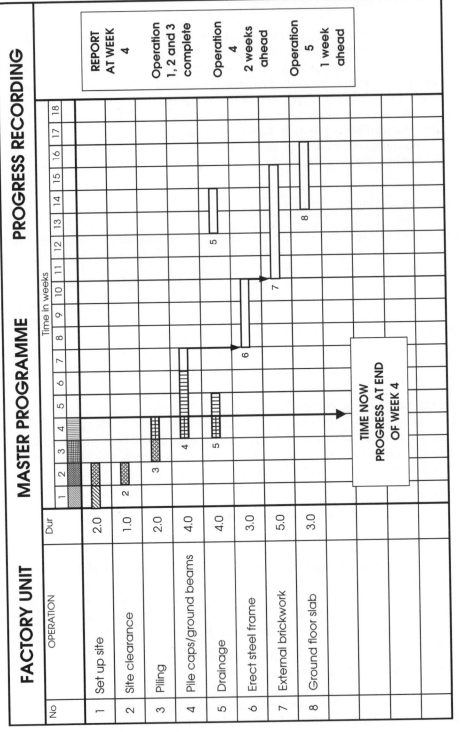

Figure 16.2

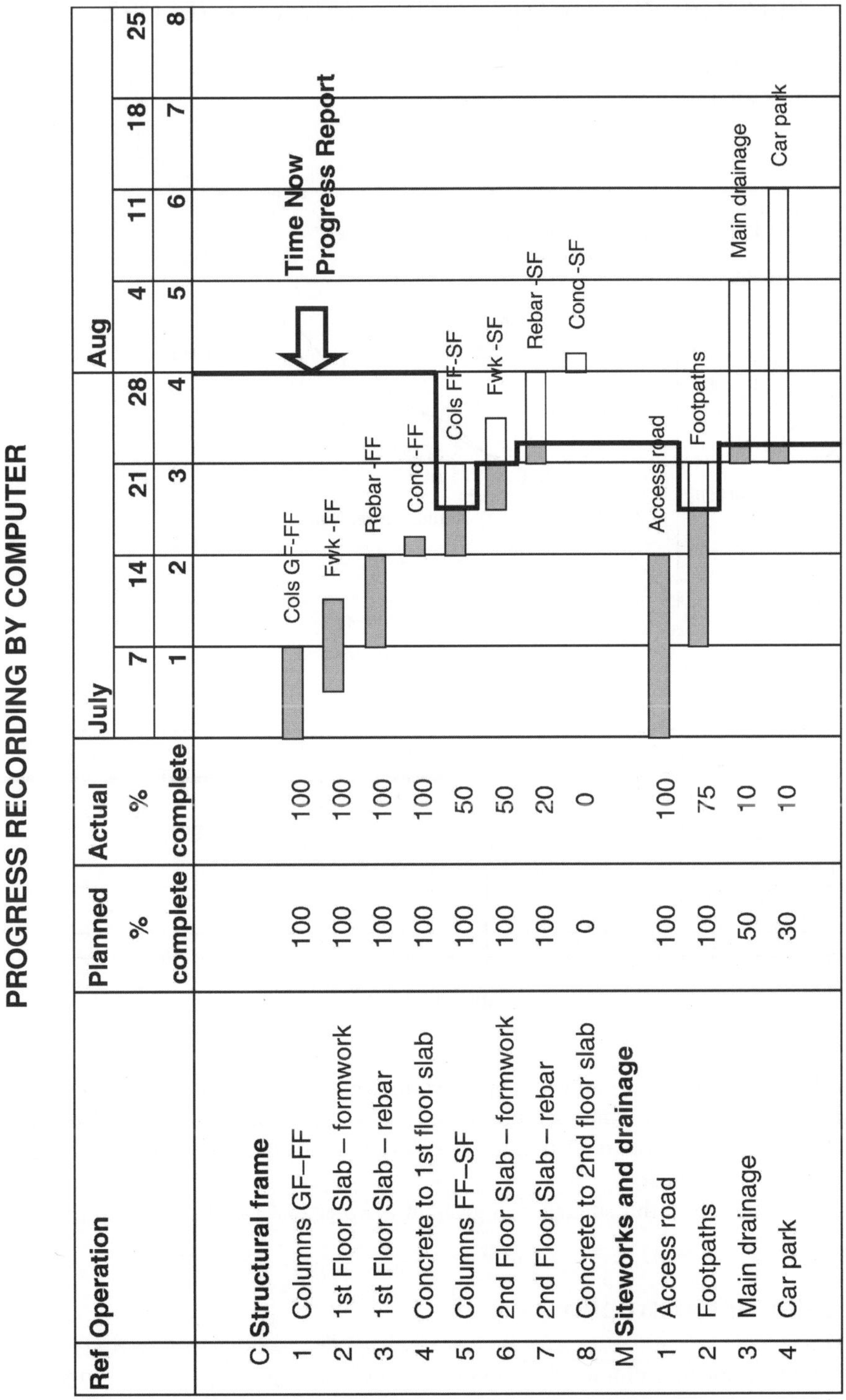
PROGRESS RECORDING BY COMPUTER
Ref
Operation
Planned % complete
Actual % complete
July
Aug
7
14
21
28
4
11
18
25
1
2
3
4
5
6
7
8
C Structural frame
1 Columns GF–FF 100 100
2 1st Floor Slab – formwork 100 100
3 1st Floor Slab – rebar 100 100
4 Concrete to 1st floor slab 100 100
5 Columns FF–SF 100 50
6 2nd Floor Slab – formwork 100 50
7 2nd Floor Slab – rebar 100 20
8 Concrete to 2nd floor slab 0 0
M Siteworks and drainage
1 Access road 100 100
2 Footpaths 100 75
3 Main drainage 50 10
4 Car park 30 10
Cols GF-FF
Fwk -FF
Rebar -FF
Conc -FF
Cols FF-SF
Fwk -SF
Rebar -SF
Conc -SF
Access road
Footpaths
Main drainage
Car park
Time Now
Progress Report

Figure 16.3

PROGRESS CONTROL USING VALUE ENVELOPE

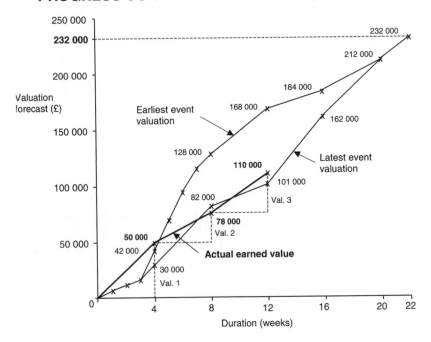

	VALUE FORECAST		ACTUAL EARNED VALUE	VALUATION PROGRESS REPORT
Week	Earliest Event	Latest Event		
4	42 000	30 000	50 000	Ahead of programme
8	128 000	82 000	78 000	Behind programme
12	168 000	101 000	110 000	On programme
16	184 000	162 000		
20	212 000	212 000		
22	232 000	232 000		

Figure 16.4

and may be produced by one of the usual project management software packages. The report highlights the progress position operation by operation. Reasons for project delays may be shown. The current position in relation to key subcontractors, materials and orders yet to be placed is often highlighted.

An extract from a monthly site report is given in Chapter 19 – City Road project.

16.8 Delay and disruption

Delay and disruption are common features of construction projects and this was highlighted in both the Latham and Egan reports – Rethinking Construction (1998) said that *Projects are widely seen as unpredictable in terms of delivery on time.* However, whilst improvements have been made since those reports were written, Constructing Excellence maintains that around 40% of construction projects still finish late.

Reasons for delay and disruption

Most projects that finish late are likely to be delayed by a few weeks or months, but there have been a number of notable high profile examples over recent years, including the Scottish Parliament building and the new Wembley Stadium, which were handed over years later than planned. Finding plausible reasons for such lengthy delays on major projects is not easy, and all the participants will probably have a different view of things. However, delays in the flow of design information, over-optimistic programming, contractual disputes and unusual procurement methods are some of the many reasons cited in the official enquiries into these projects (NAO 2003, the Scottish Parliament 2004).

On 'normal' projects, the reasons for delays and disruption may include:

- Where the employer/client is at fault
  - Incomplete design at commencement on site
  - Delays in the flow of design information
  - Variations and changes in the scope of the works
  - Discrepancies in contract documents
- Where the contractor is at fault
  - Poor quality of workmanship
  - Inadequate planning
  - Under resourcing of site operations
  - Accidents and incidents on site
- Where no one is at fault
  - Unforeseen physical conditions
  - Strikes affecting site labour or the supply of materials
  - Fire, explosion or act of God (*force majeure*)
  - Bad weather

Projects that finish late have implications for both the employer and the contractor. For instance, the employer or client will not be able to enjoy his new facility when anticipated and this may lead to loss of income or utility. Alternatively, the contractor will not be able to release key managers and other resources for new projects as planned leading to delays on other projects or there may be extra costs to bear for running the site establishment for longer than initially planned.

Definitions

'Delay and disruption' is a phrase which is commonly used in construction just like 'loss and expense' or 'health and safety'. The words go together quite naturally but in each case the individual words have quite separate and distinct meanings. As far as 'delay and disruption' is concerned, both words relate to issues of time and completion of the works but 'delay' is quite distinct from 'disruption'. For instance, a project might have been delayed without any disruption taking place, and activities on site may be disrupted without causing any delay to completion of the works.

- **Delay** relates to the act or state of being late – failing to complete the works by the contract completion date for instance.
- **Disruption** is concerned with loss of efficiency due to lower than expected productivity or some interference with normal progress.

Admittedly, delay may become a consequence of disruption, although not necessarily, whereas delay to progress may have nothing to do with the disruption of work on site.

Delay

Delays are caused by events or circumstances which lead to failure to start or complete on time. However, delays to progress and delays to contract completion must be carefully distinguished.

Sometimes delay occurs when programmed activities do not start or finish on time, or both, but this may not necessarily affect the overall completion date of the project. For instance, the employer may defer the site possession date causing a delayed start to a project but the project may still complete on time because there may be spare time in the programme or the contractor may speed up construction so as to make up for lost time. Consequently, if the contractor is delayed, this will not automatically result in an extension of time.

On the other hand, should there be a delay which impacts on the contract completion date, the cause of the delay must be determined in order that an extension of time can be considered or, alternatively, liquidated and ascertained damages may be charged. Delays do not always lead to entitlement to financial reimbursement, but conversely, delays to non-critical activities might result in a claim for loss and expense with no entitlement to an extension of time because the completion date is not affected.

Disruption

Disruption is where the contractor is involved in duplication of effort, out of sequence working or where he has to bring subcontractors or other resources back to site to do more work. Disruption to site activities may not lead to a delay but invariably will result in a claim for loss and expense under the contract due to the extra effort required to do the work.

Types of delay

There are several meanings to the word 'delay':

- **Delay event** – this refers to something that happens, or a set of circumstances that arise, that lead to a delay. The delay event may be the responsibility of the employer or the contractor or may be beyond the power of either party.

- **Culpable delay** – where the delay is the fault of the contractor. This is also referred to as **non-excusable delay** or **contractor delay**.
- **Non-culpable delay** – where the contractor is delayed through no fault of his own. The delay may be due to the fault of the client (or his agents) in which case it is commonly called **excusable delay** or **employer delay**.
- **Neutral delay** – this is where the delay is caused by factors beyond the control of either of the contracting parties but does not mean that neither party bears the risk:
  - Neutral delay at contractor's risk:
    - bad (but not exceptional) weather
    - late delivery of materials
    - delay by subcontractors
  - Neutral delay at employer's risk:
    - exceptionally bad weather
    - discovery of ancient artefacts on the site
    - strikes affecting site work, off-site manufacture or transport
- **Concurrent delay** – where two or more delay events arise at the same time and at least one of the delays has been caused by the client and another by the contractor.

16.9 Extensions of time

The issue of delay is expressly dealt with in standard construction contracts through specific clauses dealing with:

- Extensions of time for specific events beyond the contractor's control
- Procedures for the contractor to follow when making an application for an extension of time
- Procedures for the contract administrator to follow when giving consideration to the contractor's application

There are usually no similar provisions dealing with disruption and therefore the contractor may have to rely for his entitlement upon terms likely to be implied into the contract by a court. Such implied terms may include the employer's undertaking not to hinder or disturb the contractor's execution of the works. The contractor may also seek to rely upon express contract conditions entitling him to loss and expense by applying for additional payment should the regular progress of the works be affected by specified relevant events.

In any event, the job of the contract administrator is to monitor progress and make decisions if there are delays. For example, under the JCT 05 Standard Building Contract, the architect is required to:

- Grant the contractor an extension of time where completion of the works is likely to be delayed due to a 'relevant event' (Clause 2.28.1)
- State the extension of time granted for each relevant event – there may be several (Clause 2.28.3.1)
- Review the situation later on in the light of any further relevant events or instructions which affect the completion date

- Where the contractor is not proceeding regularly and diligently or if the works are suspended, give notice to the contractor (Clause 8.4.1)
- Finally determine the completion date once the works are complete taking everything relevant into account (Clause 2.28.5)

Under JCT contracts, all of the contract administrator's decisions must be 'fair and reasonable' and subject to the contractor endeavouring to prevent delay and do all that may be necessary to proceed with the works. Under Clause 2(7) of ICE7, the engineer is to act impartially and consider whether the delay *fairly* entitles the contractor to an extension of time under Clause 44(3). The NEC3 contract is silent on this matter, but Eggleston (2006) reminds us of the well-established implied term in *Sutcliffe* v. *Thackrah* [1974] that *the architect* (and the like) *will act in a fair and unbiased manner.*

Whilst there would seem to be no real science to all this, in most cases the architect/engineer and contractor will see 'eye-to-eye' and there will be no need for a dispute or any complex analysis of cause and effect should there be any delays. However, agreement may not be so easy in some cases where, perhaps, the causes of delay may be complex and especially so where large sums of money are involved.

In the case of *John Barker Construction Ltd* v. *London Portman Hotel Ltd* [1996], the Judge concluded that the architect did not carry out a logical analysis in a methodical way of the impact which the relevant matters had, or were likely to have had, on [the contractor's] planned programme. The architect was criticised for making an impressionistic, rather than a calculated, assessment of the time which he thought was reasonable. This seems to be a somewhat unfair criticism especially as most contracts do not spell out how extensions of time are to be decided nor do they stipulate the information to be supplied by the contractor to support his application:

- JCT 05 – the architect/contract administrator *estimates* what is fair and reasonable
- ICE7 – the engineer makes an *assessment* of the delay
- NEC3 – the project manager *assesses* the compensation event (delay)

In practice, it is probable that the contract administrator will base the extension of time decision on:

- the contractor's latest programme
- the information submitted by the contractor
- awareness of progress to date
- careful consideration of available facts at the time
- experience

Often there is a tendency to assume that financial recompense automatically follows an extension of time. However, whilst the extension of time and loss and expense clauses are clearly related, it is not necessarily the case that time and money claims are linked. This often comes as disappointing news for contractors leading to lengthy and costly disputes. To avoid this happening, it makes sense to employ a fair and equitable means of determining who is responsible for the effects of any delay events that might arise.

In order to do this, it is necessary to establish what happened and why. However, from a practical standpoint, this is likely to be a complex matter because, unless the project is a simple one, there is unlikely to a single cause of delay – indeed, there may be concurrent delays and some delay events may be more significant than others (i.e. a dominant cause of delay). Consequently, any delay analysis will be dependent for its success on the quality of information and records available and the skill, experience and objectivity of the delay analyst.

Standard forms of contract allow for extensions of time to be granted for delay which is not the fault of the contractor. However, individual contract forms treat the issue of extensions of time differently. For instance, under JCT 2005, the contractor is entitled to an extension of time for relevant events where the contract completion date is *likely to be delayed*, whereas under ICE 7, extensions of time are granted *after the cause of any delay has arisen*. The various contractual provisions vary from contract to contract, as shown in the following examples.

JCT 05 Standard Building Contract

Extensions of time are granted when the architect or contract administrator decides that a 'relevant event' has occurred which will delay the contract completion date.

Clause 2.29 of the contract lists 13 relevant events, including:

- Exceptionally adverse weather conditions
- Compliance with architect's instructions
- Omission or default of the architect (which could be implied to mean failure to comply with an information release schedule)
- Deferment of the possession date by the employer

Clause 2.29.2.1 allows for giving extensions of time where special objects of interest are found (i.e. antiquities).

JCT 05 Design and Build Contract

The provisions of this contract are similar to those of the Standard Form. There are 13 relevant events listed but some of these are slightly different. For example, Clause 2.26.12 allows the contractor an extension of time when a delay in the statutory approvals process affects the contractor's progress.

ICE Conditions 7th Edition

Clause 44 lists six grounds on which the contractor can apply for an extension of time:

- Variations ordered by the engineer
- Increased quantities
- Any cause of delay specified in the conditions of contract
- Exceptional adverse weather conditions

- Delay or default of the employer
- Other special circumstances (unspecified)

Possible causes of delay to be found in other clauses include:

- Clause 7 – delay in the issue of drawings and other instructions
- Clause 12 – adverse physical conditions or artificial obstructions
- Clause 40 – suspension of work
- Clause 42 – failure to give possession

Engineering and Construction Contract

Time is dealt with under Core Clause 3 but this mainly concerns the use and updating of the programme as a management tool, possession of the site, instructions to start or stop work and acceleration.

Extension of time provisions is covered under Core Clause 6 – compensation events. The clause defines 18 compensation events, which include:

- Project manager's instructions
- Employer's failure to give possession
- Project manager's failure to reply to communications in time
- The contractor encounters physical conditions
- An employer risk event occurs

Concurrent delay

Where there is a single cause of delay, there is usually little problem in dealing with prolongation claims from the contractor. However, in practice, the situation is usually far more complex as several causes of delay frequently occur at the same time. Where there is concurrent delay (i.e. where the contractor is also in delay at the same time as an employer's risk event), the contractor's entitlement will be affected. Therefore, extensions of time for instructions issued by the architect after the contract completion date when the contractor is in culpable delay must be calculated on a net basis following the judgment in *Balfour Beatty Building Ltd* v. *Chestermount Properties Ltd* (1993).

Mitigation

An important consideration with regard to both extensions of time and delay and disruption is the long-standing legal principle of mitigation. Consequently, where there is delay and/or disruption, an implied term of the contract will require the contractor to mitigate his loss. This means that the contractor in non-culpable delay cannot just sit back, rub his hands and wait for the money to pour in from the employer but must do something positive, within reason, to reduce the impact of the circumstances in question.

Some standard contracts also include the express term that the contractor will 'constantly make his best endeavours' to prevent delay.

Float

Float, sometimes called 'slack', may be defined as the amount of 'free time' available in the contractor's programme. This free time may arise for a number of reasons:

- It may be the difference between the time for completion stated in the contract and the contractor's target programme. This is considered to be 'free' time because the contractor's obligation is to complete the works on or before the contract completion date. The question as to whom this time 'belongs to' has long been debated in the industry.
- It may be simply be a contingency allowance made by the contractor where the duration of a particular activity is longer than actually necessary to complete the work required.
- It may be shown on the contractor's programme as a contingency allowance for the entire project.
- It may be the latitude available for carrying out an activity which is not on the critical path meaning that the activity may be started or completed earlier or later than shown on the programme without affecting other activities or overall completion (i.e. activity float).

Activity float is the most common type of float where total float represents the amount of time which can be used up by delaying either the start or completion of an activity, or both, without delaying project completion.

The different types of activity float are illustrated diagrammatically in Figure 16.5 and are explained below:

- **Total float** indicates the extent to which an activity may be delayed without compromising the contract completion date. Most modern project management software programmes will calculate total float and this is often shown on the bar chart display as an extension to the activity bar line. Some software packages will also show free float.

  Example of total float = latest start time of activity B − earliest start of activity A − duration of activity A

  In Figure 16.5, this is given by $21 - 10 - 5 = 6$ days.

Total float, however, gives a false impression of how much spare time there may be in the programme because there are a number of components to total float:

- **Free float** – which is that part of total float which can be used without affecting the start of any succeeding activities.
- **Independent float** – which is that part of total float which can be used without affecting any preceding or following activities.
- **Interfering float** – which is that part of total float which if used will 'interfere' with either the start or completion of dependent activities.

The critical path links all critical activities together and is the longest route through the programme. By definition, therefore, critical activities have no float because they must be started and completed on time so as not to delay completion of the project. In the delay situation, critical activities can become sub-critical or even non-critical and other activities

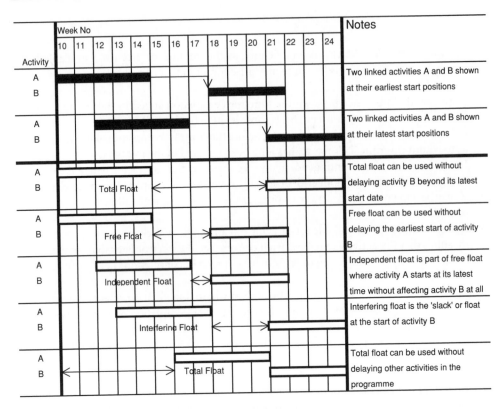

Types of Float

Figure 16.5

can assume critical status in their place. It should be noted that there may be more than one critical path in the programme.

A further point to bear in mind about float is that it is possible that critical activities could have float 'built in' even though they appear on the critical path. This may be because the contractor has included a contingency allowance in his time calculations, or it may simply be the case that estimates of activity durations are not scientifically calculated and are therefore bound to be imprecise. However, any such 'float' would not be evident when undertaking an analysis of the critical path, as this would reveal only those activities with calculable float.

16.10 The 'as-planned' programme

The as-planned programme represents the contractor's view of the project at the outset of the construction phase. Programming software (such as Project Commander, Microsoft Project and Asta Powerproject, etc.) calls this the 'baseline' programme because the programme is 'frozen' and used as a fixed reference against which actual progress can be measured.

The **baseline programme** usually shows:

- Broad bars representing the original planned sequence with logic links hidden
- Thin bars underneath which will remain in place when the programme is updated later on (this is the original 'baseline') – see Figure 16.7 for an example.

Note: In project planning software, the baseline is created by saving the original plan as a baseline. Future changes to the programme will not affect the baseline which will remain in place for change control purposes.

In delay analysis terms, the as-planned programme may have shortcomings such as:

- It may not show all activities included in the project, either by mistake or because, at the time the programme was prepared, such activities were not highly significant
- It may not show some activities in enough detail – e.g. a single bar may be shown to represent a subcontract package which contains several sub-activities within it
- There may be no actual logic or the logic links may be incorrect – there may be activities which have hanging or incomplete logic or they may simply be wrong
- The critical path(s) may be incorrect due to either wrong calculation or because the links are wrong

This may seem inconceivable but programming (especially where the project is large or complex) is not easy and every planner will have a different view of how to put the project plan together. Consequently, the delay analyst may have to 'tidy-up' the as-planned programme and this may lead to accusations that bias or hindsight may have influenced the analysis.

16.11 The 'as-built' programme

The as-built programme is a representation of what actually happened on the project so that this can be compared to what was planned. This programme is not simply the contractor's updated progress programme but is based on the complete project record database including:

- Site diary entries
- Progress records – usually based on percentages complete
- Interim valuations – i.e. monthly financial progress (but this may prove to be unreliable data due to under or over valuation)
- Labour and plant returns such as allocation sheets, daywork sheets and plant hire sheets
- Subcontractors' accounts and claims
- Progress photographs (especially useful when memories begin to fade or management staff move on)
- Correspondence – hopefully all significant events and instructions will have been confirmed in writing and all contractual notices sent in on time
- Minutes of site meetings and the contractor's internal progress meeting

16.12 Delay analysis

Under the common standard forms of contract, it falls to the contract administrator, architect or engineer, to determine whether or not the contractor is entitled to an extension of time under specific circumstances. For instance, under the JCT 05 Standard Building Contract, a decision has to be made whether a 'relevant event' has happened and what the contractor's entitlement will be. Such decisions need to be made when it becomes apparent or likely that the contract completion date will not be met.

When delays occur on a project, the contracting parties will be keen to ensure that they recover their due entitlement under the contract such as an extension of time, loss and expense or liquidated and ascertained damages. It is frequently the case that agreement cannot be reached and adjudication, arbitration or even litigation may follow. Not surprisingly, each of the parties will want the best settlement and to this end will 'argue their corner' as best they can. This often results in a jaundiced view, based more on fiction than fact, which does not stand up to independent scrutiny.

The question is a simple one:

- 'what was the cause of the delay and
- who bears the time and money risk?'

Finding the right way to get the answer is not so easy though because, whilst there are a number of appropriate methodologies available, much depends on the rigor of the investigations undertaken and the skill of the analyst who must determine why a project was delayed.

According to Farrow (2006), the techniques used may be categorised as either **theoretical** or **actual**, although it is recognised that a degree of theory is present in both theoretical and factual techniques due to the extent of personal judgement and interpretation of facts required in most analyses.

When delay or disruption occurs, the particular conditions of contract for the project must be referred to in order to determine whether any relief is available for the event in question.

16.13 Delay and Disruption Protocol

The Society of Construction Law (SCL) Delay and Disruption Protocol (2002) is a non-mandatory guide to good practice for dealing with delay and disruption issues. The Protocol provides guidance on compensation and extensions of time, the preparation and maintenance of programmes and records and proposes a number of methodologies for analysing delays and their effects on project completion. It recommends that, for all but the simplest of projects, the programme should be based on a critical path network using commercially available software, i.e. a programme, whereby the activities are linked with logic.

It is further suggested that where the contractor is proposing to complete the project in a faster time than stated in the contract, the employer should be made aware of his intentions so that the contractor may be granted fair and reasonable extensions of time and/or costs where delayed by the employer and thereby avoiding arguments such as those which arose in the *Guinness* v. *Glenlion* case.

The protocol has its critics and some delay analysts, including Lowsley and Linnett (2006), feel that its overly prescriptive insistence on the use of critical path analysis is costly and places too much emphasis on delay analysis and claims to the detriment of a common sense approach to project planning.

16.14 Delay analysis methodologies

In an excellent series of articles, Farrow (2002) summarises the various methods as follows:

(1) **theoretical methods** – a model of what was planned is produced ignoring how the project was actually constructed and actual events are imposed on the model to see how they may have influenced the end date
(2) the **'but-for' methods** (i.e. as-planned but-for/as-built but-for delay events) – create models of either planned intentions or the as-built project and seek to address causation on the basis of assumption
(3) the **as-planned versus as-built** approach, which overlays two models of the project in order to explain the causes of variances, an approach seemingly favoured by lawyers
(4) **update methodologies**, which relive the project in incremental stages, consider events at the time they arose, and offer the most extensive analyses

Farrow maintains that none of the methodologies are perfect because they all include an element of assumption, subjective assessment and theoretical projection. Farrow also explains that there is no preferred delay analysis methodology, that the more theoretical the methodology the weaker the analysis, and that it is always better to demonstrate what actually occurred.

From a delay analysis point of view, the SCL Delay and Disruption Protocol (2002) suggests that a distinction needs to be made between current and completed contracts and that different delay analysis techniques may be applicable during the project and after completion of the project. Consequently, the protocol categorises the available methodologies as:

- Prospective – looking forward or anticipating future events
- Retrospective – looking back at past events

The above methodologies are fully explained by Lowsley and Linnett (2006).

A further delay analysis methodology called 'windows analysis' is referred to by Farrow (2004), Lowsley and Linnett (2006) and others. It is also called 'time-slice analysis' or the 'snapshot' or 'update' method because it views the project in time slices (e.g. monthly/weekly intervals). The time slice is updated for progress, then the delay events are inserted one by one and the impact of delay events are analysed sequentially. The time delay (if any) at the end of one window is used at the beginning of the next window so the delay (if any) is accumulated.

16.15 Delay analysis in practice

At a practical level, delay analysis is not easy and is a professional discipline in its own right. There are accepted techniques available to help in the process and modern project management software is helpful for speeding up iterations, developing 'what if?' scenarios and producing colourful and professional displays. However, Lowsley and Linnett (2006) maintain that all of the various techniques have *difficulties and shortcomings*, and it is difficult to find *an approach that can withstand robust examination.*

Where a project is currently ongoing, extension of time judgements will be based on the known facts to date (i.e. the latest updated programme and current site records) together with assessments about what might happen later on in the project. When a project is completed, on the other hand, a 'forensic' assessment of what actually happened might be more appropriate. However, the reliability and acceptability of this approach will be heavily dependent on the quality of the records available, such as site diary entries and progress records, to build up the 'as-built' view of the project.

Showing delay on the programme

From a programming point of view, the difficulty is how to show the effect of delays on the contractor's programme. This is not as easy as it sounds because:

- additional activities and/or events have to be introduced into the original programme
- where a delay event is complex a sub-net or frag-net may have to be introduced into the programme
- new logic links have to be made and these need to be correct and without bias
- existing activities may have to be 'split' so as to introduce the delay events and associated logic links (not all software allows this)

Figure 16.6 shows three bar chart displays:

- An original programme comprising three activities A, B and C
- A delayed programme where a 2-week delay is introduced into the programme by increasing the duration of the affected activity (activity B)
- A delayed programme which shows activity B split at the point where the delay event occurred and an intervening 2-week delay event

Note: All logic links are retained.

Delays are rarely straightforward and more often than not there are a number of events or activities making up the delay as a whole. The sequence of delaying events could therefore be put together into a mini-programme or sub-network in order to build up a picture of the delay. The sub-net would have its own logic links and this can then be incorporated into the main programme.

Software

As far as project management software for delay analysis is concerned, in order to successfully determine the effects of delay and disruption, the software must be capable of:

- activity linking, e.g. start–finish, start–start or finish–finish
- producing a baseline of the contractor's original master programme

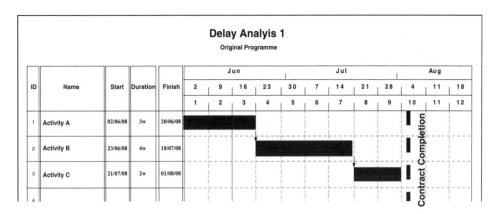

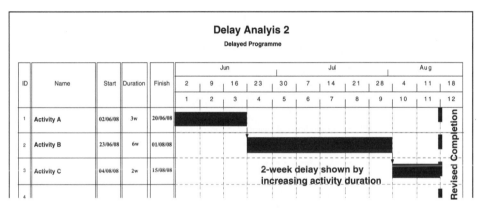

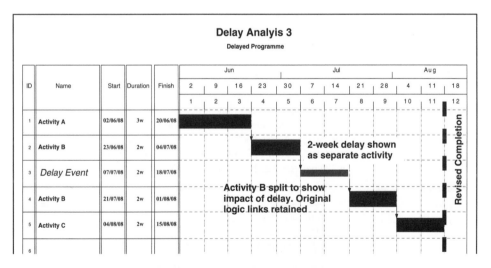

DELAY ANALYSIS
(Displays using Project Commander)

Figure 16.6

- regular updating with actual progress, activity start and finish dates and percentage completion for incomplete activities
- rescheduling the programme to see the effects of delays

Farrow argues that, ideally, the software should also have an integrated database for forensic analysis of project records and that whilst some of the more popular packages are critical path analysis tools, they are weak as regards data processing and vary enormously in terms of cost and the extent of training and experience required to use them effectively for delay analysis purposes.

Worked example

The following worked example is a very much simplified example of a time-impact analysis.

Figure 16.7 shows two bar chart displays for a groundworks project.

Master programme	• Critical activities (activities 1, 2, 4 and 5)
	• Non-critical activities (activity 3)
	• Available float
	• Logic links showing dependencies
	• Contract completion date
Baseline programme	• As-planned programme
	• Baseline bars beneath activity bars
	• Logic links hidden (a facility on most software packages)

Both programmes are effectively the contractor's as-planned programme, but the baseline programme is more likely to be the contractor's internal control document.

Figure 16.8 shows the same project with programme status indicated at week 4.

Progress at week 4	• The baseline programme is shown
	• Progress is shown by a 'jagged' line at week 4
	• Piling is 70% complete (behind programme)
	• Pile caps are 10% complete (behind programme)
	• Other activities have not yet commenced
Reschedule at week 4	• The programme is updated for progress
	• Incomplete and yet to be started activities are moved to the week 4 progress line
	• The baseline remains in place
	• A delay of 4 days is indicated
Note: The delays shown in Figure 16.8 are **contractor risk events** where the cause of delay is the contractor's responsibility under the contract, e.g. bad weather, slow or inefficient working, late delivery of materials or lateness on the part of subcontractors.	

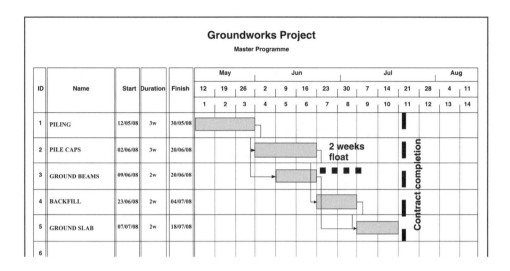

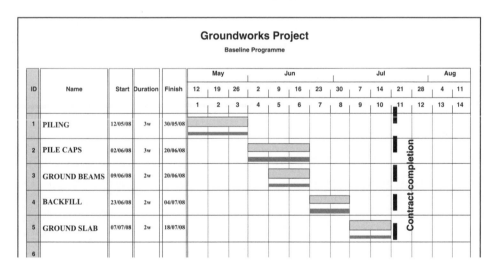

DELAY ANALYSIS

(Displays using Project Coomander)

Figure 16.7

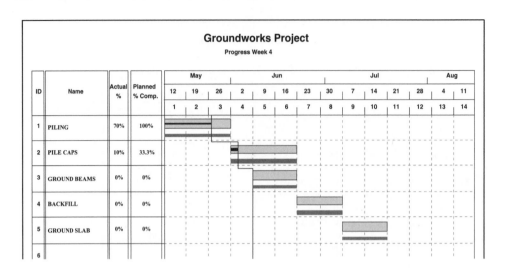

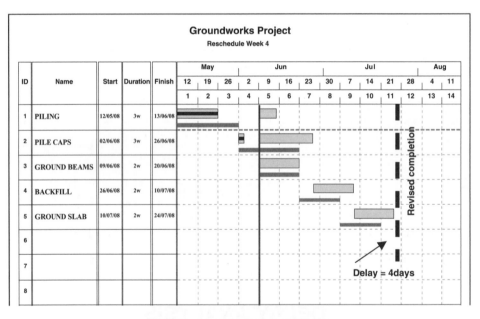

DELAY ANALYSIS
(Displays using Project Commander)

Figure 16.8

Figure 16.9 shows two further displays for the groundworks project, this time at week 8 of the project.

Revised programme at week 4:	• Updated with progress to week 4
	• Logic links unhidden to show dependencies
	• Including 2 employer risk events impacted after the week 4 update:
	   ○ 10 days delay for the re-design of the rebar to the ground beams
	   ○ A further 8 days delay for off-site fabrication of the new rebar

Progress at week 8	• Updated for progress to week 8 • Showing revised completion to week 12 + 3 days • Overall delay to date = 8 days

Note: The delays in Figure 16.9 are partly contractor risk events and partly employer risk event. **Employer risk events** are where the cause of delay is the employer's responsibility under the contract, e.g. delayed instructions, exceptionally inclement weather, unforeseen ground conditions or, as in this case, design variations to the ground beams.

The contractor's extension of time entitlement would therefore be:

Overall delay to completion	8 days	Contract completion delayed from week 10 to week 12 + 3 days
Less contractor delay events	4 days	Contractor was in culpable delay for the piling and pile caps activities
Therefore employer risk events	4 days	The employer risk events followed the contractor's culpable delay so there was no concurrency of delay

Even though the employer risk events amounted to some 18 days, the contractor's extension of time entitlement is only 4 days. This is because the overall delay was only 8 days because the ground beams activity was non-critical and had 10 days float. The contractor was also in culpable delay which amounted to 4 days.

16.16 Project acceleration

Where a project has been delayed for one reason or another, the contractor may wish to bring forward the completion date or may be asked to do so by the client (employer). If the contractor is not at fault in these circumstances, he should be paid for his trouble and effort through a separate agreement with the client. Some standard conditions of contract, such as the ICE conditions, provide for this.

There are several types of acceleration, which have different legal meanings:

• **Pure acceleration** – where the contractor speeds up work on site so as to finish earlier than scheduled at the request of the client

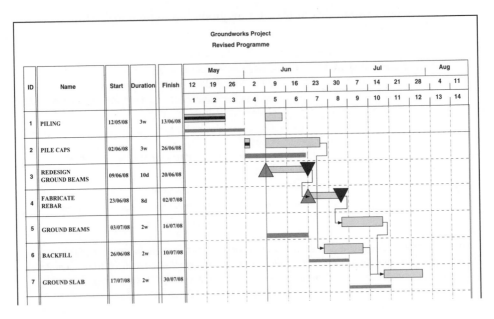

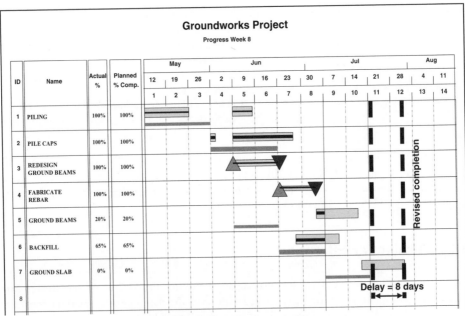

DELAY ANALYSIS
(Displays using Project Commander)

Figure 16.9

- **Constructive acceleration** – where the contractor is effectively forced to work at a faster rate because the contract administrator has delayed or refused a legitimate application for an extension of time
- **Expedite** – where there is culpable delay and the pace of work on site has to be speeded up so as to get back on programme (see ICE conditions for example)

Acceleration, or speeding up the work, can be achieved in three main ways:

- Re-organising the work more efficiently
- Increasing the resources on site
- Both

The work may be re-organised, for instance, by increasing the concurrency of site operations. This may be successful or may introduce other problems – inefficiencies or accidents for instance – where operatives and package contractors may be working on top of each other.

By increasing resources, work may be speeded up initially but here again problems may arise. For example, inefficient gang sizings may be introduced or the site may become congested with plant and operatives. There may also be problems of quality and supervision and extra costs may be incurred due to the need for additional site management personnel.

The contractor, therefore, needs to balance these and other considerations and decide to what extent it is physically and financially viable to speed up the work. There is a break-even point here beyond which diminishing returns may set in.

One method of seeking out this break-even point is to use the management technique of time–cost optimisation. This attempts to balance the direct cost of doing the work with the indirect costs of managing the process. The resultant analysis provides an optimum time and cost solution. The principles of time–cost optimisation, together with worked examples, are covered in Chapter 9.

References

Eggleston, B. (2006) *The NEC 3 Engineering and Construction Contract.* Blackwell Publishing.

Farrow, A. (2002) *Delay Analysis – Methodology and Mythology – Part 1.* Trett Consulting Digest, Issue 27, Article 1.

Farrow, A. (2004) *Delay Analysis – Methodology and Mythology – Part 2.* Trett Consulting Digest, Issue 28, Article 1.

Farrow A., (2006) *Assessing Extensions of Time.* Trett Consulting Digest, Issue 31, Article 5.

Lowsley, S. & Linnett., C. (2006) *About Time – Delay Analysis in Construction.* RICS Books.

The Society of Construction Law (SCL) Delay and Disruption Protocol (2002) The Society of Construction Law.

Web references

Trett Consulting: http://www.trett.com

National Audit Office (2003), The English national stadium project at Wembley: http://www.nao.org.uk/

The Scottish Parliament, The Holyrood Inquiry (2004): http://www.scottish.parliament.uk/

17 Controlling money

17.1 Introduction

The control of money in contracting and development is a complex subject but may be simply considered under the headings shown in Table 17.1.

17.2 Reporting procedures

Reporting procedures on the financial position of a contract must facilitate comparison of what is actually happening during the project with what was planned. It is essential that realistic budgets are prepared at the pre-contract stage and that they are monitored during the progress of the contract.

Monitoring the forecast during the project is an essential part of the project control process. The procedure for preparing cumulative value forecasts based on the master programme is outlined in Chapter 14 and illustrated in Figure 14.10.

Small firms

Within the small contracting organisation, little is done regarding cost and value reporting during the project. The principal is more concerned about his cash flow position and keeping his bank at bay. Cash flow is vital for the survival of the small firm because of the shortage of working capital and the pressures coming from the bank and creditors (including HM Revenue and Customs).

The principal must keep the money flowing despite whatever crisis he may have on the project and comparison of value with cost is only really made at defined stages of the contract, if at all. It may be undertaken at the completion of the substructure or superstructure work, or at the completion of the building.

The majority of small contractors do not operate any form of cost control system. They may simply look at the actual cost and value situation, perhaps at the final account settlement stage, or they may be content to have their overdraft under control. This rather ad hoc approach to the management of cost information may be satisfactory for the small contractor, but as the business expands, there is a greater need to report in a more formal manner.

Table 17.1 Control of money.

Cost	Cost = money out Wages, salaries, general costs and monies paid or owing to builders' merchants, suppliers and trade subcontractors	Control must be exercised over **creditors** (to whom money is **owed**) by checking invoices carefully, making sure that the goods or services have been supplied and not paying the account too early
Value	Value = money in The value of work carried out as certified usually based on the quantity surveyor's interim valuation	Control must be exercised over **debtors** (from whom money is **owing**) to maximise **cash flow**. 'Value' excludes the value of **work in progress** (work carried out which has not been invoiced)
Cash flow	The **difference** between **money in** and **money out**	Sufficient **working capital** must be available to avoid **overtrading** (taking on too much work with insufficient liquid funds) and liquidity problems

Medium-sized firms

Within the medium-sized organisation, which may be undertaking contracts in the order of £500 000 in value, the reporting of financial information takes on more importance. Projects of this size cannot be allowed to drift along with management unaware of the financial position, and procedures must be implemented for reporting on contract profitability as the work proceeds.

The majority of medium-sized contractors undertake some degree of cost reporting during the progress of work. They also have cost–value reconciliation (CVR) procedures as an integral part of the monthly valuation process and these are conducted as one of the surveyor's normal monthly responsibilities. Whether or not senior management seriously consider the information to be reliable is another matter. A detailed account of the principles and practice of CVR procedures is described by Barrett (1981).

Large firms

As a company expands further, emphasis on the analysis of project performance becomes more important and consideration has to be given to more reliable methods of reporting. Within large organisations, there is also a legal requirement for formal reporting procedures.

In the 1970s, a number of large construction firms operated reporting systems which collected data on every single site operation in progress. This practice appears to be out of favour these days and a more global approach is taken to the collection and analysis of data. Companies have at last begun to realise that collecting data for data's sake is an expensive luxury they simply cannot afford.

Financial control systems

As a general rule, 80% of the value of a contract is contained within 20% of the items in the bills of quantities. This is known as the Pareto principle. It is therefore worth considering setting up a control system to deal with the 20% of bill items which ultimately affect the contract's profitability.

Changes in the industry resulting from the introduction of information technology, the greater use of subcontractors and the use of work packages have resulted in the collection and analysis of more selective data. Reductions in profit margins and staffing levels have led to a serious review of existing control systems in order to make the business a little leaner and to gain a more competitive edge. Old, outdated systems have been abandoned due to changes in the nature of construction work practices.

17.3 Monthly cost–value reporting

CVR is the comparison of the project value with the project cost at predetermined intervals during the progress of a project. This interval is normally monthly and tends to tie in with the company's valuation and accounting procedures. The purpose of CVR is to allow management and statutory accounts to be prepared on a more meaningful basis. The basic principles of CVR are illustrated in Figure 17.1.

It is worth noting that a novel approach to understanding accounts has been suggested by a number of Manchester academics as part of an accounting system which attempts to put a 'brave face' on company figures. It has been devised as a way of assessing a company's health by using cartoon faces. This approach should appeal to those responsible for interpreting cost–value reports in construction and should assist in providing instant feedback to senior management. Figure 17.2 illustrates the possible application of 'expression management' to cost–value reporting.

17.4 CVR terminology

The procedure for reconciling cost and value at a cut-off date at the end of each month during a project is outlined in Figure 17.3 and this should be read in conjunction with the following definitions.

Forecast value (cumulative)

This is a forecast of the cumulative value based on the contract master programme. It is obtained by allocating money to the bar chart or schedule of operations and presenting the monetary figures in the form of a cumulative line graph or cumulative value forecast. It must, however, be based on a realistic assessment of the sequence of work or contract programme (refer to Chapter 14, Figure 14.8). The forecast may be presented in tabular or graphical format.

PRINCIPLES OF COST–VALUE REPORTING

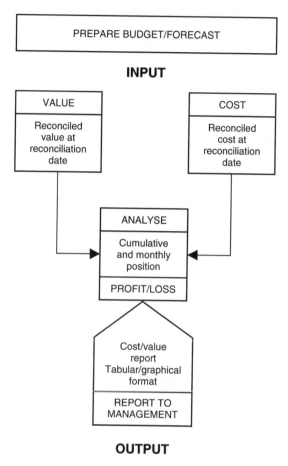

Figure 17.1

Reconciled value (cumulative)

This is the project value assessed at the cut-off date, which is often referred to as the contract 'CVR date'. This may be somewhat different to the value agreed at the monthly valuation date with the professional quantity surveyor and the contractor. This value, for the purposes of comparison with the project costs, must be the reconciled value adjusted for the time variance between the valuation date and the reconciliation date. This may involve assessing the value of the work undertaken in the intervening period between the actual valuation date and the cut-off date. Depending on valuation date, the adjustment may be positive or negative. Figure 17.4 indicates a typical format for presenting the assessment of the reconciled value.

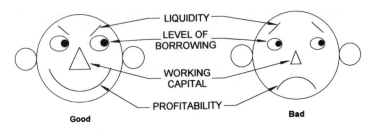

EXPRESSION MANAGEMENT

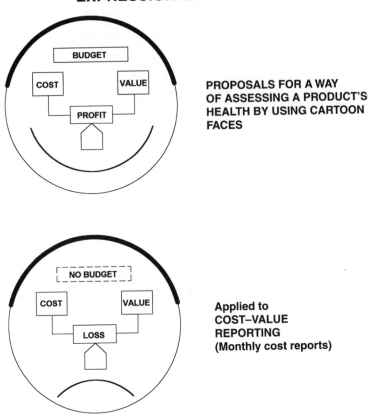

PROPOSALS FOR A WAY
OF ASSESSING A PRODUCT'S
HEALTH BY USING CARTOON
FACES

Applied to
COST–VALUE
REPORTING
(Monthly cost reports)

Figure 17.2

Reconciled cost (cumulative)

This is the cost expended at the date of reconciliation. It is the cost in the contractor's cost
ledger, adjusted to the cut-off date or reconciliation date. The cost assessment must include
all accruals for materials and subcontractors which have not been included in the cost ledger
at the date it was closed. Assessment of the reconciled cost is one of the major areas of error in
the reconciliation process, especially in the assessment of the subcontractor accruals. Similar
errors also occur in the assessment of the material costs. Materials accruals represent the cost

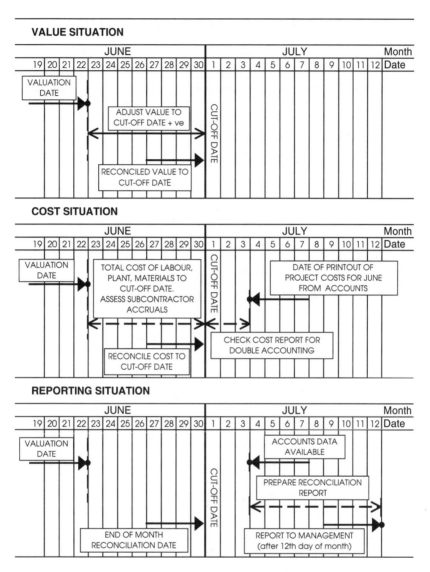

RECONCILIATION PROCEDURE

Figure 17.3

of materials delivered to site but not yet included in the cost ledger. For this purpose, it may be necessary to put a value to goods received records and material delivery notes.

Reconciliation date (or cost–value cut-off date)

This is the date agreed by management when the comparison or reconciliation of cost and value is to take place. It is usually the date when the monthly accounting period is closed and is

ASSESSMENT OF RECONCILED VALUE

GROSS VALUE		VALUATION DATE May		
Gross Value Certified by the Professional Quantity Surveyor				176 000
		ADD	OMIT	
ADJUSTMENTS TO RECONCILIATION DATE May				
PLUS or MINUS				
Under valuation +ve				
Over valuation –ve			5 000	
Adjusted value to date of reconciliation +ve		15 000		
Variations issued (not yet included in valuation)		2 000		
Dayworks – ditto.		1 000		
Remeasured work sections		3 000		
Preliminaries adjustments – under valuation +ve or over valution –ve			5 000	
Materials on site adjustments		1 000		
		22 000	7 000	15 000
RECONCILED VALUE				**£191 000**

Figure 17.4

frequently referred to as the **cut-off date**. This may be the last Friday in each month, the 30th of each month or simply the last day in the month.

Date of report to management

The report date in principle should be as close to the cut-off date as possible but will depend on when the project cost figures are available. Figure 17.3 illustrates this as being achieved by the end of the 12th day of the month.

17.5 Cost–value reports

Figure 17.5 illustrates a typical format for a CVR report in a large contracting organisation. Various adjustments to the cost are indicated in order to match cost with the adjusted value.

COST–VALUE RECONCILIATION REPORT

Contract _____   Valuation No.   _____

Contract No. _____   Date of valuation   _____

Contract duration _____   Month   _____

VALUATION ASSESSMENT	CUMULATIVE	THIS MONTH
Value of certificate to / /		
ADJUSTMENTS :		
Adjustment to valuation date		
Preliminaries adjustment		
Overvaluation		
Variations		
ADJUSTED VALUATION TOTAL		

CONTRACT COST ASSESSMENT	CUMULATIVE	THIS MONTH
Contract costs to / /		
ADJUSTMENTS TO COST (ACCRUALS)		
Plant		
Materials		
Subcontractors		
Inter-site costs		
PROVISIONS		
Subcontractor liabilities		
Future losses		
Maintenance/defects costs		
Cost of delays		
Liquidated damages		
ADJUSTED COST TOTAL		

PROFIT (LOSS) As a value		
Percentage		

Date of reconciliation   /   /   Prepared by :

Figure 17.5

Accruals are costs which have been incurred but for which invoices have not been received by the accounts department. Accruals represent provisions recognising future costs which have to be assessed and taken into account in order to report true value.

The comparison of the cumulative value with the cumulative cost is often referred to as the monthly cost report and a suitable form for this is shown in Figure 17.6. The comparison

MONTHLY COST REPORT
COST—VALUE RECONCILIATION

Date	Val. no.	Certified value	Recon. value	Cum. cost	Cumulative Profit	%	Monthly Value	Cost	Profit	%
30/5	1	27 000	29 000	25 000	4 000	16.0	29 000	25 000	4 000	16.0
30/6	2	64 000	67 000	59 000	8 000	13.5	38 000	34 000	4 000	11.7
28/7	3	110 000	112 000	102 000	10 000	9.8	45 000	43 000	2 000	4.6
26/8	4	170 000	175 000	163 000	12 000	7.4	63 000	61 000	2 000	3.3
25/9	5	270 000	280 000	265 000	15 000	5.6	105 000	102 000	3 000	2.9

Forecast cumulative value
(based on programme)

Val. no.	Forecast cumulative value	Val. no.	Forecast cumulative value
1	30 000	5	310 000
2	70 000	6	410 000
3	130 000	7	480 000
4	200 000	8	560 000

Figure 17.6

may be presented in such a table or alternatively in graphical format. Figure 17.7 indicates the principles of the relationship between the cumulative value forecast, actual reconciled value and reconciled cost presented in graph form. The value variance and time variance have been highlighted.

Variance analysis which highlights the difference between actual and expected figures forms an essential part of the cost–value reporting procedure, but it should be noted that it is

PRINCIPLES OF COST–VALUE RECONCILIATION ANALYSIS PRESENTED GRAPHICALLY

Value–time/Cost–time relationship

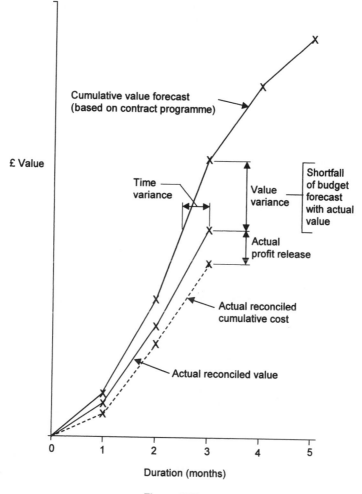

Figure 17.7

important to review the contract's progress when considering variances as there must be reasons for any shortfall in the project value.

Data relating to a project has been indicated on the monthly cost report in Figure 17.6, together with the forecast cumulative value for the project. The relationship between forecast value, actual value and cost has been presented in graphical form in Figure 17.8 and the percentage profit release situation can be seen in both cumulative and monthly terms in Figure 17.9.

VALUE–TIME/COST–TIME RELATIONSHIP

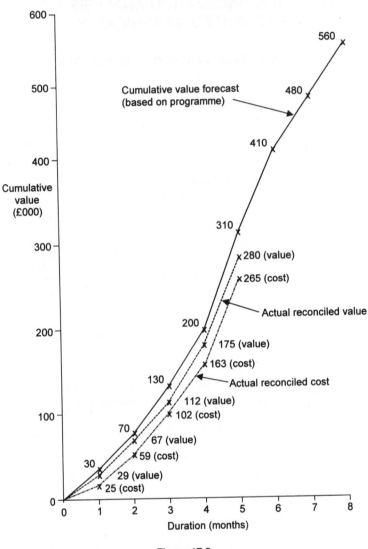

Figure 17.8

Management tends to react more to data presented graphically, which clearly indicates the relationship between forecast, actual value and cost.

Commentary on Figure 17.9

The cumulative profit release has been slowly declining each month of the project. In 5 months, profit has fallen from 16 to 5.6%. During this period, the monthly profit has declined from 16

PERCENTAGE PROFIT RELEASE

Date	Validation No.	CUMULATIVE PROFIT	%	MONTHLY PROFIT	%
30/5	1	4 000	16	4 000	16
30/6	2	8 000	13.5	4 000	11.7
28/7	3	10 000	9.8	2 000	4.6
26/8	4	12 000	7.4	2 000	3.3
25/9	5	15 000	5.6	3 000	2.9

Profit forecast at tender stage 10%

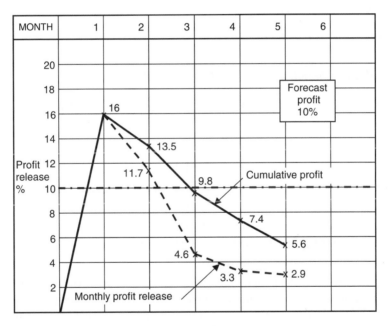

Figure 17.9

to 2.9%. During months 3, 4 and 5, the average monthly profit release has been some 3.5%. At month 5 of the 8-month contract, the cumulative profit release of 5.6% is well below the forecast profit of 10%. It is doubtful that the contract will achieve its forecast margin without some drastic action by senior management.

Part of the cost–value reporting procedures within the larger contracting organisations includes preparing a forecast of the project's cost and profitability to completion. This will involve consideration of the profit level capable of being achieved on the remaining operations to be completed (between month 5 and the end of the contract).

17.6 CVR case study

Contract brief

Table 17.2 indicates the contract value and cost position at the end of month 3 of a 4-month project.

Table 17.2 Contract value and cost position.

Month	Cumulative value forecast (£)	Actual cumulative (£)	Actual cumulative cost (£)
1	30 000	22 000	20 000
2	90 000	65 000	60 000
3	140 000	100 000	92 000
4	180 000		

The forecast project profit margin is 10%.

Analysis of the variance at the end of each month

An analysis of the variance at the end of months 1–3 has been outlined in the form of a report to senior management, as follows.

Contract variance analysis at the end of month 1

	£		£
Forecast value	30 000	Actual value	22 000
Actual value	22 000	Actual cost	20 000
Value variance	**8 000**	Profit variance	**2 000**

The value variance of £8000 represents a time variance of approximately 1-week delay in the progress of the works. This position must be verified by contract progress reports and the reasons for the 1-week delay at this stage of the project must be established. Is the delay due to the contractor, his subcontractors or delay in the receipt of information from the architect? All these questions must be answered.

The actual profit release to date is 10% and the contract is therefore within the forecast margin.

Contract variance analysis at the end of month 2

	£		£
Forecast value	90 000	Actual value	65 000
Actual value	65 000	Actual cost	60 000
Value variance	**25 000**	Profit variance	**5 000**

The value variance of £25 000 represents a time variance of approximately 2-week delay in the progress of the works. The reasons for the 2-week delay must again be investigated and actual progress checked against the master programme. The effect on the project completion date must also be assessed and, if appropriate, requests for an extension of time should be considered.

The cumulative profit release to the end of month 2 is 8.3% and the monthly profit release is 7.5%. Both of these figures indicate a need for concern at this stage of the project.

It is recommended that the work rate be increased during the next 8 weeks in order to bring the contract back on programme. Intensive short-term planning procedures should be implemented at site level.

Contract variance analysis at the end of month 3

	£		£
Forecast value	140 000	Actual value	100 000
Actual value	100 000	Actual cost	92 000
Value variance	**40 000**	Profit variance	**8 000**

The value variance of £40 000 is now giving rise for concern as the project is now some 3 weeks behind programme. In order to complete the programme on time, some £80 000 value of work will require to be completed next month. Realistically, this is not possible considering the project's past performance.

The action recommended at the end of month 2 has not been fully implemented. A serious overview of the reason for the delays to date must be undertaken. A realistic project completion date must now be assessed and the client informed of any delay to the project completion date.

The cumulative profit achieved to date is 8.7% and the monthly profit release is 9.4%. Possible liquidated damages to be levied at contract completion will affect the final profit release. A cost and value forecast to the completion of the project should be prepared.

Figure 17.10 indicates the relationship between forecast value, actual value and actual cost presented graphically. At the end of month 3, the total value variance is £40 000 (negative) and the time variance shows the project approximately 3 weeks behind programme.

17.7 Interim valuations using S curves

Case study

Project particulars

- The construction of a 60-bed hotel and adjacent leisure facilities. The six-storey hotel building is of timber-framed construction with traditional brick finishes
- The contract has been awarded as a design and build project as part of a partnering arrangement with the contractor. To date, 20 similar projects have been successfully completed under this arrangement
- Project value is some £2.1 million with a contract period of 28 weeks. The contract payment terms are every 2 weeks and payment is to be made within 21 days. The contract is fast-track

VARIANCE ANALYSIS

MONTHLY AND CUMULATIVE REPORT

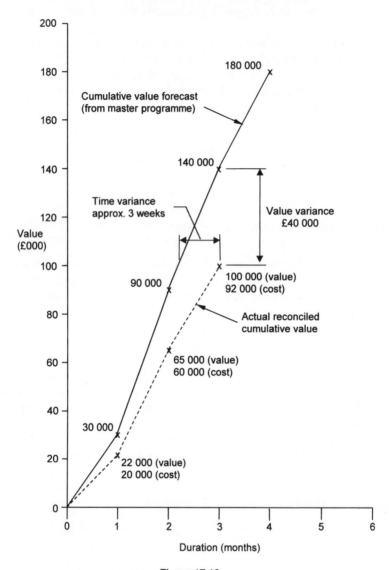

Figure 17.10

using named subcontractors, all of whom have worked with the contractor on previous similar contracts
*   The project is under the control of an executive project manager employed by the client and extensive trust has been developed between the parties over the partnering period

Table 17.3 Abstract from S curve.

Week number	Cumulative value forecast (£)	Week no.	Cumulative value forecast (£)
2	69 250	16	1 311 000
4	180 600	18	1 506 000
6	326 500	20	1 683 000
8	499 300	22	1 835 000
10	691 800	24	1 953 000
12	896 200	26	2 031 000
14	1 105 115	28	2 100 000

The valuation process

Interim payments (every 2 weeks) are based on a forecast S curve agreed between the client and contractor at project commencement. The S curve is based on similar principles to formulae used by some government departments. An abstract from the S curve printout is given in Table 17.3.

Release of valuation monies

The client's project manager allows the contractor to self-certify up to the end of week 14 (7th payment stage). At this point in the contract, a check is made on the progress of the works. If the contract is on programme, self-certification continues up to week 24 (12th payment stage) of the project. At this stage, a further progress check is made and the value of contract variations to date assessed and added to the next payment.

At the end of the contract, the contractor submits his final account within 6 weeks of contract handover.

Note: This is a practical example of partnering at its best – where complete trust exists between client, contractor and subcontractors.

Reference

Barrett, F.R. (1981) *Cost Value Reconciliation.* Chartered Institute of Building.

18 Controlling resources

18.1 Introduction

Most people in the construction industry would acknowledge that the contractor's site manager has one of the toughest jobs in the business. Armed with a set of contract documents, a programme and a health and safety plan, the site manager has to use his knowledge, experience and management skills to bring his project to a successful conclusion – on time, on budget and zero accidents – as well as making sure that the client is satisfied with the quality of the finished product. Oh, and by the way, there is usually a contract profit target to achieve and sales handover targets for speculative developments.

The site manager is not alone of course, and has the full backing and resources of the company at his disposal. In most companies this will include planning, health and safety, quantity surveying, subcontractor procurement, purchasing and accounts backup. There will also be support from experienced middle and senior management who will play a 'hands-off–eyes-on' role in the project.

Clearly, a project needs to be adequately resourced in relation to the scope and content of the work required and the contractor's programme and this has to be achieved within the cost budget allowed by the contractor at tender stage. If the costs increase, and the scope of the work remains unaltered, the contractor's profit margin will be eroded, possibly to the point where the contract makes a financial loss.

In order to achieve a successful project outcome, the site manager will need to marshal the resources at his disposal to best effect including site labour, subcontractors, plant, materials and site supervisory staff. However, these resources are scarce and/or expensive and therefore cannot be considered to be available 'on tap' or capable of being used wastefully or indiscriminately. Consequently, resources have to be budgeted, planned, mobilised and controlled and resource availability has to be a key consideration in drawing up the master programme.

The control of project resources requires consideration of:

- Labour
  - Directly employed labour
  - Subcontract labour
- Materials
  - Control of waste at site level
- Plant

- Subcontractors
- Preliminaries

18.2 Labour control

Many contractors prefer to have little or no site-based labour at all and they package up the entire project to a variety of subcontractors. Other contractors like a combination of directly employed (cards-in) labour and self-employed labour for general building operations such as concrete work, drainage, brickwork, carpentry and joinery, etc., and sublet the remainder of the work. Some contractors, on the other hand, employ the entire workforce directly. There is nothing typical about construction!

Directly employed labour

Despite the predominance of specialist subcontractors, there are some contractors who employ their own labour force and some companies that employ no subcontractors at all. This is often the case with contractors who specialise in particular markets such as pub refits. In this market, the contractor has to be 'in and out fast' and therefore needs to employ all the trades directly to ensure that they will be available when required.

There are also examples of very large contractors who employ their own workforce – both general building trades and specialist trades – because their business model consists of general construction, plant hire, precast concrete manufacture and specialist trading companies (such as demolition, piling, mechanical and electrical).

On large projects, the contractor is more often concerned with keeping the overall labour expenditure within the estimated allowance. The monitoring of labour expenditure involves recording the actual man-weeks expended each week on the project and matching this with the forecast (see Chapter 15). It is important to relate the analysis to the actual contract progress situation, because the reason for an apparent overspend can simply be due to the project being ahead of programme.

Self-employed labour

It has always been common practice in construction for self-employed people looking for employment to call at a new site and ask for 'a start'. However, the site manager cannot take people on in this way any more because of health and safety competence issues and also because of income tax legislation.

One of the problems with self-employed labour is the New Construction Industry Scheme (New CIS) which is designed to prevent abuse of the income tax system (see Chapter 6). The intention is not to prevent self-employment but to prevent the employment of self-employed people on a permanent basis thereby avoiding the PAYE and National Insurance system.

Self-employed labour therefore has to be engaged on a contract-by-contract basis on terms that clearly indicate a contract *for* services rather than a contract *of* service (which implies direct employment under a 'master–servant' arrangement). The issue is one of 'control'. If the operative organises his own work, provides the necessary tools for the job, makes his own decisions as to how the work is to be done, and there is a clear agreement as regards the price to be paid, then this implies self-employment. The operative is thus classified as a 'subcontractor' under tax rules and if his tax status with HM Revenue and Customs has not been 'verified' he will suffer a 30% reduction in the labour element of his payments.

The widespread use of subcontractors in construction has shifted the emphasis away from the need for contractors to employ sophisticated labour control procedures, as was the case in the days of the general contractor who directly employed his own workforce.

Resource histogram

When planning a project, it cannot be assumed that any number of resources will be available just as required on the contractor's programme and it may be the case that the programme will have to be adjusted to suit the resources available.

A resource histogram is a useful visual presentation of labour demand which shows how much labour is required and when. The histogram shows where the peak labour demands will be and, from this, the site manager can see whether he will have sufficient labour available to meet the demand. It is usual to have a separate histogram for each type of labour or plant.

Resource histograms can be easily prepared using standard project management software such as Powerproject, Microsoft Project or Project Commander, etc. The procedure is as follows:

(1) Create the resource categories, e.g. bricklayers, joiners, general operatives, etc.
(2) Decide on the numbers of operatives available in each 'pool' or category
(3) Allocate the labour demand from the method statement to each activity on the master programme
(4) Ask the software to give you a resource graph for each of the labour categories
(5) Look at the histograms in the same view as the bar chart programme in order to see which activities on the programme as causing the peak demands
(6) Inspect the resource histograms to see where resources are 'overloaded', i.e. where the peak demand exceeds the resources available

Once the resource allocation procedure has been followed, activities can be inspected for criticality and perhaps re-scheduled to overcome over-allocation problems. This might result in a longer construction period however. Failing this, the site manager might get round the problem by considering working extra hours each day or working overtime at weekends. To do this, simply ask the software to change the default calendar for individual activities to create a longer working day or change weekends from 'non-working' to 'working' for that activity.

Resource levelling

An additional problem concerning resources is that of 'peaks and troughs' in resource demands. It might be the case that the contractor is able to resolve resource over-allocation problems for bricklayers and joiners, etc., but this does not necessarily ensure efficient resource usage in itself.

Consequently, the labour configuration may still fluctuate such that six joiners are required on site one day and four the next, none the next day and then six the following week. This is not an efficient use of available resources and ideally there should be a gradual increase in resource demand in the early stages of a project which builds to a peak and then gradually slows down towards the end of the job.

Sorting out this type of problem is called Resource Smoothing or Levelling. This is not easy to do and an ideal solution is difficult to achieve in practice. However, by looking at float in the programme, and working overtime where possible, the demand may be 'smoothed' to some extent. The choice is whether to level within the slack (float) available or do so irrespective of delay to the project completion time. This would make the project either 'time critical' or 'resource critical'.

The software will level the resources for you and this can be done automatically or manually. Automatic levelling is best avoided as the computer will make decisions which a human would probably not make. For instance, the computer might decide to 'split' an activity into two parts with a time gap in between so as to level resource demands. This might be OK in theory but is probably impractical in reality as it may not be feasible to split a large concrete pour or a slipform operation for example. Also, the computer does not optimise its decisions and does not consider and 'weigh-up' all the options as a human would. Manual levelling gives the option to select specific activities to be levelled and allows more control than automatic levelling.

It should be noted that it may be difficult to smooth or level more than one resource because revising the programme again for another resource may reverse the initial smoothing decisions and cause further resource overloading. The resource to smooth will be either the most used, or the most expensive, or the least flexible or the least available. The eventual outcome will inevitably be a compromise.

Worked example

Figure 18.1 illustrates a 45-day project with nine activities A–I. Activities A, C, D, F, H and I are on the critical path (i.e. there is zero float) and activities B, G and E have 15, 24 and 9 days of float respectively.

The contractor has allocated resources based on the earliest start for the programme activities. This is shown as a histogram in Figure 18.2 which indicates that there is a peak demand for joiners between day 5 and 15. The problem for the contractor is that he only has six joiners available to work on this project.

If the project completion date is not to be extended, the only solution to the problem is to level the labour demand within the programme according to the number of joiners available. This means adjusting the work activities within the available float (but obviously not the critical ones) to reduce the peak labour demand situation. This is shown in Figure 18.3 which also indicates improved continuity of work for the joiners. The solution is not ideal, however, because there is a fall in demand during days 17–24. This gives the site manager a problem from a labour control point of view because the joiners may find work on another site and decide not to come back.

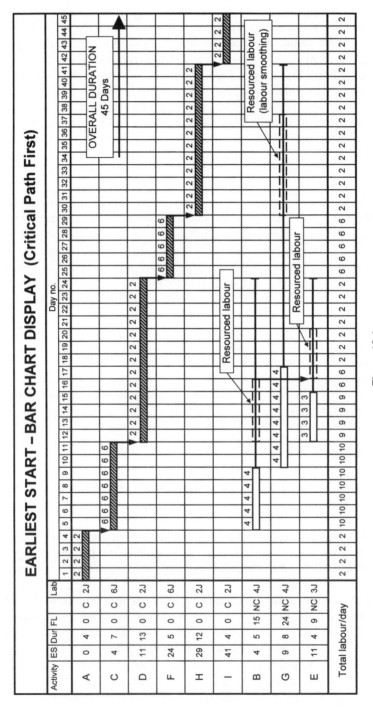

Figure 18.1

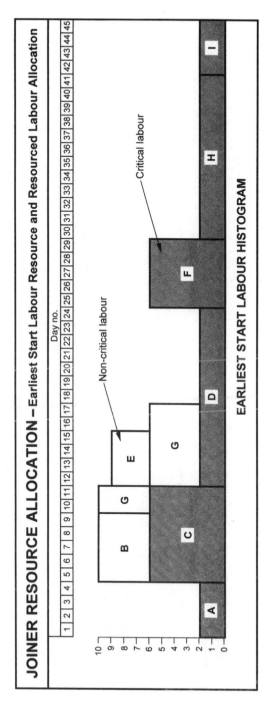

Figure 18.2

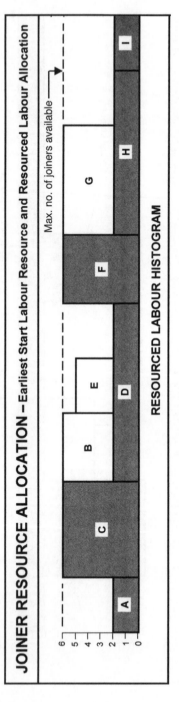

Figure 18.3

18.3 Materials control

Materials expenditure represents a major proportion of contract value. Therefore, the control of purchasing, scheduling, delivery and handling of materials on site is an essential part of the control process.

Within small organisations, the responsibility for all aspects of material control lies with the principal. He is the estimator, buyer, surveyor and contracts manager all rolled into one. In medium-sized companies, responsibility for the purchase of materials may rest with the estimator/buyer or may be part of the surveying function. Large companies, on the other hand, usually have departments responsible for buying and procurement, estimating, surveying, contracts and administration, and therefore a formalised approach to the procurement and management of materials is necessary.

Good communications are important and the buying and contracts sections need to liaise closely in order to ensure that materials arrive on site on time. Contact needs to be established between the contract buyer and the site manager in order to ensure that the material call-off schedules clearly tie in with the programme of work.

Responsibilities also have to be clearly defined. The responsibility for buying materials within the estimate allowances lies with the buyer and any resulting savings created by efficient buying contribute to the profitability of the contract.

Materials management

Once materials have been delivered to site, they become the sole responsibility of the construction site manager who has to make arrangements for handling, distribution around the site and fixing them in position. He is also responsible for material loss and accounting for excessive waste.

An essential component of materials management begins with site layout planning. Designated areas must be allowed for materials storage in order to reduce waste and damage to materials and these areas should be located away from site traffic routes where materials could be splashed with water and mud.

There is a wide variety of mechanised materials handling equipment available to the site manager. Some enable materials to be offloaded and placed into the exact position required for final fixing, such as rough terrain fork lift trucks and telescopic handlers, and others facilitate the direct placing of ready-mixed concrete and screeds using concrete or screed pumps. Elevating materials from ground level may be accomplished with telescopic elevators, tile and block loaders as well as scissor lifts, mast climbers and cranes with various lifting and mobility capabilities.

Effective materials handling requires planning and organisation and, by choosing the right piece of equipment for the prevailing conditions, site efficiency can be increased and costly waste reduced. There is really no excuse for inefficient materials handling but on many sites numerous examples of the mismanagement of materials can be observed:

- Excessive waste left under scaffolds, including bricks, blocks, skirting boards, fascia boards, drainage fittings, etc.
- Expensive facings and engineering bricks being bulldozed into the ground and then covered over with topsoil to provide 'instant brick gardens'

- Materials being stored on uneven ground, adjacent to unprepared access roads, allowing the material to become contaminated with mud and water
- Pallets of bricks and blocks unloaded directly on to unprepared ground, away from the workplace
- Damage to materials while unbanding the packs
- Roof trusses being stacked on unprepared areas, allowing them to distort and twist
- Lack of covering and protection to internal timber floor joists, door frames and finishing joinery items. Structural timbers left unprotected in the rain
- Excessive thickness of ready-mixed concrete to in situ concrete kerb beds
- Out of sequence working, resulting in the excessive waste of stone filling materials, bricks and blocks, etc.
- Commencing foundation work with no provision for adequate access to the works. This resulted in chaos with respect to the storage of materials around the work area

Waste

Historically waste has always been an 'on-cost' in construction (i.e. a percentage added to the net cost of materials) but new thinking suggests that waste should be seen as a resource. By improving on-site practices, the industry has the opportunity to benefit the environment and improve its performance and 'bottom line'.

According to DEFRA (2006), controlled waste from construction and demolition amounts to 100 million tonnes per year and waste removal accounts for around 4.5% of a construction company's profit. DEFRA (Department for Environment, Food and Rural Affairs) also estimates that around 13% of all the solid materials delivered to construction sites are wasted and up to one-third ends up in landfill. Around a third of all fly tipping waste is from the construction, demolition and excavation sectors.

It is also claimed (DEFRA, 2006) that studies show the typical cost of a construction skip is made up of skip hire (6%), labour filling the skip (12%) and materials in the skip (82%). If it costs say £150 to hire a skip, this implies that the labour cost is £300 and that there is £2050 worth of materials in the skip making a total cost of £2500. The materials probably would not be reusable, of course, because they would most likely be timber offcuts, broken bricks and offcuts of plasterboard, etc. – an inevitable part of the building process. The materials might be recyclable, however, and a local licenced waste management contractor claims that less than 20% of all their waste is sent to landfill and that they have processes to separate wood, bricks/concrete, plastics and cardboard, etc., for recycling. Site managers should check their skips anyway for excessive waste.

Duty of care

The Construction and Demolition (C&D) 'waste stream' has been prioritised by the Environmental Agency due to the amount of waste it generates, its hazardous nature and the potential for recycling. 'Waste streams' refer to the process by which waste is grouped together to identify where it comes from and how it should be treated and disposed of safely without risk to human health. C&D waste consists of metals, asphalt, tar and tar products, concrete, bricks, tiles, ceramics, gypsum-based products and soil and stones.

C&D waste is normally classed as 'commercial' or 'industrial' waste and therefore can be designated as a controlled waste subject to waste-related legislation. When waste has properties that make it especially hazardous or difficult to dispose of it is referred to as 'Hazardous waste' which requires a pre-consignment note system for recovery or disposal. Hazardous wastes, such as asbestos, chemicals, oils or contaminated soils, may require a detailed waste transfer consignment note.

Everyone who produces or handles waste from construction activities has a legal duty of care for its safe keeping, transport and disposal. Failure to comply can result in an unlimited fine under Section 34 of the Environmental Protection Act 1990.

The duty of care requires the principal contractor, and therefore the site manager, to ensure that:

- The waste is properly cared for whilst on site
- The waste is passed on to an authorised body (registered waste carriers, holders of a waste management licence)
- A waste transfer note is made out when the waste is handed over
- All reasonable steps are taken to prevent unauthorised handling or disposal of the waste

From 30 October 2007 all waste sent to landfill must be treated beforehand, unless the waste is inert or would not benefit from treatment, and a written declaration should be included with the waste transfer note. This should state who has treated the waste, how it has been treated and what amount has been sorted out for recovery or alternative treatment. Treatment is anything which changes the volume of the waste, reduces its hazardous nature or makes the waste easier to handle or recover, e.g. collecting waste streams separately for recycling, composting or anaerobic digestion or incineration.

The production of waste and its eventual treatment is a process that the site manager should manage from the outset of a project and he should consider at each stage whether any waste stream can be eliminated.

Site waste management plans

Following the voluntary Waste Management Code of Practice launched by the DTI in 2004, DEFRA has developed proposals for compulsory SWMPs (site waste management plans) for works involving construction or demolition waste. The Clean Neighbourhoods and Environment Act 2005 (Section 54 of Part 5) is the enabling legislation for the introduction of regulations requiring a SWMP on projects worth more than £250 000 effective from April 2008.

Figure 18.4 illustrates a pro forma approach to preparing a SWMP. There are many other ways of preparing a SWMP and all contractors will have their own preferred method. Whichever form they take, SWMPs provide a structure for the systematic management and disposal of waste created in the course of construction or demolition work including the design and pre-construction stages. The aim is to encourage contractors and others to predict how much waste will be created on a project and how it will be recycled or disposed of responsibly. If a project is started without an SWMP, the person in charge of the project, and the principal contractor, will both be guilty of a serious criminal offence and individual directors or managers may also be found guilty.

SITE WASTE MANAGEMENT PLAN

Contractor		Prepared	
Project		Updated	
Waste coordinator			

Waste management goals

Waste prevention measures

Reuse and salvage items

Communication plan

Contamination prevention measures

Documentation

Project waste estimate			Design/Demolition/Constructionphase	
Material	Quantity	%	Waste management contractor	Site handling procedure

Waste summary			
Total C&D waste			
Total recycled			
Total to landfill			

Signed	Date

SITE WASTE MANAGEMENT PLAN

Figure 18.4

Every SWMP must include:

- a description of the construction works proposed including the location of the site and estimated value
- the contractor's identity or, if there is more than one contractor, details of the principal contractor
- the person in charge of the project and details of the person who drafted the SWMP
- a record of decisions made before the SWMP was drafted concerning the nature of the project and its design, construction method or materials employed.
- the types of waste that will be produced and an estimate of the volume of each
- the waste management action proposed for each different waste type including reusing, recycling, recovery and disposal.

C&D waste should be properly managed in accordance with the waste hierarchy shown in Figure 18.5 which indicates best and worst environmental options for waste. Figure 18.6 shows a sample waste data sheet which can be employed for estimating and planning waste arisings and for comparing actual waste figures with target. Each time waste is removed from site, the SWMP must be updated with further information, including the type of waste removed, where the waste is being taken to and the identity of the waste management contractor removing the waste. The principal contractor must also monitor the plan during the project and record any lessons learnt from the process. An action plan to address lessons learned is also required along with a comparison of actual quantities of each waste type compared to the estimated quantities. Further rules apply to projects over £500 000 in value.

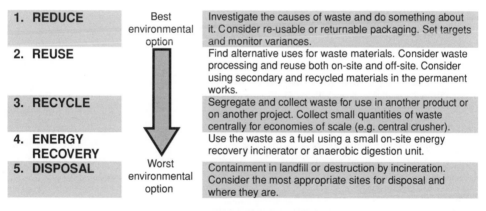

WASTE HIERARCHY

Figure 18.5

Improving waste management

In years gone by construction waste was controlled simply by comparing the actual wastage on site with the estimator's standard waste allowances in the tender. However, it is clear from

SITE WASTE MANAGEMENT PLAN – Data Sheet								
Project			Report no.					
Stage			Date					
Address			Responsible person					
Main contractor								
TYPES OF WASTE ARISING								
	Quantity m³							
	Reuse		Recycle		Sent to recycling facility	Sent to WML exempt site	Disposal to landfill	Waste Transfer Notice completed?	
Material	On-site	Off-site	On-site	Off-site				Y	N
Inert								
Non-hazardous								
Hazardous								
Totals	m³							
Target	Tonnes							
Actual	Tonnes							

WASTE DATA SHEET

Figure 18.6

current thinking and legislation that construction site managers will have to do more to develop their understanding of the impact that the industry has on the environment and they will need to use their knowledge and skills to find ways to improve recycling and waste reduction.

Site managers will need to be much more proactive in the way that they plan and manage materials ordering, delivery, handling and storage and they will need to be more environmentally aware and think of ways to manage waste more efficiently. Waste on construction sites is caused by a number of factors which may be conveniently categorised under four headings:

- Design waste – where the design results in wasteful cutting on site
- Take off/specification waste – where additional materials are ordered 'to be on the safe side' or where materials are over-specified for the job in hand
- Delivery waste – materials damaged in transit or during offloading
- Site waste – waste resulting from the production process or where incorrect materials are used

Figure 18.7 illustrates a proposed strategy for improving site waste management by focusing on three issues – management procedures, procurement and contracts and site practices. By

Management procedures	Procurement and contracts	Site practices
Notify site personnel and subcontractors of company SWMP	Client and subcontractors to sign the SWMP.	Keep the site tidy to encourage the right attitude.
Observe correct environmental procedures.	Include Site Waste Management Plan in all subcontracts.	Protect materials from the weather
Allow an adequate budget in the tender for good site layout planning. Give environmental induction and toolbox talks (including subcontractors). Improve site security and materials storage facilities.	Make subcontractors contractually responsible for waste: • Contra-charge subcontractors for waste removal. • Give labour-only subcontractors a waste allowance and contra charge for exceeding it. • Make a charge for collecting and separating waste.	Improve site materials management: • Provide bins/enclosed areas for loose materials • Provide clean hard standings for lay down areas and materials storage • Use lockable secure containers for components, e.g. windows, sanitary fittings • Use storage racks for keeping materials off the ground.
Change work practices to reduce waste. Identify waste streams and recyclable materials.	Partner with suppliers for 'just-in-time' deliveries.	Mechanise materials handling and distribution to reduce damage and waste • Tele-handlers • Rough terrain forklift trucks • Conveyors • Mast climbers and hoists.
Tighten up site procedures for materials deliveries and checking and signing delivery tickets.	Evaluate materials procurement procedures to avoid over-ordering and site waste.	Remove and organise waste regularly.
Adopt simple procedures for reconciling materials orders and usage.		Collect and recycle timber. Take more care in disposing of construction waste.
Designate a waste segregation area on the site.	Build materials and packaging return provisions into purchase contracts.	Separate different wastes into separate skips clearly labelled: • Cardboard • Plastics • Metals • Wood • Plaster/Plasterboard • General waste, etc.
Conduct regular waste audits.		Reduce number of skip collections: • Make sure skips and waste disposal containers are full • Use Roll Packer or similar drum compactors for removing air pockets in skips.

SITE WASTE MANAGEMENT STRATEGY

Figure 18.7

integrating waste management into everyday site management thinking, site waste can be better controlled and a positive impact made on both the environment and the contractor's profit margin.

Delivery and site waste can be reduced by considering site layout planning at the pre-contract stage and by allowing the site manager to become involved in the decision-making at this stage of the project. Site managers should be encouraged to manage their materials properly – they can be assisted by establishing in-company courses on materials management. The company may offer some degree of incentive to managers who achieve the minimum waste targets on their projects.

All staff should be made aware of the company materials management policy, and of the allowances built into the estimate, and senior managers could be made responsible for the losses occurring on their projects as well as site managers. It is no use blaming the site manager when his superior is oblivious to the losses when he visits the site.

The company may also consider producing a simple site guide on its waste management policy and recommendations could be included on good practice regarding materials handling and control. This could be issued to subcontractors and the company's directly employed labour.

Poor control of material waste can create health and safety problems – nails in discarded timber offcuts, discarded bricks/blocks causing a trip hazard, etc. It is very noticeable that many construction sites have very poor standards of site housekeeping and untidy sites set the wrong tone for good health and safety management. Procedures should be included in the contractor's safety management system to eliminate this problem.

18.4 Plant control

Contractors rely heavily on all sorts of construction plant and equipment for the efficiency of site operations. This requires careful site layout planning and budgetary control if the plant is to work safely and effectively and the budget is not to run out of control. Plant may be categorised as:

- **Small tools and tackle:** hand tools, small power tools, lifting tackle, slings
- **Mechanical:** excavators, cranes, hoists, mobile elevated working platforms (MEWPs), concrete placing and finishing equipment
- **Non-mechanical:** scaffolding, formwork, patented formwork systems, falsework (for supporting formwork, e.g. suspended slabs, bridge decks)

The site manager must plan, organise and control the plant required for the project so as to ensure that:

- the estimator's plant allowances (i.e. the plant budget) are not exceeded
- plant is on site and available for use as and when needed
- plant is properly inspected and maintained according to statutory requirements

- operatives are trained and competent to use the equipment
- plant is immediately 'off-hired' when not needed (the dreaded plant budget again!)

Legislation

The way in which plant is used on site is controlled by legislation and in particular by:

- **Provision and Use of Work Equipment Regulations 1998 (PUWER)**
  General legislation that applies to all tools, plant and equipment from hammers and chisels to tower cranes. PUWER may be supplemented by other legislation.
- **Lifting Operations and Lifting Equipment Regulations 1998 (LOLER)**
  Legislation over and above PUWER that applies specifically to lifting equipment and operations such as cranes, MEWPs, mast climbers, hoists, etc.
- **Construction (Design and Management) Regulations 2007 (CDM2007) Part 4**
  Applies over and above to PUWER with respect to the inspection of excavating equipment and cofferdams, etc., the use and loading of vehicles (e.g. wagons), site lighting, task lighting, etc.
- **Work at Height Regulations 2005 (amended 2007)**
  Applies specifically to access scaffolding, safety barriers, fall arrest equipment (harnesses, air bags, nets, etc.) and the like.
- **Management of Health and Safety at Work Regulations 1999**
  Requires risk assessments and appropriate health and safety arrangements together with a trained and capable workforce, etc.

Organisation

Contractors commonly allocate major items of plant to the master or stage programme which shows when plant is expected to arrive on site and how long it will be there. This helps with entitlement claims when there are delays to the programme. Plant is often delivered to site outside working hours and this needs to be accommodated in the site access and safety planning arrangements for the job. Other issues that must be anticipated include noise suppression (hydraulic breakers, compressors, etc.) and dust control where abrasive saws and chasing equipment is used.

To ensure that operatives are suitably trained and/or aware of the risks of using site equipment, the contractor should hold regular toolbox talks. They will cover issues such as abrasive wheels, cartridge-operated tools, hand tools, dumpers and lifting gear, etc. Plant operators are trained to either Certificate of Training Achievement (old card) or Construction Plant Competence Scheme (new card) and Construction Skills Certification Scheme standard, and this is essential for safe operation of the equipment.

Significant items of plant, such as excavators, MEWPs, cranes, etc., are usually supplied to the contractor under Construction Plant Association (CPA) terms and conditions and crane hire companies also work to BS 7121 – The Safe Use of Cranes. CPA conditions include requirements for the hirer (i.e. the contractor) to provide suitable access and a stable working area. Some items of plant come with a trained operator according to the hire agreement.

Planning

Under LOLER, every lifting operation where lifting equipment is used must be properly planned by a competent person, appropriately supervised and carried out safely. In cases where two or more items of lifting equipment are used together, such as a tandem crane lift, a written plan should be prepared. Where crane facilities are required on site the site manager has two choices with regard to crane procurement:

(1)   a CPA Hire
(2)   a CPA Contract Lift

When a crane is simply hired, the contractor must provide a qualified and experienced 'Appointed Person' to plan and take responsibility for the lifting operation. Method statements and risk assessments must be provided for the crane operator. Where the contractor chooses a contract lift, the crane hire company takes full responsibility for the planning and management of the lift and provides the Appointed Person/Crane Supervisor and all necessary fully trained personnel.

In the case of scaffolding – and depending upon the complexity of the scaffold layout – an assembly, use and dismantling plan must be prepared by a competent person in accordance with the Work at Height Regulations.

Inspection and maintenance

Plant must be regularly inspected and maintained according to statute and scaffolding must be checked before use. It is common practice to use a tagging system to say that the scaffold is safe and has been properly passed for use. Tagging may also be used for mobile towers, ladders, safety harnesses, hoists and MEWPs, etc., as well as for portable equipment and lifting equipment.

18.5 The control and coordination of subcontractors

The planning stages involved in the control and coordination of subcontractors follow similar stages to those that the main contractor goes through, i.e. pre-tender, pre-contract and contract planning. The decision-making process in respect to subcontractors and their involvement in the project will therefore be dealt with under these headings. Subcontractor input into the various planning stages is illustrated in Figures 18.8 and 18.9.

The pre-tender planning stage

Assuming a traditional JCT 05 main contract, the subcontractor selection process follows the following stages:

- Stage 1 – qualification
- Stage 2 – compilation of tender list

SUBCONTRACTOR INPUT INTO THE PLANNING PROCESS

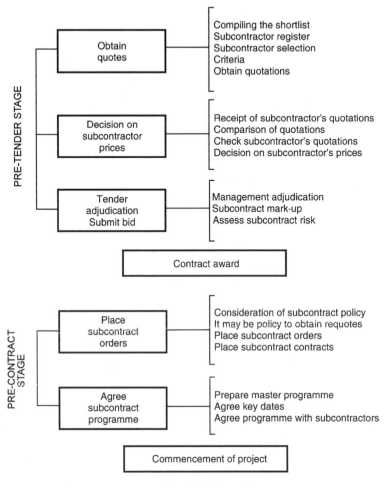

PRE-TENDER STAGE

Obtain quotes
- Compiling the shortlist
- Subcontractor register
- Subcontractor selection
- Criteria
- Obtain quotations

Decision on subcontractor prices
- Receipt of subcontractor's quotations
- Comparison of quotations
- Check subcontractor's quotations
- Decision on subcontractor's prices

Tender adjudication Submit bid
- Management adjudication
- Subcontract mark-up
- Assess subcontract risk

Contract award

PRE-CONTRACT STAGE

Place subcontract orders
- Consideration of subcontract policy
- It may be policy to obtain requotes
- Place subcontract orders
- Place subcontract contracts

Agree subcontract programme
- Prepare master programme
- Agree key dates
- Agree programme with subcontractors

Commencement of project

Figure 18.8

- Stage 3 – tender information and submission
- Stage 4 – tender assessment
- Stage 5 – tender acceptance

Each of the above stages will now be dealt with in the form of a checklist as outlined in the CIB Code of Practice for the Selection of Subcontractors (CIB 1997) and recommended in the CIOB Code of Estimating Practice (2008).

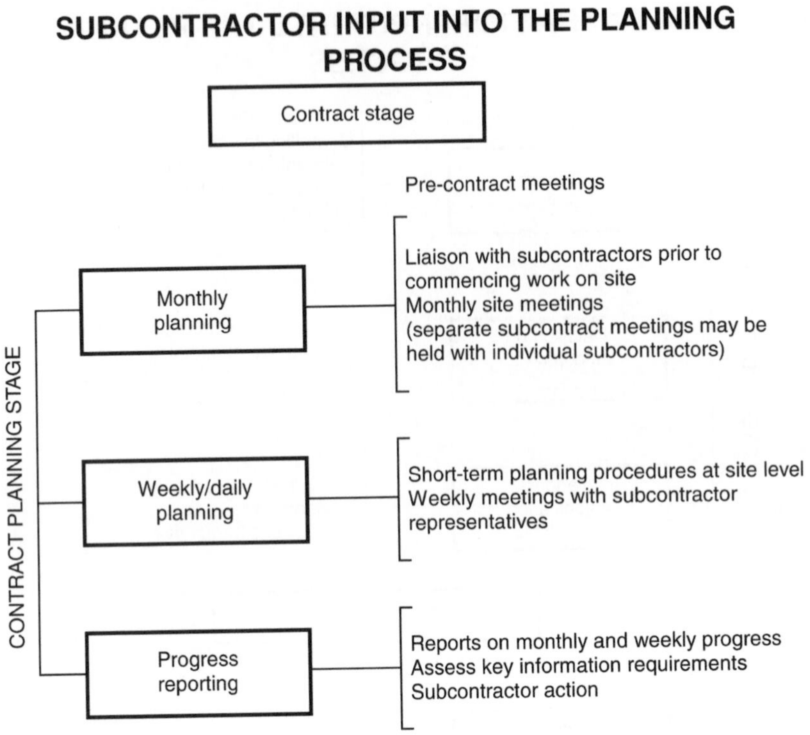

SUBCONTRACTOR INPUT INTO THE PLANNING PROCESS

```
┌──────────────────────────┐
│      Contract stage      │
└──────────────────────────┘
```

Pre-contract meetings

CONTRACT PLANNING STAGE

Monthly planning	Liaison with subcontractors prior to commencing work on site Monthly site meetings (separate subcontract meetings may be held with individual subcontractors)
Weekly/daily planning	Short-term planning procedures at site level Weekly meetings with subcontractor representatives
Progress reporting	Reports on monthly and weekly progress Assess key information requirements Subcontractor action

Figure 18.9

Stage 1 – qualification

Checklist for subcontractor selection:

- Previous experience with the subcontractor
- The subcontractor's ability to manage his resources and liaise with the main contractor's staff. Good relationships between parties are an essential requirement to developing a team approach to a successful project
- Financial standing of the subcontractor. His ability to wait to be paid
- The expertise that the subcontractor can bring to the project
- The subcontractor's reputation and his standing with the client
- The current commitment of the subcontract organisation. Their current workload with other contractors should be determined and serious consideration given to their ability to cope with the increased work. A large number of subcontractors just cannot say 'no' when it comes to taking on more work. They often pull and push their limited labour force between sites hoping the main contractor will not notice that they are stretched to the limit
- The acceptability of the subcontractor to the client. On many contracts the contractor is required to name his subcontractors at the tender stage
- The competitiveness of the subcontractor's price. The price must be right otherwise the subcontractor will never win any work. Price discounts which may be applicable and the subcontractor's response to negotiation may be an important factor

- The contractual risk which the main contractor takes on the subcontract item. Low-risk subcontract operations may be let to the more risky subcontractor and hence the main contractor may include a lower subcontract price in his estimate. Subcontract operations that are critical to the success of the contract require careful consideration. For example, a high-price-risk subcontract operation could be external brickwork, while site demolition work could be low risk
- The ability of the subcontract organisation to meet quality assurance criteria as laid down by the main contractor or as specified by the client
- References available from the subcontractor, including trade and bank references. The willingness of the subcontractor to allow previous contract work to be inspected

It is important that good relationships are established between the main contractor and subcontractor as early as possible in the planning process. This is especially important where the main contractor intends to sublet all the work on a particular project.

In the process of choosing a subcontractor, it is very important to select a firm that can liaise amicably with both the contractor's head office and site staff and satisfy different interests. A further significant consideration is the acceptability of subcontractors to clients and their consultants.

Stage 2 – compilation of tender list

As soon as it has been decided to select a subcontractor by competitive tender, a shortlist of those considered suitable to be invited to tender should be compiled. It is often the case that the main contractor has a list of subcontractors of established skill, integrity, responsibility and proven competence. This should be considered as a matter of policy within the contractor's organisation.

Main contractors should review their lists periodically so as to exclude firms whose performance has been unsatisfactory and to allow the introduction of new subcontract firms.

In some cases, the main contractor may find difficulty in compiling a short list due to the shabby way he has treated his subcontractors in the past. In this case he may have to revert to the pin-sticking approach using the building section in Yellow Pages in order to obtain quotations. There are dangers in this practice and it is much more preferable to cultivate good subcontractor relationships as recommended in the Latham Report. Many contractors are now benefiting from the wisdom of this approach.

The cost of preparing tenders is a significant element of the overheads both of the main contractor and his subcontractors. Tender lists therefore should be kept as short as practicable. Enquiries should be kept to between three and six depending on the type of subcontract work and size of project.

Perhaps the main contractor should consider adopting a policy of creating good relationships with a small group of reliable subcontractors whose businesses can expand as the main contractor becomes more established. Often the main contractor's reputation may rely solely on the excellence of his subcontractors' performance.

The use of a register of subcontractors at the selection stage is recommended good practice and this may lead to establishing records of subcontractor performance criteria which can be used to review what happened on previous contracts. It is important that such records are regularly updated and consideration given to adding new subcontractors to the register. It may be advisable to consider using a computer database, such as Microsoft Access, for compiling the subcontract register. This may prove advantageous to a medium-sized contracting organisation in order to speedily access subcontract information.

It is important that close relationships are encouraged between the main contractor and his subcontractors for the survival of both parties in the long term.

Stage 3 – tender information and submission

Both the CIB Code of Practice (CIB 1997) and CIOB Code of Estimating Practice (CIOB 2008) identify what information should be contained in the contractor's enquiry to a subcontractor. The main recommendations are summarised here as a checklist.

Details of main contract works:

- Job title and location of site
- Name of employer
- Names of architect, supervising officer, quantity surveyor and other consultants including the CDM coordinator
- General description of the works

Subcontract works:

- Relevant extracts from bills of quantities and specification
- Extracts from the contract preliminaries section
- Copies of relevant drawings
- Details of where original documents may be inspected
- Time period for completion of subcontract work (if known)
- Approximate dates when subcontract work will be undertaken
- Names of adjudicator (in case of dispute)

Subcontractor's responsibility for site arrangements and facilities:

- Watching and lighting
- Storage facilities
- Unloading, hoisting and getting in materials
- Scaffolding
- Water and temporary electrical supplies
- Safety, health and welfare provisions
- Licences and permits
- Any additional facilities

Conditions of subcontract:

- Form of subcontract agreement
- Period of interim payments and payment terms including whether 'pay when paid' will apply
- Discount applicable to the payments

- Fluctuations or fixed price tender
- Other special conditions

Particulars of the main contract conditions:

- An extract from the appendix to the form of contract will assist in providing a summary of the contract particulars. This should contain the following information:
  - Form of contract
  - Fluctuations provisions
  - Method of measurement
  - Main contract period and completion date
  - Defects liability period
  - Liquidated and ascertained damages
  - Period of interim certificates
  - Basis of dayworks
  - Insurance provisions
  - Deletions or amendments to standard contract clauses

Type of quotation required from the subcontractor:

- Lump sum quotation
- Schedule of rates

Other information:

- Date for the return of the tender
- Person in the contractor's organisation to contact
- Period for which the tender is to remain open for acceptance
- Extent of the phasing of the works and number of anticipated visits to undertake the works
- Reference to any relevant attendances likely to affect the subcontractor

Stage 4 – tender assessment

The CIOB Code of Estimating Practice (CIOB 2008) makes recommendations for analysing subcontractor quotations in the form of a domestic subcontractor register. Subcontractors' quotations are not straightforward to compare. Some do not price all the items in the enquiry, often there are mistakes and some subcontractors price net while others offer discounts, typically $2^1/_2$%. Therefore a register or spreadsheet is a useful device to enable quotations to be matched, discrepancies identified and discounts adjusted.

A typical checking procedure when comparing subcontractors' quotations should include consideration of the following:

- Does the work described in the quotation comply with the specification?
- Have all items been priced and if not are they included in other rates?
- Are unit rates consistent throughout the quotation?

- Check that the quotation does not form a counter-offer and that the subcontractor has accepted the terms and conditions of the enquiry.

Stage 5 – tender acceptance

Adjudication of the main contractor's estimate and its conversion into a tender requires management decision-making whatever the size of firm. Where the tender includes a large proportion of subcontract work, it is critical to decide carefully on the markup on the subcontractors' quotations selected for inclusion. It will also be necessary to carefully scrutinise the estimator's allowances for attendance on the subcontract works at the adjudication stage.

Consideration needs to be given to any late quotations received which will directly affect the competitiveness of the overall bid and this is where the subcontract comparison sheet or register facilitates last minute lump sum adjustments to be made to the tender.

The percentage markup on subcontractors' quotations often varies widely according to the contractor's desire to win the contract. Percentages will probably be in the range $2^1/_2$–15% depending on the contractor's view of the risk attached to the subcontract element of the contract.

Despite all the foregoing considerations, there often appears to be no rhyme or reason as to how contractors arrive at their tender adjustments.

Pre-contract liaison

Following the selection process, prospective subcontractors or works package contractors should be informed that their tenders have been accepted and that a contract will follow. However, before contracts are signed several matters need to be clarified so that disputes do not arise during the works. At this stage, it is useful to follow a checklist approach to ensure that all the pre-construction issues are covered. These will include:

- Confirm order or issue letter of intent
- Prepare contracts for signing
- Agree key dates, sequence of works and programme including integration with the project programme
- Agree timing for issue of drawings or approval of design information
- Confirm requirements for insurances
- Agree or obtain client/client representative approval of subcontracts or packages where there is a contractual requirement to do so
- Agree the provision of samples, sample panels and mock-ups to be submitted for client approval
- Agree the facilities to be provided by the subcontractor/package contractor, for example, cabins, stores, offloading, scaffolding, removing rubbish, etc.
- Agree dates for interim certificates and procedure for requesting payment
- Agree notification requirements for instructions, variations, dayworks, etc.
- Sign contract

Figure 18.10 indicates the relationships between the work package master programme, stage programme and 6-weekly short-term programme as applied to the piling and substructure

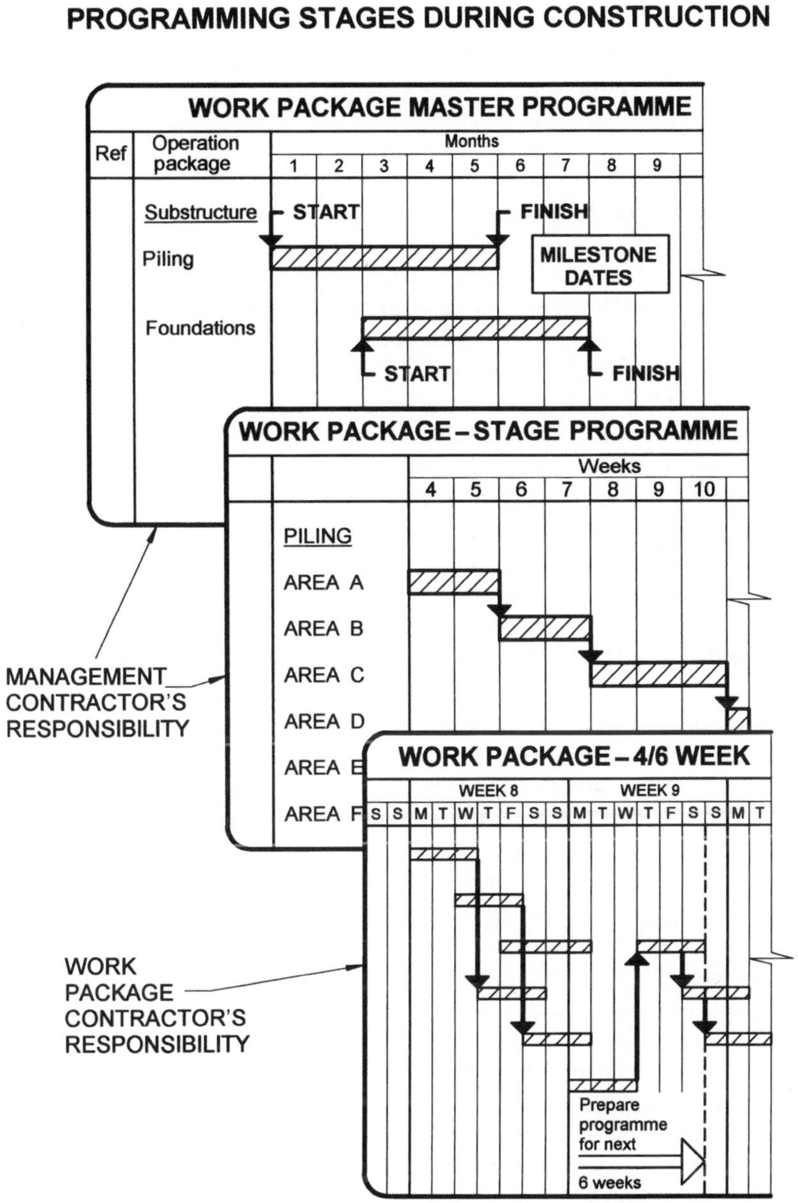

PROGRAMMING STAGES DURING CONSTRUCTION

Figure 18.10

work package. It is becoming the norm on major projects for subcontractors to be responsible for the planning of their own work, and they must be willing to include for this risk at the tender stage.

Before any work commences on site, it is vital that the main contractor or the management contractor/construction manager maintains contact with the works contractor in order to keep him fully informed of the contract progress position. For instance, where there is a likelihood of the commencement of the work being delayed as much notice as possible should be given. It is essential to maintain this contact and to ensure that notification procedures agreed at the pre-contract stage are observed by all parties.

Liaison during the contract period

The coordination and control of subcontractors and package contractors is crucial to the success of the project and lends itself to a standardised approach. Many of the large contracting organisations and construction managers have developed their own procedures for this. What is the secret to success? Well, by 'meetings – bloody meetings' as the John Cleese management training video recommended!

Many major contractors adopt a policy of ruling works contractors with a rod of iron, but this is not the way to ensure cooperation and teamworking on site. The success of a project depends on the performance and quality of those who carry out the work on site and such a short-sighted policy will ultimately fail.

Good liaison and mutual respect must be established as early in the contract period as possible and maintaining contact with subcontractors or works package contractors during the pre-contract period helps to build up an early working relationship. During the project, this is developed by regular contact at weekly and monthly progress and coordination meetings.

Figure 18.11 illustrates the meetings likely to be held during the contract stage where the works contractors will discuss problems and information requirements and iron out any difficulties which may be affecting the progress of operations on site. A checklist of points to be considered at such meetings will include:

- Review progress and quality of work
- Review the programme including relationships between the main programme operations and those of other works activities
- Action to maintain progress
- Investigate site problems and hold-ups
- Review labour situation
- Review the plant and material supply situation
- Overview of site organisation and supervision requirements
- Consider health, safety and welfare situation
- Confirm the situation regarding the issue of site instructions and variations to contract
- Review the valuation and payment situation to date

A system of short-term planning may be implemented in order to keep works contract progress under constant review. This will involve preparing weekly or 2-weekly programmes which will be discussed at the progress meetings, and preparing work plans for the next short-term period. Figure 18.12 illustrates the principles of a 2-weekly short-term planning system.

It is important that the works contractors' site representatives participate in the short-term planning procedures and a good working relationship is established at site level.

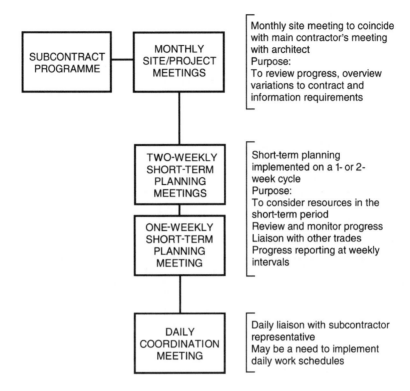

SUBCONTRACTOR LIAISON MEETINGS

Figure 18.11

With future projects in mind, it is a good idea to record the performance of the works contractors on site. This will be useful feedback for reviewing tender lists so as to ensure that works contractors are competent and adequately resourced and that there is an up-to-date database for new projects. One possible approach to this is to list the key performance criteria and to give each one a score or rating on a scale of 1–5. The performance criteria might be:

- Price
- Quality and workmanship
- Health and safety
- Standard of cooperation
- Time–programme performance

The maximum score would be 25 but a rating of, say, less than 10 might lead to exclusion from the tender list. Certain criteria, such as health and safety for instance, could be weighted for importance where appropriate.

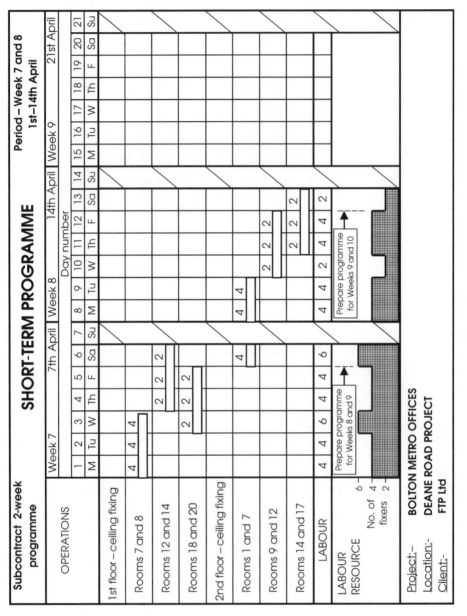

Figure 18.12

References

CIB (1997) *Code of Practice for the Selection of Subcontractors.* CIB.
CIOB (2008) *Code of Estimating Practice.* CIOB.
DEFRA (2006) *Developing a Strategic Approach to Construction Waste.* BRE.

Web reference

www.netregs.gov.uk
www.dti.gov.uk
www.defra.gov.uk
www.envirowise.gov.uk

Part D Case Studies

19 City Road project

19.1 Project description

This case study is based on the City Road project which is a development of three blocks of luxury riverside apartments complete with private basement car parking, roadworks and landscaping.

The project involves the construction of three seven-storey buildings on a restricted site in the centre of a major city in the north-west of England. A two-level basement (Basement B) is to be constructed under Block 1. Blocks 2 and 3 are to be constructed over a single-level basement (Basement A). The multi-storey blocks are of steel frame construction incorporating metal deck floors and brick external facades.

The site is bounded on the north side by the River Med where there is an existing retaining wall which is to be demolished and rebuilt. The south side of the site is bounded by City Road which is a busy main thoroughfare.

The general layout of the project is shown on the site location plan (Figure 19.1), and Figure 19.2 shows the basement location plan.

19.2 Project details

Contract value is £14 million and the agreed contract period is 70 weeks. The contract agreement is to be executed under the JCT 05 Design and Build Contract and the employer's scope designers are to be novated to the contractor to create a single point of responsibility for design and construction.

The relationship between the client's design team and the contractor is shown in Figure 19.3.

Buildability is seen as an important issue on this project and the contractor has identified a number of key areas which warrant particular attention, including basement construction proposals and the steel frame erection sequence.

The contractor proposes to let out the work as a series of work packages and some of these will be delegated with design responsibility.

19.3 Site constraints

The following factors are considered important in developing the construction methodology and sequence of work:

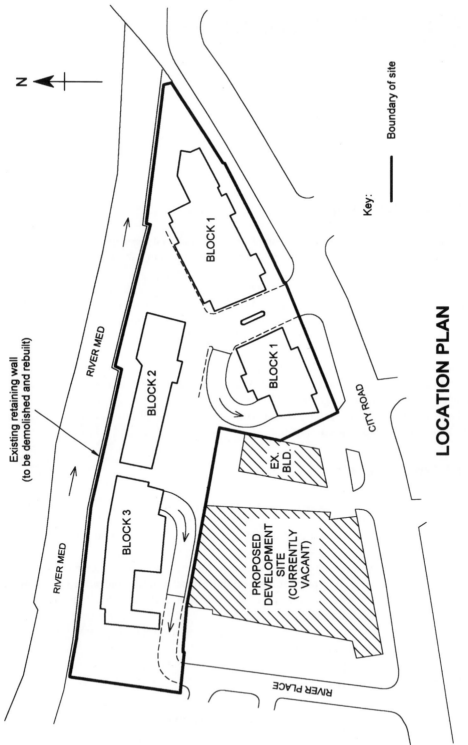

Existing retaining wall
(to be demolished and rebuilt)

RIVER MED

RIVER MED

RIVER MED

BLOCK 1

BLOCK 1

BLOCK 2

BLOCK 3

EX. BLD.

PROPOSED DEVELOPMENT SITE (CURRENTLY VACANT)

CITY ROAD

RIVER PLACE

N

Key:

Boundary of site

LOCATION PLAN

Figure 19.1

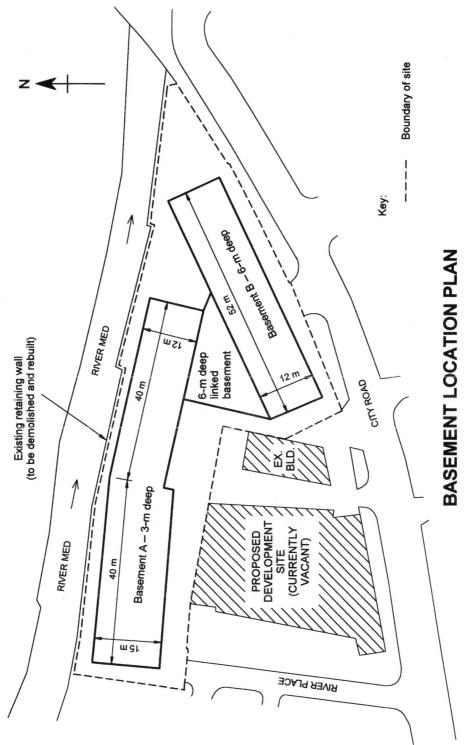

BASEMENT LOCATION PLAN

Figure 19.2

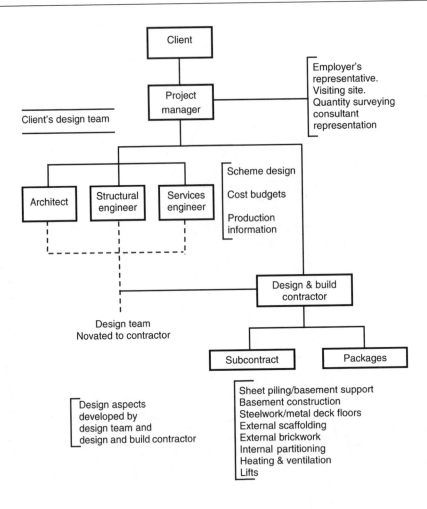

CLIENT, DESIGN TEAM, CONTRACTOR RELATIONSHIP, CITY ROAD PROJECT

Figure 19.3

- The only access in and out of the site is via River Place
- No access is available on the north side of the site due to the presence of the River Med
- Limited access is available on the south side due to City Road.
- Working space on site is limited and there is no space available adjacent to the site due to adjoining developments
- Limited site space for the storage of materials
- Basement B under Block 1 is 6 m deep and is located at the rear of the footpath on City Road
- Extensive work is to be undertaken in rebuilding the river retaining wall. This work is to be completed in sequence with the basement construction under Blocks 2 and 3 (Basement A)

- There are legal problems concerning cranes oversailing adjacent properties and the client might be asked to negotiate an easement with adjoining owners. Alternatively, luffing jib cranes could be used which can raise their jibs to avoid problems of trespass

19.4 Tender stage

At tender stage, the contractor needs to consider how the project will be undertaken and how long the work will take. There are many factors to think about but there is not much time to go into a lot of detail and so a risk assessment approach is often taken. This means that the estimator will identify those key factors that he thinks will influence the price and the programme, and the tender adjudication panel will decide on the extent of the risk involved and how much should be allowed in the tender to cover them.

In the case of the City Road project, several issues are of vital importance in putting together a winning bid for the contract and giving the contract team a realistic budget and programme to carry out the job and make a reasonable profit. These include:

- Site security
- Access and site layout
- Excavation and earthwork support
- Basement construction sequence
- Retaining wall demolition and reconstruction
- Choice and layout of cranage requirements
- Materials storage and handling
- Site accommodation and supervision

Pre-tender programme

At tender stage, there is insufficient time to go into all the issues listed above in detail and the contractor may well have to 'take a view' in deciding his tender price. Of vital importance here is the knowledge and experience of the tender team (estimator, planner, contracts manager, etc.) and the skill and judgement of the tender adjudication panel in assessing market factors and risk issues.

At the tender stage of a project, the pre-tender programme is usually presented in bar chart format using computer-based planning software. In the medium-sized organisation, the responsibility for preparing the programme will normally be undertaken by the contracts manager, whereas the larger companies will usually have the luxury of a tender planning engineer.

Figure 19.4 illustrates a typical pre-tender programme for the City Road project, showing the main activities and the overall construction period. A realistic assessment of time periods (in weeks) for each activity is determined on the basis of the experience of the contracts manager and planner.

Example – Steel frame and floors

Construction periods for the erection of multi-storey buildings may be based on the construction cycle time per floor which will depend on the plan size and complexity of the building

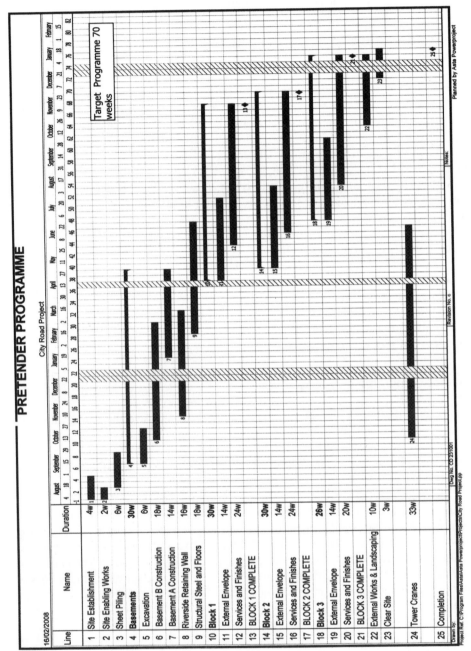

Figure 19.4

layout. The overall duration of the activity will depend on how many cranes and steel erection gangs will be used and the degree of concurrency possible on the site.

Allow 2.5 weeks per floor
Overall duration = 2.5 × 7 floors = (say) **18 weeks**
Example – Riverside retaining wall (overall wall length = approximately 150 m)

The time for the construction of the riverside retaining wall might be based on a construction cycle per 30 m length, i.e.

Form temporary access = 1 week
Construct wall – 5 lengths × 3 weeks = 15 weeks
Total duration = **16 weeks**

An overall construction period of 70 weeks is shown for the tender stage construction period.

Site management costs

On a project of this scale, responsibility for assessing the contract preliminaries would rest on a full tender team, including the business unit or contracts manager, the planning manager and the estimator.

A team approach is essential to ensure a successful contract and decisions need to be made in respect of the main phases of work in order that an appropriate organisation structure may be developed for the project.

Site-staffing assessments will be required to cover engineering, quantity surveying, administration and supervision functions during the project. Company policy will dictate the extent of full-time on-site staff and which staff would only be required on site part-time.

All too often, insufficient monies are included for site management costs at tender stage, especially when the tender adjudication panel has been looking for savings in order to make the bid more competitive. The build-up of site management costs involves consideration of:

- A realistic contract period
- The proposed site organisation structure
- Realistic pricing of staff salaries and on-costs using a spreadsheet or a standard preliminaries build-up sheet

While there would be insufficient time to prepare a formal organogram for the project at tender stage, Figure 19.5 indicates the principles in arriving at a proposed site organisation structure for the preliminaries assessment. The time in weeks allocated to each member of the construction team is indicated.

Figure 19.6 demonstrates the use of a simple spreadsheet to assess the total site management costs inclusion in the bid. The rates used for staff costs include all employment expenses (on-costs) and company cars. A total sum of £469 400 is to be included in the contract preliminaries to cover site management costs.

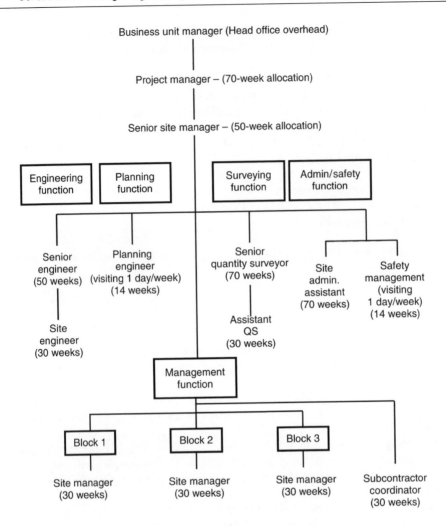

Business unit manager (Head office overhead)

Project manager – (70-week allocation)

Senior site manager – (50-week allocation)

| Engineering function | Planning function | | Surveying function | Admin/safety function |

Senior engineer (50 weeks)

Planning engineer (visiting 1 day/week) (14 weeks)

Site engineer (30 weeks)

Senior quantity surveyor (70 weeks)

Assistant QS (30 weeks)

Site admin. assistant (70 weeks)

Safety management (visiting 1 day/week) (14 weeks)

Management function

| Block 1 | Block 2 | Block 3 |

Site manager (30 weeks)

Site manager (30 weeks)

Site manager (30 weeks)

Subcontractor coordinator (30 weeks)

PROPOSED ORGANISATION STRUCTURE FOR PRELIMINARIES ASSESSMENT

Figure 19.5

Tender strategy for basement excavation

At tender stage, the estimator would need to develop a strategy for the construction of the basement. This might take the form of a tender method statement.

This would consist of a simple written description of the scope of the works with the suggested construction approach for pricing purposes. Figures 19.2 and 19.7, respectively, depict the basement layout and typical construction details.

PRELIMINARIES SCHEDULE

Site management costs

Ref.	Staff	No. off	No of weeks	Cost/weeks £	Total
A	Project manager	1	70	1200	84 000
B	Senior site manager	1	50	1000	50 000
C	Site manager	3	90	900	81 000
D	Subcontractor/ coordinator	1	30	900	27 000
E	Senior engineer	1	50	900	45 000
F	Site engineer	1	30	800	24 000
G	Planning engineer	1	14	800	11 200
H	Senior surveyor	1	70	1000	70 000
J	Assistant QS	1	30	800	24 000
K	Site administrator	1	70	600	42 000
L	Safety manager	1	14	800	11 200
	TOTAL Site management costs for inclusion in preliminaries				**£469 400**

Figure 19.6

Example

Scope of works:

- Basement A (80 m long × 14 m wide) is 3.5 m deep and is to be constructed alongside the new river retaining wall
- Basement B and the linked basement, (52 m long × 12 m wide) is 6.5 m deep and is to be constructed at the rear of the footpath on City Road
- Both basements are connected via a one-way ramped traffic system.
- Volume of excavation = approximately 10 400 m³
- All excavated material must be removed via the River Place entrance/exit
- No earthwork support required along the river boundary due to the retaining wall reconstruction

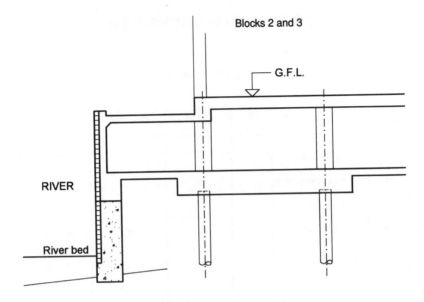

SINGLE-STOREY BASEMENT A
(Under Blocks 2 and 3)

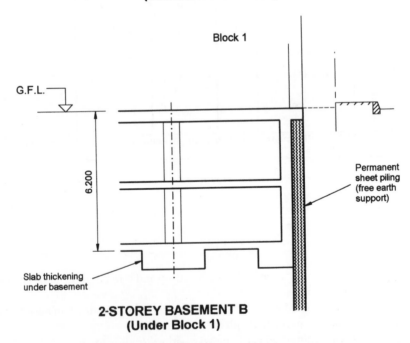

2-STOREY BASEMENT B
(Under Block 1)

BASEMENT CONSTRUCTION DETAILS

Figure 19.7

Method:

- Install permanent steel sheet piles to Basements A and B using free earth support (cantilever) design
- Stage 1 excavation from ground level to a depth of −3.5 m throughout Basements A and B
- Form ramp for access to Basement A
- Install large diameter tubular steel top struts to Basement B to support sheet piles
- Stage 2 excavation to Basement B and linked basement from level −3.5 m to level −6.5 m
- Load wagons and remove excavated material from site

See Figure 19.8 for sheet-piling layout plan and Figure 19.9 for details of the excavation sequence.

Cranage strategy

The allowances included in the contract preliminaries for major items of heavy plant will have a direct impact on the success of the tender bid.

Major decisions will have to be made in relation to the type and number of cranes to be included at the estimate stage. This is particularly important on sites with restricted access and limited storage space.

At the pre-tender planning stage, a number of key questions will need to be answered:

- What materials and components are being handled?
- What type and lifting capacity of crane(s) is required?
- At which stage of the project will the cranes be required and for how long?
- What facilities in the form of hardstandings, track, bases and power supplies are required?
- What size of crane would be most suitable in respect of location and reach?

The work in connection with the basement and the building superstructure poses different problems. The use of a single tower crane to serve the basement and frame may not lead to a satisfactory solution. The possibility of bringing in mobile low-pivot jib cranes for the basement construction and tower cranes for the steelwork and external envelope needs careful consideration.

Two proposals for cranage arrangements could be considered, as follows.

Proposal A

The basement construction is to be undertaken using mobile low-pivot jib cranes working from an access road formed around the basement perimeter. Figure 19.10 illustrates the crane locations relative to Basements A and B. The cranes will be used for handling reinforcement and formwork and some in situ concrete wall pours. The larger concrete pours for the foundations, basement slabs and walls will be placed using concrete pumps.

Once the basements are completed up to ground floor slab level, a tower crane will be located on each block to erect the multi-storey steel-framed structure as shown in Figure 19.11.

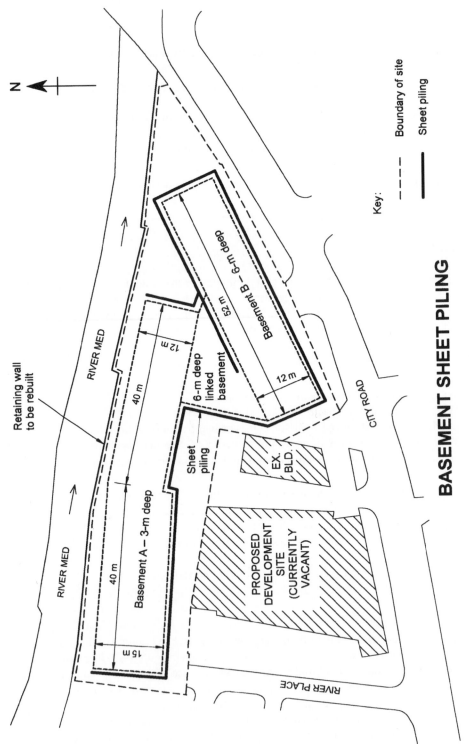

BASEMENT SHEET PILING

Figure 19.8

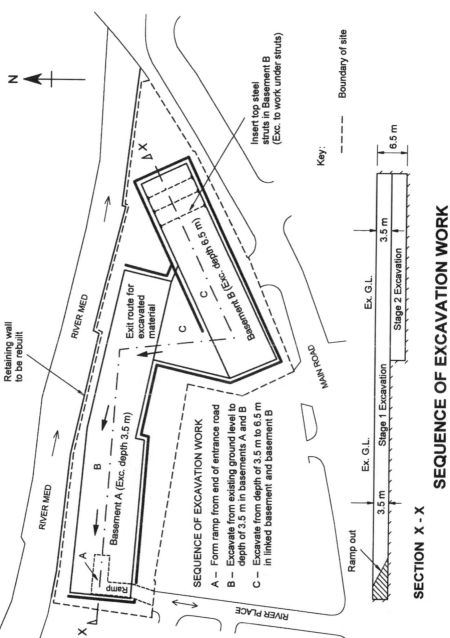

N

Retaining wall
to be rebuilt

RIVER MED

RIVER MED

RIVER PLACE

MAIN ROAD

Ramp out

Ramp

A

B Basement A (Exc. depth 3.5 m)

C

C

Basement B (Exc. depth 6.5 m)

Exit route for
excavated
material

Insert top steel
struts in Basement B
(Exc. to work under struts)

SEQUENCE OF EXCAVATION WORK

A – Form ramp from end of entrance road

B – Excavate from existing ground level to
depth of 3.5 m in basements A and B

C – Excavate from depth of 3.5 m to 6.5 m
in linked basement and basement B

Key:

– – – –   Boundary of site

Ex. G.L.

Ex. G.L.

Ex. G.L.

3.5 m

3.5 m

6.5 m

Stage 1 Excavation

Stage 2 Excavation

SECTION X - X

SEQUENCE OF EXCAVATION WORK

Figure 19.9

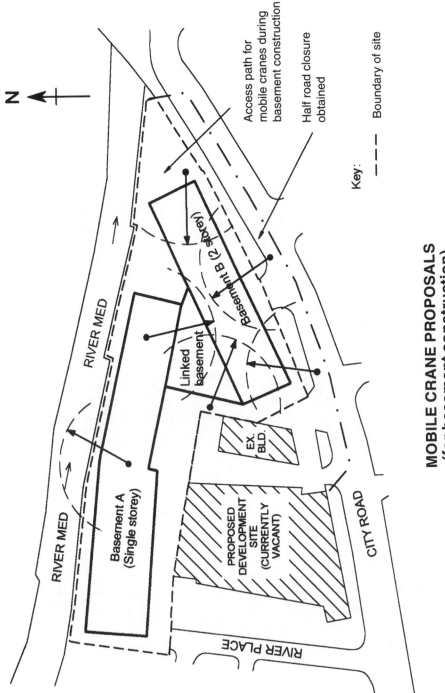

MOBILE CRANE PROPOSALS (for basement construction)

Figure 19.10

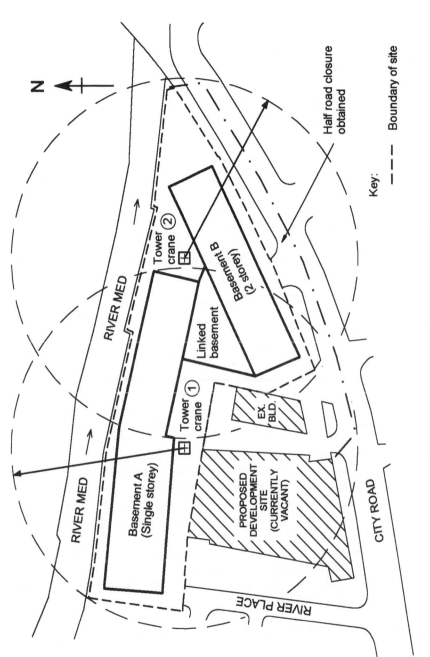

TOWER CRANE PROPOSALS
(for basement and frame construction)

Figure 19.11

- Duration for 2 mobile cranes = **28 weeks**
- Duration for 2 tower cranes = **18 weeks**

Proposal B

Two large 40 m radius tower cranes to be utilised to serve Basements A and B, as shown in Figure 19.11.

The tower cranes will be erected on completion of the basement excavation works and will be used for all work in connection with the basement construction and erection of the multi-storey buildings.

- Duration for 2 tower cranes = **36 weeks**

Price build-up for proposals A and B will be based on the fixed and time-related costs. The cost of providing temporary access roads for the mobile cranes in Proposal A needs to be included.

Cost studies of Proposals A and B should be undertaken to establish the most economic solution on which to base the estimate.

19.5 Pre-contract stage

At the pre-contract stage, the contractor will know that his tender has been accepted and much more time will be invested in the planning and organisation of the project.

Detailed proposals with regard to the layout of the site need to be formulated in order to ensure that adequate space is available for site accommodation, materials storage (or lay down) areas, mortar silos, scaffolding, transport around the site and materials handling arrangements.

The pre-tender programme must now be developed into a construction (or master) programme and this needs to be supported with detailed considerations of sequence and method. The master programme will be prepared in conjunction with the contractor's project manager and a senior planning engineer on the basis of the contract period for the project. In effect, each activity on the pre-tender programme will be expanded into its constituent operations showing their dependency links in relation to each other.

Finally, it is becoming increasingly common for the contractor's tender to be judged not only on the basis of cost and programme but also in terms of quality and health and safety management proposals. This needs to be developed at the pre-contract stage into detailed arrangements to ensure that standards of quality and health and safety can be ensured on site.

The following worked examples of pre-contract proposals are given below:

- Site layout planning and materials handling
- Sequence study for basement construction
- Project quality plan

Site layout planning and materials handling

Access

- Apply for a partial road closure to City Road to provide additional space for access and site accommodation. This road would become a one-way street during the period of the contract
- Form a temporary ramp into Basement A from River Place. This allows crane and vehicular access to all basement areas and to the adjoining riverside retaining wall construction
- Five access and storage areas have been indicated on the site layout plan in Figure 19.12. These are designated as Areas A, B, C, D and E

Location of site accommodation

- Two-tiered fully serviced site accommodation is to be located on City Road, as shown on the site layout plan. Two access gates are to be provided in the hoarding to allow access to Area A and Area D
- Basement A is sufficiently wide to allow subcontractors' cabins to be located here for a temporary period until the installation of the basement piles and floor slab
- Following erection of the steel frame to Blocks 1, 2 and 3, site accommodation for subcontractors will be provided in the ground floor areas of each block

Materials storage areas

- Tower cranes A and B are to be erected in the locations indicated on the layout plan. Area A on City Road has been designated an unloading area for main materials as vehicles can reverse into this area and be unloaded by tower crane B. Once Basement A is completed, the area marked E can be used for unloading materials with tower crane A. This area will no longer be required for access
- Arrangements for materials storage to be as follows:
  Area A   Main vehicle unloading area for key materials
  Area B   Five-metre-wide area in front of blocks 2 and 3 to be used for palletted brick storage
  Area C   Ground floor storage area part over linked basement
  Area D   Key storage area with access to City Road via gates in the hoarding. Site silo for mortar mixing and supply to be located in this area
  Area E   Unloading area
- Due to lack of storage space, phased deliveries for the following key materials is to be considered:
  - Bricks and blocks
  - Stud partitioning
  - Internal joinery
  - Heating and ventilation equipment
- Consideration might also be given to the pre-loading of the floors of the building during frame erection with pre-packaged materials and components boxed-up individually for each apartment type:
  - Door linings
  - Pre-cut architraves

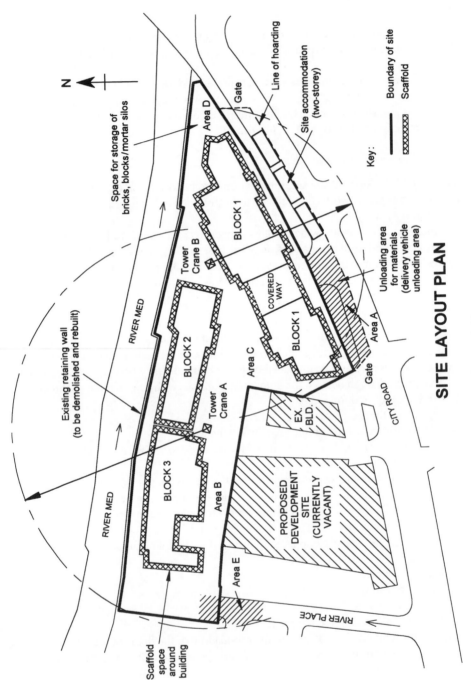

SITE LAYOUT PLAN

Figure 19.12

   - Ironmongery
   - Electrical wiring sets
   - Heating packs

Sequence study for basement construction

The following diagrammatic sequence study has been developed for the construction of Basement A.

Description of works

The work involves the construction of a single basement 80 m × 15 m × 3.5 m deep comprising reinforced concrete floor and thickening beams, associated piling works and reinforced concrete walls.

The basement is to be permanently sheet piled on three sides on the basis of the free earth support method (cantilever). The long side of the basement is adjacent to the new riverside retaining wall. See Figures 19.2 and 19.7 for details of the basement layout and typical cross sections.

The sequence study is to include all operations to the underside of the ground floor slab level. The suspended in situ concrete waffle slab forming the ground floor is not included.

Development of construction sequence

The planner or construction manager needs to carefully consider all the stages involved in the construction sequence, including intermediate activities and any temporary works required. This can be gained only from experience on contracts – it cannot be learnt from a book. The following is offered as a guide to the thought process:

- Consider the work sequence or stages involved:
   - excavation work
   - forming the basement slab
   - constructing the basement walls
   - links between fixing reinforcement
   - erecting formwork, etc.
- Consider overlapping operations
- Think about start-to-start relationships:
   - When can the floor construction commence in relation to the excavation works?
   - How will the floor be poured? How many bays?
   - When can the basement walls start relative to the floor construction?
- Consider continuity of work for operative gangs:
   - Steelfixers
   - Carpenters
   - General labourers for concreting
- Think about use of formwork (shuttering):
   - How many lifts?

- o Is formwork to be full height?
- o How many uses for wall shutters?
- Make allowance for the striking times of formwork:
  - o Beam sides and vertical wall faces – next day
  - o Soffits to beams and suspended floor slabs – 7–10 or 14 days depending on specification and possible evidence of concrete cube strengths
- Consider utilisation of plant resources:
  - o Handling reinforcement, formwork and concrete
  - o Concrete will normally be pumped with maybe three or four pours arranged for the day when the concrete pump is on site

Sequence of operations

- Stage 1   Install permanent sheet piles (three sides of excavation)
- Stage 2   Excavate basement and form piling mat
- Stage 3   Piling to basement floor
- Stage 4   Excavate thickening beams to basement floor
- Stage 5   Cut off piles and blind excavation with concrete
- Stage 6   Fix steel reinforcement to basement floor slab and thickenings
- Stage 7   Concrete basement slab
- Stage 8   Fix formwork and reinforcement to basement walls
- Stage 9   Concrete basement walls
- Stage 10   Strip formwork to basement walls

Figures 19.13–19.17 show a sequence study of the basement construction stages described above for Basement A. Figure 19.13 shows the basement plan and cross sections and, in particular, the in situ concrete floor slab thickening details. The sequence study has assumed that the thickenings and floor slab will be poured at the same time irrespective of whether Proposal A or B is chosen for the floor pour sequence (Figure 19.17 refers). As an alternative, the contractor could pour the thickenings up to the underside of the floor slab and then pour the floor as a subsequent operation.

Project quality plan

Upon notification of the award of the contract, the construction manager will initiate the preparation of the quality plan. This will need to be completed before work commences on site. It should be noted that the following is a basic quality plan and some clients often require a more comprehensive document linked into the company quality procedure.

Contents of the quality plan

The contract-specific quality plan is based on ISO 9002 and the company's procedures manual and contains the following headings:

- Quality statement
- Contract particulars

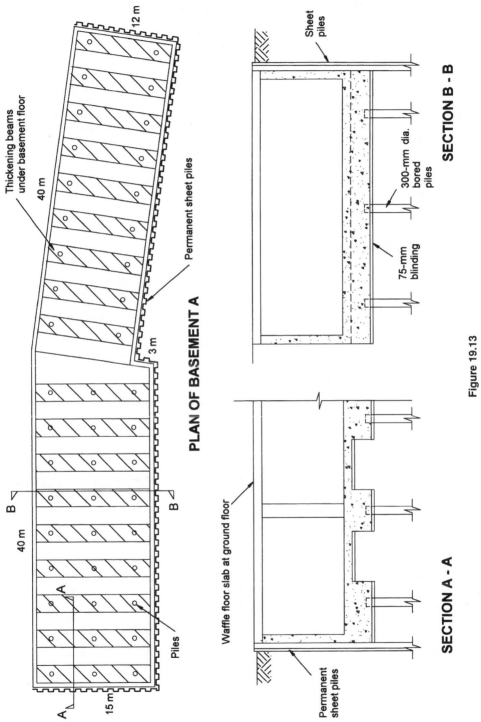

PLAN OF BASEMENT A

Thickening beams under basement floor

40 m

12 m

40 m

15 m

3 m

Permanent sheet piles

Piles

B

B

A

A

SECTION B - B

Sheet piles

300-mm dia. bored piles

75-mm blinding

SECTION A - A

Waffle floor slab at ground floor

Permanent sheet piles

Figure 19.13

DIAGRAMMATIC CONSTRUCTION SEQUENCE

STAGE 1 – Insert permanent sheet piles

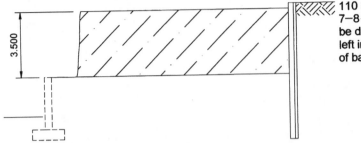

110 linear metres of 7–8-m long piles to be driven. Piles left in as rear face of basement wall

3.500

STAGE 2 – Excavate basement to depth of 3.5 m

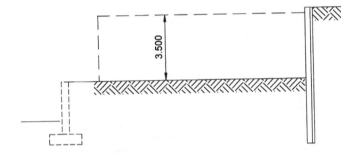

3.500

Excavation volume approx. 3800 m³ to be excavated and carted to tip-off site

Duration = $\dfrac{3800}{300 \text{ m}^3/\text{day}}$

Approximately = 12 days

STAGE 3 – Piling to basement slab

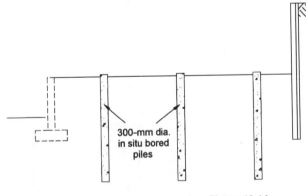

Approximately 40–300 mm dia. bored piles in basement Allow 5 days set up/dismantle equipment

8 piles/day - 5 days drilling

Approximately = 10 days overall

300-mm dia. in situ bored piles

Figure 19.14

DIAGRAMMATIC CONSTRUCTION SEQUENCE

STAGES 4 & 5
Excavate thickening beams to basement floor – Cut off piles and blind slab

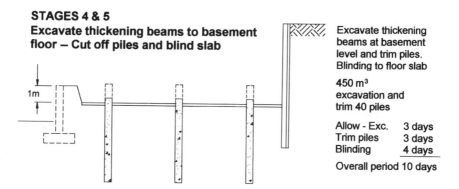

Excavate thickening beams at basement level and trim piles. Blinding to floor slab

450 m³ excavation and trim 40 piles

Allow - Exc.	3 days
Trim piles	3 days
Blinding	4 days

Overall period 10 days

STAGE 6 – Reinforcement to floor slab and thickenings

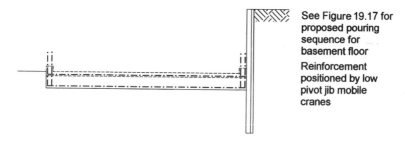

See Figure 19.17 for proposed pouring sequence for basement floor

Reinforcement positioned by low pivot jib mobile cranes

STAGE 7 – Concrete basement slab and thickenings

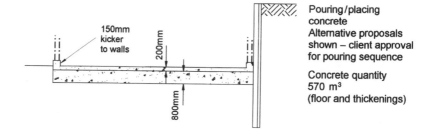

Pouring/placing concrete
Alternative proposals shown – client approval for pouring sequence

Concrete quantity 570 m³ (floor and thickenings)

Note – Thickenings may be concreted as a separate pour, prior to basement slab

Figure 19.15

- Organisation structure for the project
- Specific staff responsibilities
- Specialist services
- Control of workmanship and method statements
- Issue of drawings

DIAGRAMMATIC CONSTRUCTION SEQUENCE

STAGE 8
Reinforcement and formwork to walls
using full height formwork

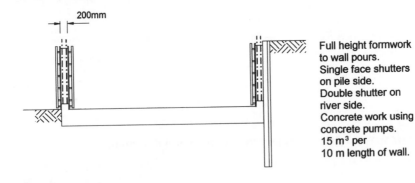

200mm

Full height formwork
to wall pours.
Single face shutters
on pile side.
Double shutter on
river side.
Concrete work using
concrete pumps.
15 m³ per
10 m length of wall.

STAGE 9 – Concrete to walls

Mobile crane to be used for handling reinforcement and formwork panels.
Maximum length of pour 10 m on plan.
Reinforcement to precede formwork as continuous operation around basement.

Concrete pours to alternate floor bays may be coordinated with wall pours for
economy of concrete pumping.

Sequence of constructing the wall and floor as shown on floor plan.
Sequence of wall pour, Figure 19.17 refers, to follow flooring sequence chosen.

Figure 19.16

- Control of subcontractors
- Control of on-site materials

Quality statement

'The company aims to meet the requirements of customers and of the industry and recognises
that quality is a constantly changing practice. The company recognises that quality is everybody's responsibility and it is committed to training and establishing an organisation that will
ensure a total understanding by all employees.

To achieve its requirements, the company has adopted a policy to operate a coordinated
quality system which conforms to the requirements of BS 7850 and ISO 9002. Mr Baker,
Regional Director, has been appointed with the responsibility and authority to manage the
adopted quality management system.'

Considerations for concreting basement slab

Proposal A

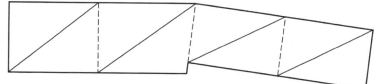

Pour floor slab in 4 pours
Approximately 140 m³ per pour
or in 8 pours at 70 m³ per pour

Proposal B

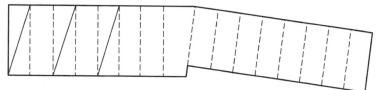

Pour slab and thickenings in 5 m wide alternative
bay construction – approximately 36 m³ per pour
three bays may be constructed per concrete pump visit – 108 m³/3 days

All floor pours to be concrete pumped from a location inside the basement area
or with the pump located on storage area B

Wall pour sequence

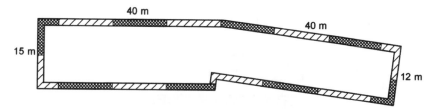

FLOOR AND WALL POUR SEQUENCE

Figure 19.17

Contract particulars

This part of the quality plan contains a summary of the conditions of contract, making reference to specific clauses in the form of contract. A description of works together with the contract commencement and completion date should be given.

A summary should also be given of the main participants involved in the contract, i.e. client, architect, design team and employer's representative. A list of the contractor's personnel engaged in managing the project should be included, indicating individual responsibility for construction management, surveying and health and safety.

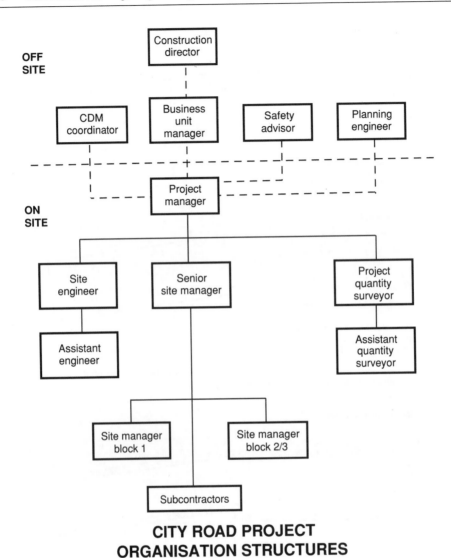

**CITY ROAD PROJECT
ORGANISATION STRUCTURES**

Figure 19.18 ABC

Organisation structure for project

A diagram showing the organisation structure for the project is commonly provided. This might be a 'family tree' diagram or perhaps a linear responsibility matrix showing roles and responsibilities, as suggested by Walker (2006).

Figure 19.18 indicates the line management structure from construction director through to site level with emphasis on safety management.

Specific responsibilities

This part of the quality plan contains a schedule of responsibilities allocated to the construction project manager, assistant site manager, site engineer and assistant surveyors and foreman. Responsibilities may be defined under the following general headings:

- Pre-start arrangements
- General site administration
- Management reports
- Materials management
- Site surveying functions

Specialist services

This section contains a checklist for the design of temporary works, formwork and the permanent works.

The parties responsible for the design of various elements of the permanent and temporary works will be identified, including the novated design team, works package contractors and suppliers of specialist components and equipment.

Control of workmanship and method statements

This section highlights operations from a specific list of activities which relate directly to the project. The activities are judged to be quality critical to the success of the project. In principle, they relate to the company's quality assurance scheme, to method statements affecting construction operations and to health and safety.

A schedule of method statements already prepared for early site operations and following operations would be provided.

Issue of drawings

This section contains instructions for the issue and distribution of drawings to all parties, i.e. head office, site and design team. The quick and efficient communication of information is vital to avoid delays and disruption on site.

Control of subcontractors

A schedule of operations relating to work undertaken by subcontractors is listed in this section. Their responsibility for developing project-specific method statements and attendance at meetings is outlined.

Control of on-site materials

This is a schedule of tests to be undertaken on site materials and relates to quality checks and destructive and non-destructive materials testing (i.e. cube tests for concrete, sonic tests for piles, etc.).

19.6 Contract stage

Subcontract procurement and management procedures

The main construction operations on site have been divided into a number of individual sub-contract work packages. A number of the packages involve some degree of design responsibility. The subcontract packages include:

- Sheet piling (earth support)
- Formwork, reinforcement and concreting of the basement works
- Structural steelwork, design and erect
- Metal deck floors
- External brickwork and blockwork
- Heating, ventilation and domestic services
- Lift installation

In this part of the case study, subcontract management arrangements are outlined taking a typical subcontract package through from the pre-contract stage of the contract to the end of contract review. The sheet-piling subcontract is used as the example in relation to subcontract procedures at the pre-contract and contract stage.

Tender procedures

At the tender stage of the project, enquiries have been sent to three sheet-piling contractors, and design proposals and firm prices have been received. Proposals from R&M Piling have been included in the tender submission.

Pre-contract procedures

These include:

- Approval of subcontract list
- Preparation of subcontract procurement schedule
- Review procurement schedule
- Subcontract pre-selection process
- Subcontract pre-placement meeting
- Place subcontract order (QS function) with R&M Piling
- Notification to commence on site

Contract procedures

These include:

- Start on site including
  - Submission of risk assessments and method statements

    ○ Submission of insurance requirements
    ○ Signed order
- Site induction procedures
- Quality inspection plan
- Agree subcontract target programme
- Subcontract weekly and monthly site meeting involvement in short-term planning
- Procedures at site management level
- Agreeing any subcontract variations and extensions of time
- Dealing with subcontract payments and final account
- Subcontract notice of completion
- End of contract review of subcontractors' performance

Subcontract management is a team effort between the project manager, the senior quantity surveyor and the planning engineer. Management of the subcontractor at the site face is undertaken by the appropriate site management personnel.

The above procedures are those implemented by the main contractor on the City Road project. This is a good example of a contractor/subcontractor relationship aimed at achieving a successful contract for all parties involved in the contract.

Procedures manual for planning and control

The following extracts are taken from the procedures manual of a large contracting organisation. They have been simplified and applied to the City Road project.

Role of the planning department

To perform the following functions:

- Initial programming for a project at the client design stage of a design and build project enquiry
- Detailed tender programme preparation
- Contract master programme preparation
- Preparation of a procurement programme and procurement schedules
- Progress evaluation at site level (including the preparation of an independent progress report on a monthly basis)
- Procurement updating
- Review of information flow
- Preparation of an as-built programme as work proceeds

Responsibilities of the site planning engineer

The planner will carry out the following tasks:

- Review project documentation
- Review the project drawings

PREPARATION OF MASTER CONTRACT PROGRAMME

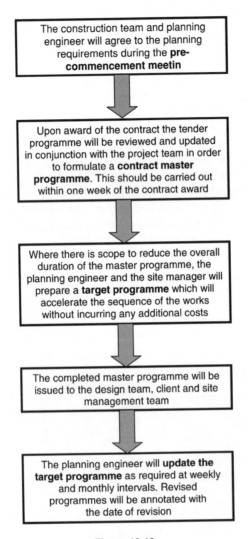

The construction team and planning engineer will agree to the planning requirements during the **pre-commencement meetin**

Upon award of the contract the tender programme will be reviewed and updated in conjunction with the project team in order to formulate a **contract master programme**. This should be carried out within one week of the contract award

Where there is scope to reduce the overall duration of the master programme, the planning engineer and the site manager will prepare a **target programme** which will accelerate the sequence of the works without incurring any additional costs

The completed master programme will be issued to the design team, client and site management team

The planning engineer will **update the target programme** as required at weekly and monthly intervals. Revised programmes will be annotated with the date of revision

Figure 19.19

- Review procurement and contractual considerations
- Review the proposed phasing and sequencing of work
- Undertake direct comparisons of proposed works against actual similar work previously completed
- Undertake a review of construction methods with the site manager
- Prepare the 2-weekly short-term programmes in conjunction with the appropriate site personnel

MONTHLY PROGRESS REPORT

Contract	City Road	Contract No	1714
Report No	5	Name	A Smith
Date	16 Feb	Contract Period	70
Weeks completed	20	Weeks remaining	50
Contract Completion Date		22 May	
Target Completion Date		4 Feb	
Anticipated completion		4 Feb	

MAIN PHASES OF WORK	PROGRESS COMMENTARY
Site establishment	Commenced on time, now 100% complete
Sheet piling	Started on time, completed 2 weeks later than planned due to work in connection with the adjacent river retaining wall. Work now 100% complete
Riverside retaining wall	Started 4 weeks later than planned due to variable ground conditions and changes in the foundation design. Currently 5 weeks behind programme Work now 60% complete
Basement A	Excavation works completed on time. Delays to the wall construction of Basement B have delayed commencement of the basement slab construction. Basement works are currently 4 weeks behind programme. Work now 20% complete
Basement B	Basement concrete and formwork started 3 weeks late due to design delays. Work currently progressing to basement walls at second level. Basement works currently 4 weeks behind programme. Work currently 80% complete

PROCUREMENT	
Subcontracts	Orders placed for steelwork which is due for erection on Block 1 at week 26. The commencement may have to be delayed by 3 weeks
Materials	Brick samples approved and orders placed

COPIES TO	Site manager ☐	Project manager ☐	Chief planning eng ☐

Figure 19.20

- Prepare a weekly and monthly report on progress for submission to the site manager and project manager with a copy to the chief planning engineer

Contract control documentation

On award of the contract, the following control documents form the basis of the link between the planner and the site construction team:

- The master contract programme
- The target programme
- The materials requirement schedules
- The subcontract procurement schedules
- Information requirement schedules

Operating procedures when preparing contract programmes

The purpose of these procedures is to produce a master contract programme to be issued to the design team and client and to provide a construction target programme for construction and monitoring purposes. Figure 19.19 illustrates the procedures to be followed.

Progress records and reports

Liaison between site-based planning engineer and site managers

- The site-based planning engineer will be responsible for following the company procedures as indicated in the procedures manual, as outlined above
- Progress on the target programme will be recorded using a computer software package at both weekly and monthly intervals
- The site-based planning engineer will work closely with the appropriate site manager in the preparation of detailed stage programmes in relation to major sections of the work. During the undertaking of the works on site, short-term 2-weekly programmes will be prepared and implemented by both the planner and site manager in charge of that particular work section

Monthly progress reports

Figure 19.20 contains an extract from a monthly site report produced at the end of week 20 of the contract. This is produced for the project review meeting at the end of each month. It is compared with the independent report produced by the head office planner. The report is presented in operational format, supported by an as-built bar chart together with the updated target programme.

Reference

Walker, A. (2006) *Project Management in Construction*, 5th edn. Blackwell Publishing.

20 Eastlands project

20.1 Project description

The Eastlands project is part of a £2 billion urban regeneration programme in a large city in the north-west of England. This particular scheme is a residential development of both high-rise and low-rise housing with car parking, garages and landscaping.

The site is roughly triangular in shape and lies between existing houses to the north and Ashton Canal to the south. A busy dual carriageway is to the west of the site and further established housing lies to the east.

Figure 20.1 shows the site layout plan of the development.

20.2 Project details

The total sales value of the project is approximately £25 million and the construction period on site is 49 weeks. Site acquisition, design, statutory approvals and sales make up the overall project duration. Block A contains 47 units and block B has 68 one and two bed apartments. Each floor level accommodates between six and seven apartment units per floor.

The site is a 'brownfield' site and prior to handover by the local authority to the developer/contractor, a programme of remediation was carried out under separate contract to a specialist company. This involved dealing with old mining shafts and contaminated ground before construction work could commence.

20.3 Construction details

The project involves the construction of two cylindrical drum-like multi-storey apartment blocks and 25 linked low-rise housing units. The apartment blocks are 24 m in diameter and blocks A and B are seven and ten storeys high respectively.

The circular 'drums' are of precast concrete cross-wall construction with pre-stressed concrete floor panels supported on internal cross-walls which radiate out of the central service 'core'. Some master bedrooms have projecting steel balconies. The external elevations are finished with a rain screen panel cladding system and in the area of the window and balcony units feature external brick panels are employed. Each block incorporates communal areas, lifts and staircase access.

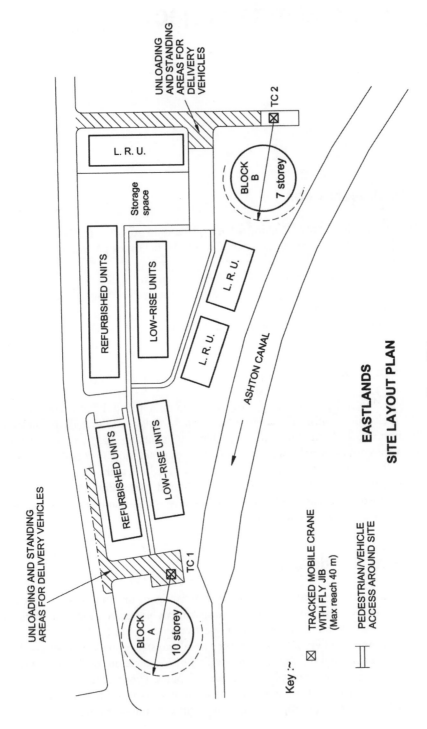

EASTLANDS

SITE LAYOUT PLAN

Figure 20.1

Key :-

⊠   TRACKED MOBILE CRANE WITH FLY JIB (Max reach 40 m)

▯▯   PEDESTRIAN/VEHICLE ACCESS AROUND SITE

Figure 20.2 shows a typical floor plan with the internal cross-wall layout and load-bearing flank walls indicated. An internal floor plan of a typical flat unit is shown, i.e. flat 5.

20.4 Organisation structure of the developer/contractor

Head office

The developer/contractor is a London-based private development company with an annual turnover of some £500 million. The residential housing division operates from three regional offices including one in the north-west of England.

Figure 20.3 illustrates the organisation structure of the north-west region. The Eastlands project falls under the direct control of the regional operations director. The company employs their own 'in-house' teams of architects, planners, quantity surveyors and sales personnel.

Site organisation

Figure 20.4 shows the site organisation structure for the project and, unusually, the site management team is led by a 'build manager' whose role and responsibilities are explained below.

The project manager is the build manager's assistant/deputy and his role is similar to that of the traditional sub-agent. He is responsible for organising the individual site managers employed to supervise the work package subcontractors on each section of the work. There are three site managers – one for each of the tower blocks and another for the low-rise units.

The title used for the person responsible for running a construction project varies from company to company and other companies may refer to the contractor's representative as the project manager, site agent, team leader, construction executive or simply the construction manager. It may seem somewhat confusing that the job titles for personnel carrying out the same roles and responsibilities are not common in the industry, but this is the nature of construction!

A materials coordinator assists the project manager and his job is to make sure that materials are called forward for delivery at the right time (concrete, bricks and blocks, windows etc.). Just-in-time deliveries avoid the need to stockpile materials and this reduces double handling and waste. A separate materials coordinator is employed by the precast frame subcontractor.

Figure 20.5 indicates the specific responsibilities of the site management personnel and support services from head office.

Duties and responsibilities of the 'build manager'

The build manager is appointed at the design/pre-contract stage of the project in order that his expertise in relation to construction matters may be called upon. His responsibilities are to:

- Coordinate the key information requirements between the design team, i.e. architect, design engineers, quantity surveyors and planners

PLAN OF BLOCK

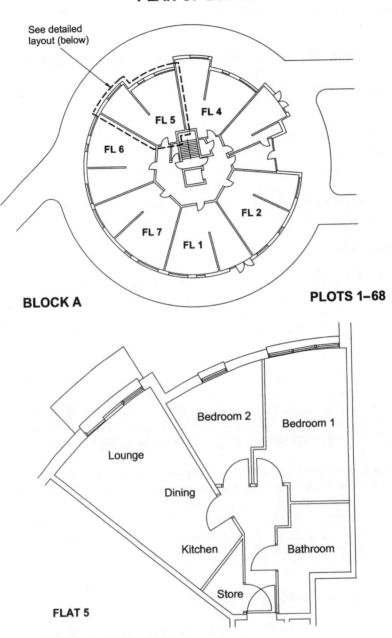

See detailed layout (below)

FL 5

FL 4

FL 6

FL 2

FL 7

FL 1

BLOCK A

PLOTS 1–68

Bedroom 2

Bedroom 1

Lounge

Dining

Kitchen

Bathroom

Store

FLAT 5

Figure 20.2

REGIONAL ORGANISATION

Figure 20.3 NORTHERN REGION ORGANISATION

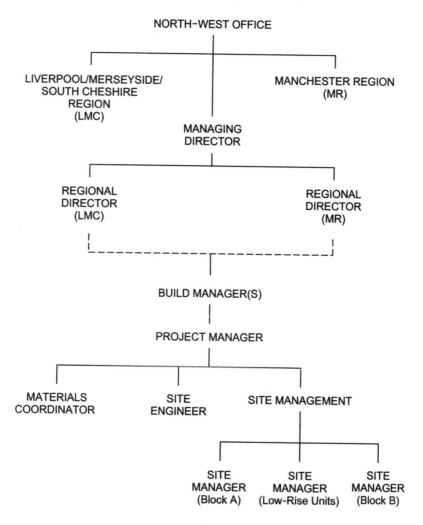

Figure 20.4 SITE/PROJECT ORGANISATION

- Assist with the preparation of the pre-contract Method Statement and Exit Plan (handover proposals) in conjunction with senior management
- Provide input to the master programme (overall project strategy) in conjunction with the operations director
- Develop a team approach to the management of the project

SITE ORGANISATION STRUCTURE

RESPONSIBILITIES OF SITE MANAGEMENT

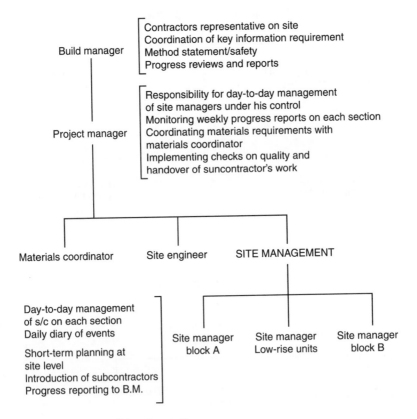

Build manager — Contractors representative on site
Coordination of key information requirement
Method statement/safety
Progress reviews and reports

Project manager — Responsibility for day-to-day management
of site managers under his control
Monitoring weekly progress reports on each section
Coordinating materials requirements with
materials coordinator
Implementing checks on quality and
handover of suncontractor's work

Materials coordinator   Site engineer   SITE MANAGEMENT

Day-to-day management
of s/c on each section
Daily diary of events

Short-term planning at
site level
Introduction of subcontractors
Progress reporting to B.M.

Site manager
block A   Site manager
Low-rise units   Site manager
block B

Visiting support staff from head office

- Quantity surveying – Valuation and payment of subcontractors
- Planning – Contract planner available for progress updates and detail planning at site level
- Safety – Visiting safety manager
- Customer care – Liaison with build manager at handover stage of both s/c work stage and final quality inspections

Figure 20.5

- Monitor and control the procurement schedules and master programme
- Encourage safe working practices on site at all levels by implementation of the company safety and care policy
- Report on the contract's progress situation to the operations director

Build manager's role

On this project, the build manager led by example. He developed a range of site information boards on aspects of safety, site layout planning and good materials handling practice. Good relationships were established with the key work package subcontractors and the contract key programme dates were successfully achieved.

The organisation structure on site fostered good relationships between site personnel and operative gangs which helped to achieve a tight programme. Excellent backup services were provided to the build manager from head office in areas such as safety, quantity surveying, planning, sales and customer care services.

20.5 Project planning

Part of the planning documentation prepared prior to the commencement of work on site comprises a Method Statement (construction operations) and Exit Plan (phasing, sales and handovers) which is a joint document prepared by the commercial manager/regional operations director together with input from the build manager, planning engineer and sales department.

Additionally, a master programme is prepared and Figure 20.6 is an extract from this showing the main operations connected with the construction of the tower blocks A and B together with site establishment and external works, etc. Logic links and float times indicate relationships between the various stages of work and the critical and non-critical activities on the programme.

The master programme is based on the following data:

Sequence of operations	Duration in weeks
Establish site	3 weeks
Piling	2 weeks for piling each block
Substructure work	9 weeks per block
Superstructure – precast frame	Block A 13 weeks
	Block B 8 weeks
Roof seal	7 weeks (both blocks)
Structural finishes	Block A 16 weeks
	Block B 11 weeks
External envelope	Block A 16 weeks
	Block B 12 weeks

20.6 Method statement – tower blocks A and B

The unusual circular nature of the design of the high-rise 'drums' poses a number of difficulties for the contractor from a planning point of view and these problems are compounded by

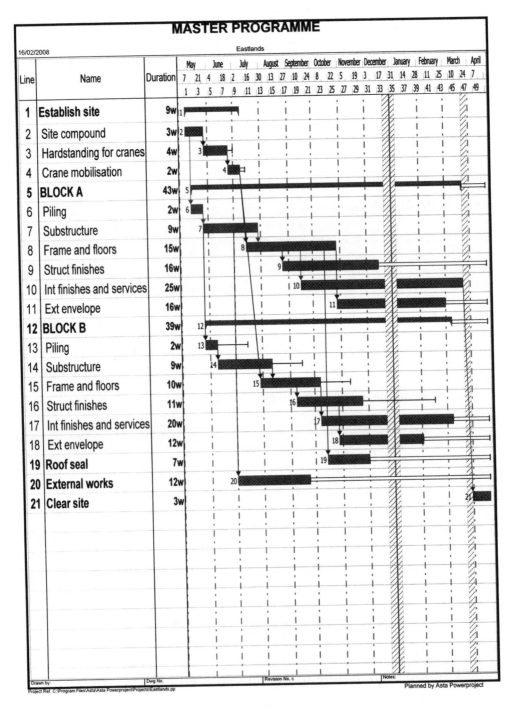

Figure 20.6

the projecting steel balconies and the two different cladding systems on the external facade. Materials handling and access are further problems posed to the contractor.

The following aspects of the contractor's method statement are now considered in turn:

- Establish site
- Piling
- Substructure
- Precast concrete frame
- Roof seal
- Structural finishes
- External envelope
- Materials handling

Establish site

Normally, the contractor's site set-up is located directly on the site of the works but, in this case, the lack of space on site requires an off-site set-up. Fortunately, a suitable area for the site offices, welfare accommodation and secure storage for materials and equipment is available about 100 m from the site. The area is secured with metal security fencing.

The site accommodation is tiered in two units which incorporate a site canteen facility, drying rooms and mess facilities for the operatives at ground level and offices above. A car parking facility for office staff and site visitors is provided.

A particular feature of the compound area is a daily site hazard board which informs site staff and operatives of current hazards on the site.

Piling

Blocks A and B are each piled with 70 flight auger bored piles commencing with block A. This takes 2 weeks. On completion of Block A, a further 2 weeks is allowed for piling to Block B.

Substructure

Excavation work to the pile caps and ground beans is followed by fixing formwork and rebar to the ground beams and ground floor slab. The 140-cm thick ground floor slab and ground beams are planned as a single pour using mobile concrete pumps. Prior to concreting, a methane barrier is to be laid under the slab. A period of 9 weeks has been allocated for the complete foundation construction for each block.

Precast frame

The precast concrete framework package for the tower blocks is awarded to a specialist manufacturer under a 'design and build' contract. The units are manufactured at a factory some 60 km from the site. Whilst being contractually responsible for the design, supply and fixing of the units, the manufacturer is to sublet the site erection work to another subcontractor who assumes responsibility for offloading, handling and fixing the units in place. This is effectively a sub-subcontract.

The precast concrete units are delivered to site on large flatbed articulated vehicles and are then offloaded and directly lifted in place using two 75-tonne lattice-boom tracked crawler cranes – one for each tower. This means that there are no precast concrete units stored on site. The main contractor is responsible for providing a firm hard standing area for the cranes to stand on. A further aspect of the contractor's 'attendance' on the specialist subcontractor is the provision of a hardcore access road for the delivery wagons. The wagons will back into the site from the road and position themselves close to the cranes for offloading. Further space is needed for delivery vehicles standing whilst awaiting offloading. Figure 20.1 is a site layout plan showing the main hard standing areas for the delivery vehicles and the crawler cranes.

A mobilisation period of 2 weeks is required by the subcontractor for assembly of the mobile cranes and this is built in to the precast concrete frame erection periods on the master programme (block A is 15 weeks and block B is 10 weeks). The frame is erected floor by floor on a 6-day construction cycle (see Figure 20.9 for the programme sequence per floor). Completion of the frame and floors for both blocks A and B is programmed for the end of week 25 – note that block B frame erection has 3 weeks float (see Figure 20.6).

Subcontractor's organisation

The frame erection is to be carried out by a specialist contractor engaged by the precast manufacturer under a sub-subcontract. The organisation structure of the specialist subcontractor and the interface with the precast concrete supplier and the main contractor is illustrated in Figure 20.7. This shows the role of the area frame supervisor who is effectively the site manager for the precast erection package. The actual site operations – lifting and placing the units, temporary propping and grouting, etc. – are under the control of the subcontractor's site foreman. A site engineer is to visit the site for setting out and checking the line and level of the precast panels.

A key member of the subcontractor's site management team is their materials coordinator (not to be confused with the main contractor's materials coordinator). He is responsible for liaising with the production manager at the precast factory and with the area frame supervisor. It is essential to ensure that the precast units are delivered to site in the sequence in which they are to be erected so as to avoid delays, double handling and damage to the units. This is a daunting task as up to four delivery vehicles are scheduled to service the site each day and arrangements have to be made to unload them in accordance with the daily sequence of erection. This is not as easy as it sounds because the vehicles have to arrive in the correct order each day, the right materials have to be on each vehicle to suit the erection sequence and they have to be stacked on the vehicles in the correct order for offloading. One mistake and the project will be delayed!

Safety hazards

A number of hazards are evident from the precast concrete erection sequence a few of which are listed below together with suitable control measures:

Hazard	Risk	Control measure
Work at height (1)	Precast concrete wall panels are delivered to site stacked vertically on a flatbed vehicle. Slinging of the panels for offloading places an operative at risk because he has to climb on top of the panels to attach the slings. The operative would be approximately 4 m above ground level.	Panels to be stacked vertically either side of a suitable steel frame which runs from back to front of the trailer. Permanent steps are incorporated into the frame to provide access for operatives.
Unstable load	There is the danger that panels could move during slinging/offloading thereby causing a risk of crushing or trapping.	Ditto.
Work at height (2)	During the placing of precast concrete floor units operatives will be working at an unprotected leading edge with the risk of falling 3 m to the floor below. Also, after installation of the floor units, but before the external wall panels are erected, an exposed leading edge could result in an operative falling over the edge of the building to the ground which could be up to 30 m below.	Safety nets (or alternatively air bags) to be installed below the work area to arrest falls. Nets to be slung from steelwork or suitable fixing points in the concrete panels. Access for fixing nets to be from a hydraulic access platform (cherry picker). A patented scaffold barrier system to be bolted to the external face of the building to provide external edge protection.
Unstable structure	Once the precast concrete wall panels are lifted into place they are unstable until the final site stitching is completed. This leaves operatives at risk of being crushed.	Temporary propping is required until final site stitching, e.g. Acrows or Powershores bolted and braced to floor units. Scaffolded perimeter rails to be installed at each floor level.
Moving plant or vehicles	Delivery wagons and mobile cranes for offloading and lifting of precast concrete units pose the risk of being trapped or crushed by moving vehicles.	Pedestrian separation wherever possible and trained banksman working with machines.
Falls of materials	Whilst the precast frame is being fixed at upper floor levels there is a danger of panels falling onto operatives while they are fixing external windows to the floors below or onto other site personnel passing by.	The establishment of exclusion zones beneath work areas and creation of safe entry and exit points away from the precast erection activities.

PRECAST CONCRETE ERECTION – SUBCONTRACTOR ORGANISATION STRUCTURE

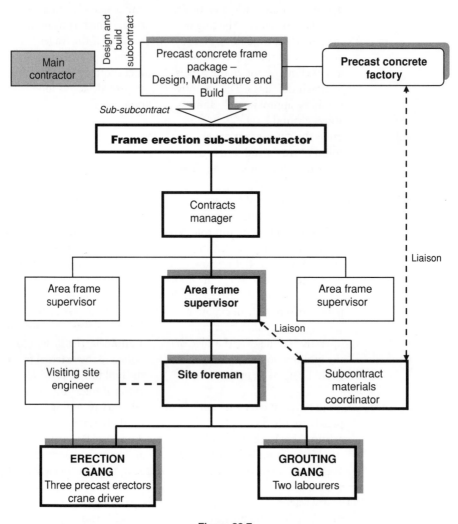

Figure 20.7

Sequence of work

The frame erection sequence is based on a construction cycle time of 6 days per floor (see Figure 20.9). Any delays would impact both the programme and the delivery schedule for the precast concrete units. The construction cycle comprises the following operations:

(1)   Erect external perimeter wall panels
(2)   Erect internal cross-walls and staircase core units

EASTLANDS			PRECAST FRAME				MAIN CONTRACTOR'S PROGRAMME								
Line	Name	Plot no.	Start	End	Dur	% Comp	2007 February				March			
							5 / 19	12 / 20	19 / 21	26 / 22	5 / 23	12 / 24	19 / 25	26 / 26
1	Site set up		05/02/07	09/02/07	1w		▨							
2	Ground floor walls		12/02/07	13/02/07	2d		▨							
3	1st floor and stich		14/02/07	16/02/07	3d		▨							
4	1st floor walls		19/02/07	21/02/07	3d			▨						
5	2nd floor and stich		19/02/07	21/02/07	3d			▨						
6	2nd floor walls		27/02/07	01/03/07	3d				▨					
7	3rd floor and stich		02/03/07	06/03/07	3d					▨				
8	3rd floor walls		07/03/07	09/03/07	3d					▨				
9	4th floor and stich		12/03/07	14/03/07	3d						▨			
10	4th floor walls		15/03/07	19/03/07	3d						▨			
11	5th floor and stich		20/03/07	22/03/07	3d							▨		
12	5th floor walls		23/03/07	27/03/07	3d								▨	

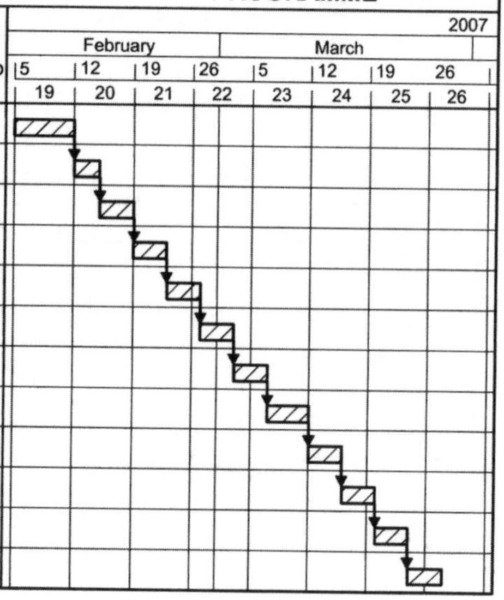

Figure 20.8

(3) Fix external perimeter protecting rail
(4) On-site stitching to the vertical wall joints
(5) Install safety netting to the floor areas
(6) Insert steel beams over corridor areas
(7) Erect floor units and precast staircase access
(8) Form in situ concrete make up strips and grout the floor slab

Short-term programme

Figure 20.8 shows an extract from the main contractor's programme for the first five floors of the frame erection. The short-term cycle programme developed by the subcontractor is illustrated in Figure 20.9 which is based on the following sequence and durations:

• External perimeter wall	1 day	⎫
• Internal cross-walls	1 day	
• Scaffold edge protection	2 days	
• Grout wall joints	1 day	⎬ Overall cycle time
• Internal floor safety nets	1 day	per floor = 6 days
• Internal steel beams	0.5 day	
• Fix precast floor units	1.5 days	
• In situ strips and grout up	1 day	⎭

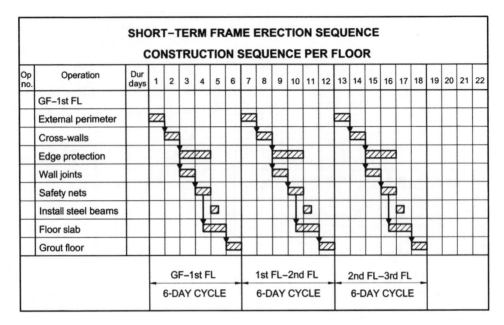

Figure 20.9

Roof seal

Roof sealing work is to commence on completion and final grouting of the precast concrete roof units. Internal finishes cannot commence above floor 5 until the roof sealing is substantially complete due to the danger of water ingress. Temporary roof sealing may become necessary in excessively wet weather conditions which would add another activity to the programme.

Structural finishes

Structural finishes include the fixing of external windows, internal timber stud work, floor screeds and dry lining and insulating the internal concrete cross-walls. This will commence at ground floor level when frame erection to the fourth floor is completed. The lining of the internal walls cannot commence until the floors have been temporarily sealed against rain penetration from above.

Subcontract labour working on the lower floors during the precast concrete frame erection activities are at significant risk from moving vehicles and heavy materials being handled overhead. Consequently, segregated pedestrian access routes into the building are of critical importance for safe working. Protected areas are therefore provided at the front of the buildings with well-signed safe access routes leading to them.

External envelope

The circular precast concrete outer walls of the blocks are finished externally with a combination of facing brickwork and rain screen cladding. The cavity between the precast frame and brickwork is insulated with a rock wool or similar material. Adjacent wall panels incorporate a rain screen cladding system fixed on a lightweight frame which is insulated prior to fixing the composite board finish.

Safe access, safe working at heights and the lifting and handling of materials for the wall surface finish are a major issue for the contractor. A further complication is the lifting and fixing of steel balcony units once the brickwork and cladding have been completed.

Two proposals are considered.

Option one – mast climbers

Mast climbers are a fast means of distributing materials and providing temporary access for operatives to the exact position required. They consist of a single or twin mast (they look similar to a tower crane mast) and incorporate a safe working platform with a perimeter barrier. They are adjustable and extendable to suit the building's contours. Mast climbers may be freestanding or braced to the external wall surface of the building and the platform moves up and down on a 'rack and pinion' system.

Using mast climbers, the external envelope would be completed in the following sequence:

(a) Erect mast climber from ground level to roof level positioned at each vertical group of flats
(b) Complete the external brickwork and cladding in the area of the mast climber
(c) Move in a predetermined sequence around the building, completing each area from ground level to roof level
(d) Adjust mast climbers to allow balconies to be fitted
(e) Fit balconies from the upper to the lower floors as the mast climbers are dismantled

The pre-manufactured balcony sections will be assembled on site and lifted into position by crane. The balconies will be fitted from the mast climbers once they have been adjusted in height.

On this project, up to ten mast climbers would be needed in order to facilitate concurrency (overlapping) of the work.

Option two – independent external scaffold

Using an external independent scaffold, the external envelope would be completed in the following sequence:

(a) Erect scaffold around complete building perimeter from ground floor to fourth floor
(b) Incorporate hoist arrangements in the scaffold for vertical movement of materials and labours
(c) Complete brickwork and cladding panels from ground to fourth floor
(d) Erect scaffold from fifth floor to roof level
(e) Complete external brickwork and cladding to roof level
(f) On completion of brickwork and cladding, dismantle the scaffold completely
(g) Lift balcony assemblies into place with a mobile crane and fix in place with the assistance of hydraulic lifters using 'top-down' construction (progress from the top of the building down to ground level)

Despite mast climbers being £20 000 cheaper than the scaffold option, the decision to use the external scaffold option is based on criteria other than cost, including:

- Difficulties envisaged with adapting the mast climbers to suit the curvature of the building using extension arms
- Mast climbers involve establishing single point loading areas at the base of each mast climber for handling bricks, mortar and the lightweight cladding panels so there would be up to ten loading areas
- The external scaffold provides an easier materials handling solution for other materials to be lifted into the building (i.e. sanitary ware, kitchen units and services components etc.)

Materials handling and access proposals

Developing an efficient materials handling system and providing safe access for operatives and site staff are essential for the successful completion of any project but high-rise buildings pose particular problems. Tower blocks A and B are ten storeys (30 m) and seven storeys (21 m) high respectively and both pedestrian access and hoisting of materials to the upper levels of the buildings are required. Finding the right solution is difficult because materials and personnel are needed *not only* at the floor levels *within* the buildings *but also* at the main scaffold platform levels *outside* the building.

The problem for the contractor here is that building floor levels are at 3-m intervals and the scaffold platforms are set at 2-m intervals – so they do not line up. Also, hoisting of materials and pedestrian access need to be segregated for safety reasons.

A system of two hoists is employed with foot access via the permanent central precast concrete staircases and also the external scaffold. Access into the building for materials from the hoist is via the large patio doors at each floor level.

Pedestrian access plan

(a) Site operatives and supervisory staff will gain access to the upper floor levels of the buildings via the permanent staircase within the central access core which has its ground level access point away from the materials hoisting area.

(b) Pedestrian access to scaffold levels is provided by a staircase access tower integrated into the main external scaffold around the building. This is fully enclosed within the scaffold with safe access points at each scaffold level.

Materials hoisting plan

Each tower block is to be provided with a twin-car goods hoist comprising two enclosed 'container' units. Each unit has different dimensions so that a variety of large, bulky and small materials can be accommodated.

The two cars are suspended either side of a central mast which is braced to the main building structure. The hoist operates using a 'rack and pinion' system and each car is independently electrically controlled.

MATERIALS HOISTING AND ACCESS PLAN
TO TOWER BLOCKS A AND B

HOIST CAR 1

- The car is positioned with its offloading door facing the building and is arranged to be able to stop at each floor level in the building and the roof
- This is a 2000-kg capacity car chosen for carrying a full pallet of plasterboard (each board is 2.4 m x 1.2 m)
- The hoist car is loaded using rough terrain forklift trucks and so the car needs sufficient headroom to allow access for loading the pallets inside
- The car is hoisted up floor by floor and the boards offloaded by hand to a central point to be distributed internally
- It is intended that the hoist will also be used by subcontractors for the distribution of their goods including kitchens, kitchen equipment, bathroom fittings, etc.

HOIST CAR 2

- The car is positioned with its offloading door facing 90° from the building onto a specially constructed scaffold tower. The tower has large boarded platforms at each scaffold working level so that materials can be offloaded and handled
- This car is also 2000-kg capacity and has been chosen for its suitability for accommodating the rain screen cladding materials and packs of external facing bricks, etc.
- The stopping positions for this car are set at 2-m scaffold height
- The car is able to stop at each scaffold level

A general arrangement of the hoisting and access plan is shown below:

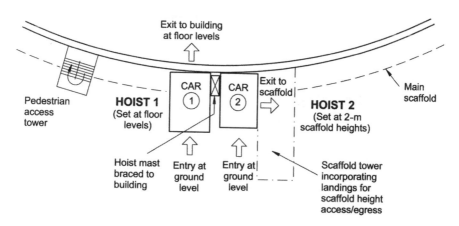

Figure 20.10

Further hoisting facilities for each tower block include the provision of separate scaffold hoists for lifting scaffolding materials into position.

Figure 20.10 shows the materials handling and pedestrian access arrangements (permanent staircases and scaffold hoists not shown) and explains how the system operates.

21 Highway project

21.1 General description

Figure 21.1 shows a proposed dual carriageway trunk road which is to be built to motorway standards under the Department of Transport Specification for Highway Works.

Under the ICE Conditions of Contract, the contractor is required to submit a Clause 14 programme to the engineer within 21 days of the award of the contract. Additionally, if requested by the engineer, the contractor is also required to submit details of his proposed methods of construction, including temporary works and contractor's equipment. A time-chainage diagram would be an appropriate display on this type of project.

A typical cross section through the road is given in Figure 21.2. This shows the permanent fencing and fence line drainage which both have to be completed in advance of bulk earthworks operations. At chainage 3 + 600, there is a proposed box culvert which would also have to be completed in advance of filling to the embankment.

The main activities for the project and estimated durations are given in Figure 21.3.

There are three access points available on to the site, one at the existing roundabout, one at Colliery Road and the other via Green Lane.

21.2 Construction strategy

Earthworks

The project is significantly influenced by the earthworks in that the sequencing of cut and fill operations largely controls the sequencing of the entire project. However, no activity can be considered mutually exclusive as the sequencing of the earthworks is itself affected by the construction of the bridge and box culvert.

On major road projects, the quantities of muck to be shifted are often very large and two or three million cubic metres is not unusual. The quantities for this case study are therefore relatively small, but there are a number of problems which face the earthworks contractor:

- The major cut and fill zones are at opposite ends of the site and the average haul distance on the project is approximately 2 km each way
- The second fill zone is bisected by a proposed bridge site
- A box culvert is required at chainage 3+600 which must precede filling to the embankment

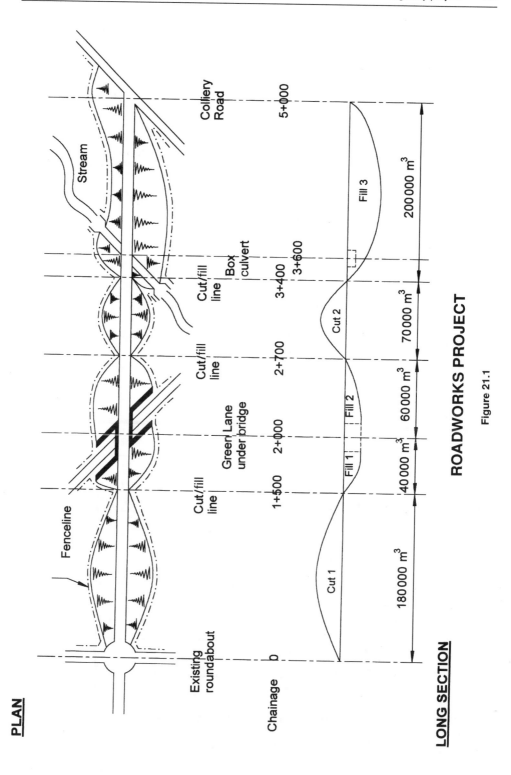

PLAN

Fenceline

Stream

Existing roundabout

Cut/fill line

Green Lane under bridge

Cut/fill line

Cut/fill line

Box culvert

Colliery Road

Chainage   0   1+500   2+000   2+700   3+400   3+600   5+000

LONG SECTION

Cut 1

Fill 1   Fill 2

Cut 2

Fill 3

180 000 m³   40 000 m³   60 000 m³   70 000 m³   200 000 m³

ROADWORKS PROJECT

Figure 21.1

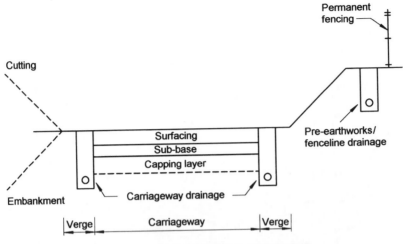

TYPICAL CROSS-SECTION

Figure 21.2

- There is a shortfall of acceptable fill material and therefore 50 000 m³ will have to be imported
- The existing road at Green Lane must be kept open to traffic during construction of the new bridge

ACTIVITY	DURATION (weeks)
Set up site	6
Fencing and site clearance	10
Topsoil strip	4
Fenceline drainage	14
Excavation and filling	20
Capping layer	11
Carriageway drainage	13
Sub-base	11
Surfacing	13
Box culvert	8
Bridge	26
Finishes	19
Tie-ins to existing roads	6

SCHEDULE OF ACTIVITIES

Figure 21.3

Earthworks plant for a project of this nature has changed over the last 30 years or so. Conventionally, major earthworks were carried out using rope-operated scraper boxes towed by caterpillar-type tractors, but motorised scrapers, which were faster and with larger capacity, became popular. Scrapers often need dozers pushing to help loading, especially in heavy ground conditions.

Nowadays, it is also popular to use large tracked back actors with articulated dump trucks (ADTs) carrying excavated materials to fill zones. Modern excavators are fast and powerful and the new breed of dumpers carry large volumes of spoil. However, where the haul distances are long, this means either the excavators being kept waiting or employing a large fleet of dumpers. A contractor may well use a combination of scrapers and excavators/dumpers, the main considerations being:

- Volume of earthmoving
- Nature of materials to be removed
- Whether material is to be used for fill or carted off site
- Cost of delivering plant to site
- Cost and availability of plant
- Use of own or hired plant
- Haul distances
- Time constraints
- Access
- Restrictions (e.g. bridges)

Drainage

The drainage works comprise:

- Pre-earthworks or fence line drainage – mainly consisting of porous rubble-filled 'french' drains with associated catchpits and connections to surface water ('carrier') drains, culverts and lined and unlined ditches
- Carriageway drainage – comprising both 'carrier' drains, manholes and gullies and 'french' drains and catchpits

The pre-earthworks drainage normally commences early in the project and follows on behind erection of the permanent fence line and the topsoil strip over the site.

When the pre-earthworks drainage is reasonably advanced, bulk earthworks can follow on behind. Earthworks may be interrupted by culverts crossing the carriageway, which will affect muck-shift sequencing and may involve temporary diversions around the site of the culvert.

Pre-earthworks drainage poses problems for the contractor such as:

- **Access** – There may be no haul roads in and plant may have to 'track' along from the main access points, e.g. access for the box culvert would have to be from Colliery Road
- **Materials** – Without proper access for road vehicles, site conditions and the weather will very much dictate whether access will be possible along the spread of the job. Drainage materials such as pipes, manhole/catchpit components, bricks and ironwork may have to

be delivered to the main compound and then double-handled by site transport such as tractor and trailer to the various locations along the job. Drain fills are delivered from the quarry by road transport and may have to be tipped at access points and reloaded on to site dumpers. Ready-mixed concrete may similarly be a problem

- **Drainage arisings** – A major proportion of the spoil from drainage excavations will not be required for filling trenches but may not be suitable for filling elsewhere on the site. The suitability of fills is the responsibility of the main contractor/earthworks subcontractor and therefore the drainage subcontractor will have to be instructed as to what is required of him. Part of the contractor's thinking could be either for the drainage subcontractor to load the arisings into wagons/dumpers or, perhaps more cost effectively, for the arisings to be left to one side to be removed by the earthworks subcontractor. This is an interface which would require managing by the main contractor, and the respective responsibilities would have to be written into each of the subcontract packages

- **Resources** – The main contractor will have to carefully manage the drainage subcontractor to make sure sufficient gangs are available so as not to slow the earthworks down. This is important because the earthworks 'season' is relatively short. Often on large projects, the drainage is split into two or more packages and different subcontractors are employed on each. Again, there will be interface problems here, particularly with regard to the control of materials, which the main contractor has to think about

Structures

One of the problems facing the contractor is the construction of structures across the carriageway, such as bridges and culverts, especially in fill areas. In this example, there is a bridge at chainage 2 + 000 and a box culvert at chainage 3 + 600, both of which would have to be constructed in advance of embankment construction. Fill can then be brought up to and behind each structure so as to bring levels up to the correct height for road construction to follow.

With regard to the bridge at chainage 2 + 000, abutment walls and foundations will have to be constructed parallel to Green Lane. This may require traffic management arrangements or possibly a temporary diversion of traffic around the bridge site. Alternatively, it may be possible to close the road during construction.

In order that bridge construction can proceed without impeding earthworks operations, a temporary haul road will be constructed around the bridge site. This will involve traffic control measures on Green Lane as heavy plant will be crossing frequently. This will also complicate the time-chainage programme as drainage works and other activities will have to be left incomplete at this location until later. This has not been shown on Figure 21.4 for simplicity.

Capping layer

Depending on the design of the roadworks, a layer of fill may be required as a foundation for road construction. This is known as the capping layer and this operation follows on from bulk earthworks.

Due to the earthworks sequencing on this particular project, the capping layer, drainage and road construction would start at chainage 5 + 000 and progress towards chainage

TIME-CHAINAGE DIAGRAM   Roadworks Project

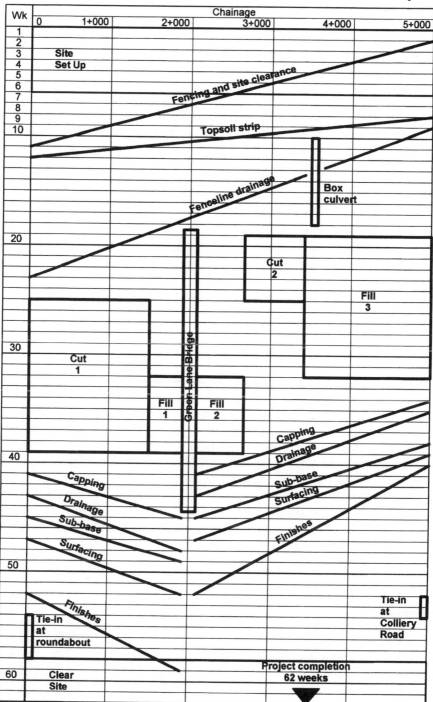

NB: Holidays and other activities omitted for clarity

Figure 21.4

2 + 000. This would be more efficient than starting from chainage 0 + 000 as capping, etc., could follow filling to zone 3.

21.3 Method statement (simplified)

Earthworks

Bulk excavation to Cut 1 assuming all material is acceptable as fill. Part of Cut 1 is used for embankment construction at Fill 3 and the remainder for bringing up embankment construction either side of Green Lane Bridge.

Quantity = 180 000 m³ of excavation

Method

Cutting to be taken out using a combination of motor scrapers push-loaded with a tracked dozer and 40 tonne 360° tracked excavators loading into articulated dump trucks.

Average haul distance = approximately 2.25 km

Filling and compaction is a separate activity.

Outputs

The average haul distance is relatively long and the scrapers will do well to achieve three cycles per hour. The scraper crews usually work a long day in the order of 10 working hours over 6 or 7 days per week in the earthworks season. A large scraper will move an average of 450 m³ each per day over the haul distances on this project.

The output of the back actors is usually governed by the haul times of the dumpers, unless the digging is hard. Therefore, the output is decided on how many trips/cycles can be achieved. Again, because of the average haul distance, this will be no more than three per hour giving around 300 m³ per dumper per day. Using three dumpers this equates to 90 m³ per hour for the excavators, which is no problem for two large machines. The gang therefore has capacity for trimming operations and dealing with obstructions, etc.

Gang

Foreman
Three Terex TS24 motor scrapers
One Cat D7 dozer (pushing)
One Akerman H10 excavator
One Komatsu PC380 excavator
Three Cat D400 dumpers
Banksman

Activity duration:

$$= \frac{180\,000\ \text{m}^3}{[(3\ \text{scrapers} \times 450\ \text{m}^3) + (3\ \text{dumpers} \times 300\ \text{m}^3)] \times 6\ \text{days}}$$

$$= \textbf{14 weeks}$$

Drainage

Assume the carriageway drainage from chainage 0+000 to 2+000 varies from 150 to 300 mm in diameter at an average depth of 1.50 m with one catchpit at approximately 100 m intervals

> Quantity = 4500 m of pipework including crossings and connections
>
> 40 No. catchpits

Method

One drainlaying gang either side of the carriageway excavating trenches, laying and jointing pipes and backfilling.

One catchpit gang following up concreting catchpit bases and placing bottom chamber ring and then next day completing remaining rings, precast slab and temporary cover.

Gullies, gully connections and brickwork and ironwork to gullies and catchpits to follow on later during roadworks.

Gangs

Drainlaying (two gangs):
        Ganger
        Komatsu PC180 tracked excavator
        Banksman
        Komatsu WA470 wheeled loader
        Three labourers
Catchpits: Ganger
        JCB 3CX
        Banksman
        Bricklayer
        Two labourers

Outputs

Drainlaying: 90 m/gang/day working a 5-day week
Catchpits: Average two per day

Activity duration:

$$\text{Drainlaying} = \frac{4500 \text{ m}}{90 \text{ m/day} \times 5 \text{ days} \times 2 \text{ gangs}} = \textbf{5 weeks}$$

$$\text{Catchpits} = \frac{40}{\text{Average 2 per day} \times 5 \text{ days}} = \textbf{4 weeks}$$

The two operations run concurrently and therefore the duration for the drainage activity is 5 weeks overall.

Structures

3000 mm × 2500 mm precast concrete box culvert at chainage 3+600 comprising 1200 mm long units laid in trench on gravel bed jointed with compressible neoprene and polysulphide sealant and backfilled with free draining material.

Quantity = 80 linear metres

One headwall at each end

Method

Excavate trench with battered sides using tracked excavator and remove material to spoil with articulated dump trucks. Lay bed of gravel and lay and joint precast concrete units to line and level. Backfill working space on completion. Headwalls to be constructed at each end of the culvert comprising gabion baskets filled with stone. The excavator will be used for diverting the stream and for all excavation, backfilling and handling bedding materials. A rough terrain mobile crane will be needed for offloading and handling the precast concrete culvert sections and for lifting and placing the gabions.

Gang

Foreman
Cat 320 tracked excavator
40 T mobile telescopic crane
Banksman
Volvo A25 dumper
Three labourers

Output

Allow forward travel of three complete precast sectional units per day

Activity duration:

Diversion of stream $= 1$ week

$$\text{Culverting} = \frac{80\text{m}}{3\,\text{units/day} \times 5\,\text{days}} = 5\,\text{weeks}$$

Headwalls $= 2$ weeks

Total $= \textbf{8 weeks}$

Capping layer

Acceptable capping layer material will be imported from a quarry located 10 miles from the site. It is assumed that the necessary quantities of material will be available to suit site production outputs.

$$\text{Quantity} = 12\,000 \text{ m}^3 \text{ of fill (chainage } 0 - 1 + 500)$$

Method

Imported material to be delivered to site in 20 tonne road wagons via the existing roundabout at chainage 0+000. The material is to be tipped and dozed into position with a dozer blade and spread in 200 mm layers with a motor grader. Compaction is to be carried out using a sit-on tandem vibrating roller. Final trim to be carried out immediately prior to laying sub-base material.

Outputs

Haulage distance and availability of wagons will largely dictate site output and the contractor will have to consider the possibility of more than one supplier for such large quantities. The gang is expected to place 300 m³ in a 10-h working day under normal conditions.

Gang

Ganger
Cat D6N dozer
Cat 120H motor grader
Bomag BW125 roller
Banksman

$$\text{Activity duration} = \frac{12\,000 \text{ m}^3}{600 \text{ m}^3/\text{day} \times 5\,\text{days}} = \textbf{4 weeks}$$

21.4 Clause 14 programme

Earthworks

The earthworks activity will occupy Cut 1 for 14 weeks between chainages 0+000 and 1+500, and this is shown plotted on the time-chainage diagram in Figure 21.4.

Drainage

Calculating the slope of linear activities requires consideration of:

- Quantity of work
- Method of working
- Gang size/make-up
- Estimated gang output (rate of forward travel)

The drainage activity is plotted between chainage $0 + 000$ commencing at week 43 and chainage $2 + 000$ finishing at week 48 (see Figure 21.4).

Structures

The culvert is a 'static' activity and is therefore plotted at chainage $3 + 600$ with a duration of 8 weeks, which is shown on the vertical axis of the time-chainage diagram (Figure 21.4).

Capping layer

A 2-week lead time has been allowed after completion of filling to zone 3 in order to give a time buffer between filling and capping to zone 2 at chainage $2 + 600$.

At week 41, capping layer and drainage, etc., are programmed to commence at chainage $0 + 000$, working towards Green Lane bridge. Note that continuity of working is preserved by using time buffers of different durations between activities (see Figure 21.4).

Index